AF696763

Lecture Notes in Physics

Editorial Board

H. Araki
Research Institute for Mathematical Sciences
Kyoto University, Kitashirakawa
Sakyo-ku, Kyoto 606, Japan

E. Brézin
Ecole Normale Supérieure, Département de Physique
24, rue Lhomond, F-75231 Paris Cedex 05, France

J. Ehlers
Max-Planck-Institut für Physik und Astrophysik, Institut für Astrophysik
Karl-Schwarzschild-Strasse 1, W-8046 Garching, FRG

U. Frisch
Observatoire de Nice
B. P. 139, F-06003 Nice Cedex, France

K. Hepp
Institut für Theoretische Physik, ETH
Hönggerberg, CH-8093 Zürich, Switzerland

R. L. Jaffe
Massachusetts Institute of Technology, Department of Physics
Center for Theoretical Physics
Cambridge, MA 02139, USA

R. Kippenhahn
Rautenbreite 2, W-3400 Göttingen, FRG

H. A. Weidenmüller
Max-Planck-Institut für Kernphysik
Postfach 10 39 80, W-6900 Heidelberg, FRG

J. Wess
Lehrstuhl für Theoretische Physik
Theresienstrasse 37, W-8000 München 2, FRG

J. Zittartz
Institut für Theoretische Physik, Universität Köln
Zülpicher Strasse 77, W-5000 Köln 41, FRG

Managing Editor

W. Beiglböck
Assisted by Mrs. Sabine Landgraf
c/o Springer-Verlag, Physics Editorial Department V
Tiergartenstrasse 17, W-6900 Heidelberg, FRG

The Editorial Policy for Proceedings

The series Lecture Notes in Physics reports new developments in physical research and teaching – quickly, informally, and at a high level. The proceedings to be considered for publication in this series should be limited to only a few areas of research, and these should be closely related to each other. The contributions should be of a high standard and should avoid lengthy redraftings of papers already published or about to be published elsewhere. As a whole, the proceedings should aim for a balanced presentation of the theme of the conference including a description of the techniques used and enough motivation for a broad readership. It should not be assumed that the published proceedings must reflect the conference in its entirety. (A listing or abstracts of papers presented at the meeting but not included in the proceedings could be added as an appendix.)

When applying for publication in the series Lecture Notes in Physics the volume's editor(s) should submit sufficient material to enable the series editors and their referees to make a fairly accurate evaluation (e.g. a complete list of speakers and titles of papers to be presented and abstracts). If, based on this information, the proceedings are (tentatively) accepted, the volume's editor(s), whose name(s) will appear on the title pages, should select the papers suitable for publication and have them refereed (as for a journal) when appropriate. As a rule discussions will not be accepted. The series editors and Springer-Verlag will normally not interfere with the detailed editing except in fairly obvious cases or on technical matters.

Final acceptance is expressed by the series editor in charge, in consultation with Springer-Verlag only after receiving the complete manuscript. It might help to send a copy of the authors' manuscripts in advance to the editor in charge to discuss possible revisions with him. As a general rule, the series editor will confirm his tentative acceptance if the final manuscript corresponds to the original concept discussed, if the quality of the contribution meets the requirements of the series, and if the final size of the manuscript does not greatly exceed the number of pages originally agreed upon.

The manuscript should be forwarded to Springer-Verlag shortly after the meeting. In cases of extreme delay (more than six months after the conference) the series editors will check once more the timeliness of the papers. Therefore, the volume's editor(s) should establish strict deadlines, or collect the articles during the conference and have them revised on the spot. If a delay is unavoidable, one should encourage the authors to update their contributions if appropriate. The editors of proceedings are strongly advised to inform contributors about these points at an early stage.

The final manuscript should contain a table of contents and an informative introduction accessible also to readers not particularly familiar with the topic of the conference. The contributions should be in English. The volume's editor(s) should check the contributions for the correct use of language. At Springer-Verlag only the prefaces will be checked by a copy-editor for language and style. Grave linguistic or technical shortcomings may lead to the rejection of contributions by the series editors.

A conference report should not exceed a total of 500 pages. Keeping the size within this bound should be achieved by a stricter selection of articles and not by imposing an upper limit to the length of the individual papers.

Editors receive jointly 30 complimentary copies of their book. They are entitled to purchase further copies of their book at a reduced rate. As a rule no reprints of individual contributions can be supplied. No royalty is paid on Lecture Notes in Physics volumes. Commitment to publish is made by letter of interest rather than by signing a formal contract. Springer-Verlag secures the copyright for each volume.

The Production Process

The books are hardbound, and the publisher will select quality paper appropriate to the needs of the author(s). Publication time is about ten weeks. More than twenty years of experience guarantee authors the best possible service. To reach the goal of rapid publication at a low price the technique of photographic reproduction from a camera-ready manuscript was chosen. This process shifts the main responsibility for the technical quality considerably from the publisher to the authors. We therefore urge all authors and editors of proceedings to observe very carefully the essentials for the preparation of camera-ready manuscripts, which we will supply on request. This applies especially to the quality of figures and halftones submitted for publication. In addition, it might be useful to look at some of the volumes already published. As a special service, we offer free of charge LATEX and TEX macro packages to format the text according to Springer-Verlag's quality requirements. We strongly recommend that you make use of this offer, since the result will be a book of considerably improved technical quality. To avoid mistakes and time-consuming correspondence during the production period the conference editors should request special instructions from the publisher well before the beginning of the conference. Manuscripts not meeting the technical standard of the series will have to be returned for improvement.

For further information please contact Springer-Verlag, Physics Editorial Department V, Tiergartenstrasse 17, W-6900 Heidelberg, FRG

QC
173.68
.I59
1992

H. Gausterer C. B. Lang (Eds.)

DISCARDED

Computational Methods in Field Theory

Proceedings of the
31. Internationale Universitätswochen
für Kern- und Teilchenphysik
Schladming, Austria, February 1992

Springer-Verlag
Berlin Heidelberg New York
London Paris Tokyo
Hong Kong Barcelona
Budapest

DISCARDED
LIBRARY
Appalachian State University
Boone, North Carolina 28608

Editors

H. Gausterer
C. B. Lang
Institut für Theoretische Physik
Karl-Franzens-Universität
Universitätsplatz 5, A-8010 Graz, Austria

ISBN 3-540-55997-3 Springer-Verlag Berlin Heidelberg New York
ISBN 0-387-55997-3 Springer-Verlag New York Berlin Heidelberg

This work is subject to copyright. All rights are reserved, whether the whole or part of the material is concerned, specifically the rights of translation, reprinting, re-use of illustrations, recitation, broadcasting, reproduction on microfilms or in any other way, and storage in data banks. Duplication of this publication or parts thereof is permitted only under the provisions of the German Copyright Law of September 9, 1965, in its current version, and permission for use must always be obtained from Springer-Verlag. Violations are liable for prosecution under the German Copyright Law.

© Springer-Verlag Berlin Heidelberg 1992
Printed in Germany

Typesetting: Camera ready by author/editor using the TEX/LATEX macro package from Springer-Verlag
58/3140-543210 - Printed on acid-free paper

Preface

This volume contains the final versions of the invited lectures presented at the "Internationale Universitätswochen für Kern- und Teilchenphysik" in Schladming, Austria, February 1992, which was held under the topical title "Computational Methods in Field Theory". A subject like the one discussed at our school would have been impossible at the first school in the series, 30 years ago. Field theory at that time was limited to analytic attempts for a solution such as Feynman graphs perturbation series; computational techniques were clever applications of the theory of special functions.

Today we have a wealth of new insight into problems in the context of quantum field theories gained by the computational approach. The first field to make heavy use of computer simulations for the solution of its problems was statistical physics. When the deep connections between the critical behaviour of spin models and the properties of regularized quantum field theories became clear, it was only a question of time and computer power before these computational techniques were applied to models in elementary-particle physics, starting in the late 1970s.

Meanwhile this field is becoming a discipline of its own. With the advent of powerful computers the spectrum of applications in physics has widened enormously and cannot be covered by any single conference or workshop. However, at our school in Schladming we always concentrated on problems in elementary-particle physics; thus we put the emphasis on those topics and methods in (quantum) field theory which have been particularly influenced by the computational approach.

John Klauder's lecture reviews in a tutorial-like fashion the issue of random variables and stochastic processes in field theory, including the stochastic approach to field quantization, which has been utilized directly in simulations of bosonic and fermionic systems.

In Oscar Lanford's lecture the impact of computers on mathematics and in particular theorem proving is analyzed. Emphasis is put on rigorous quantitative estimates.

In the process of applying the methods of statistical physics to the problems of gauge field theories one has to learn how to deal efficiently with the new types of systems. Computer studies are necessarily limited in (CPU) time and space (finite-size). Since one is interested in the behaviour of the system at large correlation lengths, finite-size effects and critical slowing down are important issues.

The lectures of both Ulli Wolff and Kurt Binder deal with these problems, the first concentrating on critical slowing down and efficiency of algorithms and the second with finite-size effects and their taming and utilization. For certain spin systems new algorithms have evolved, allowing an efficient simulation of bosonic theories. Gauge systems have an essentially non-local structure, and improvements of the simulation algorithms are hard to achieve. The close methodological interrelationship between particle physicists working in the lattice-field approach and solid-state physicists working with spin systems has proven useful in many instances. One example is the problem of learning about surface tension and bubble formation in systems with a first-order phase transition. This is one of the topics dealt with by Kurt Binder in his lecture, demonstrating how to extract information from the finite-size dependence of distribution functions.

Thomas DeGrand surveys the situation in lattice QCD, for particle physicists perhaps the most challenging topic. Early promises turned out to be too optimistic and recent years have been characterized by hard and detailed work aimed at understanding the limitations, aspects and prospects of this approach.

The microscopic theory may look very different on a macroscopic scale. The appearance of a theory may change with the scale of observation, as illuminated by the renormalization process. Gerhard Mack discusses ideas for identifying effective theories. His contribution is co-written by T. Kalkreuter, G. Palma and M. Speh and covers new ideas in methods (such as the multigrid approach) and conception.

Fred James reviews the impact of computers on high-energy physics experiments and gives an overview of design, running and analysis. He puts particular emphasis on the (superficially) most trivial and at the same time most subtle and important issue: how to get good random numbers. In the last few years there has been enormous progress and his lecture provides "computer field theorists" with a good update and background.

In his lectures at the school Alexander Migdal discussed attempts to study quantum gravity with help of computer models. Most of the work presented has been published elsewhere and he decided not to contribute to this volume.

At the winter school there were many excellent seminars closely related to the topics of the lectures. We cannot publish them in this collection but want at least to give credit to the speakers. Thus there is a list of seminar contributions at the end, and the authors may be contacted directly for related information and material.

The organizers of the school (the editors of this volume) wish to thank the lecturers for their excellent contributions. The school was supported generously by the Austrian Ministry of Science and Research, the Styrian Government and other sponsors. The contributions were prepared by the authors and written in TEXor LATEXwith the help of Springer's macro packages. Mrs. E. Neuhold helped in producing the final version.

July 1992
Graz, Austria

H. Gausterer
C.B. Lang

Contents

Computers in the Design and Analysis of HEP Experiments

By F. James (with 5 figures)

Contributors

Binder, K.

Institut für Physik, Johannes Gutenberg Universität Mainz
Staudinger Weg 7, D-6500 Mainz, Germany

DeGrand, T.

Department of Physics, University of Colorado
Boulder, Colorado 80309, USA

James, F.

CERN, Data Handling Division
CH-1211 Genève 23, Switzerland

Klauder, J.R.

Departments of Physics and Mathematics, University of Florida
Gainesville, FL 32611, USA

Lanford III, O.E.

Mathematics Department, ETH-Zürich
CH-8092 Zürich, Switzerland

Mack, G., Kalkreuter, T., Palma, G., Speh, M.

II. Institut für Theoretische Physik, Universität Hamburg,
Luruper Chaussee 149, D-2000 Hamburg 50, Germany

Wolff, U.

CERN, Theory Division
CH-1211 Genève 23, Switzerland

A Stochastic Primer

John R. Klauder

Departments of Physics and Mathematics, University of Florida, Gainesville, FL 32611

1. Random Variables

1.1 Basic Definitions

Moments. The basic notion of a random variable is intuitively clear, namely, it is a variable, say X, that can, each time it is measured, assume one or another of a set of possible values with a given probability. For example, if the possible values are integers, then the probability of their occurrence is given by

$$P(X = n) = \langle \delta_{nX} \rangle \,,$$

where $\langle \cdot \rangle$ denotes an average over the ensemble and δ_{nX} equals 1 (or 0) whenever $X = n$ (or $X \neq n$). More generally, the outcome, or measured value, is some real number, and other ensemble averages are frequently used to study such variables. Certain aspects of the distribution of real values of the random variable X are determined by the moments

$$\langle X^n \rangle \,, \qquad n = 1, 2, 3, \ldots \,,$$

averaged over the ensemble and which we temporarily assume exist. Other functions of X are also important such as

$$M(a) \equiv \langle e^{aX} \rangle = \sum_{n=0}^{\infty} \frac{a^n}{n!} \langle X^n \rangle,$$

defined for $|a| < \alpha$, for some $\alpha > 0$, that serves as the *moment generating function.*

Besides the ordinary moments $\langle X^n \rangle$ of a distribution it is also useful to introduce the *connected moments* or *cumulants* $\langle X^n \rangle_c$, defined below, which also go by other names such as linked moments or truncated moments. In brief, the cumulants are defined by the power series expansion in the relation

$$\langle e^{aX} - 1 \rangle_c = \ln \langle e^{aX} \rangle$$

or, correspondingly,

$$\langle e^{aX} \rangle = e^{\langle e^{aX} - 1 \rangle_c} \,.$$

It follows, for example, that

$$\begin{aligned}\langle X\rangle_c &= \langle X\rangle\,,\\ \langle X^2\rangle_c &= \langle X^2\rangle - \langle X\rangle^2\,,\\ \langle X^3\rangle_c &= \langle X^3\rangle - 3\langle X^2\rangle\langle X\rangle + 2\langle X\rangle^3\,,\\ \langle X^4\rangle_c &= \langle X^4\rangle - 4\langle X^3\rangle\langle X\rangle - 3\langle X^2\rangle^2\\ &\quad + 12\langle X^2\rangle\langle X\rangle^2 - 6\langle X\rangle^4\,,\end{aligned}$$

etc. The first, $\langle X\rangle$, is termed the *mean*, while the second, $\langle X^2\rangle_c$, is called the *variance*. We note especially the cases

$$\begin{aligned}\langle X^2\rangle_c &= \langle X^2\rangle\,,\\ \langle X^4\rangle_c &= \langle X^4\rangle - 3\langle X^2\rangle^2\,,\end{aligned}$$

which hold when the mean $\langle X\rangle = 0$.

There are several classic distributions that may be easily described by their generating functions for the moments and for the connected moments.

Poisson Distribution:

$$\begin{aligned}\langle e^{aX}\rangle &= e^{\mu(e^a-1)}\\ &= e^{-\mu}\sum_{p=0}^{\infty}\frac{\mu^p}{p!}e^{pa}\end{aligned}$$

which is easily seen to imply that X assumes the values $0, 1, 2, \ldots$, with probabilities

$$P(X = p) = \frac{\mu^p}{p!}e^{-\mu}\,,$$

and moreover it follows that $\langle X\rangle = \langle X\rangle_c = \mu$. Observe in this case that the connected moments $\langle X^p\rangle_c = \mu$ for all $p \geq 1$.

Gaussian Distribution:

$$\langle e^{aX}\rangle = e^{a\langle X\rangle_c + \frac{1}{2}a^2\langle X^2\rangle_c}$$

which is easily seen to imply that only the first two cumulants are allowed; all higher-order connected moments *vanish*. If $\langle X\rangle = 0$ then

$$\langle e^{aX}\rangle = e^{\frac{1}{2}a^2\langle X^2\rangle}\,,$$

which we shall denote as a *normal distribution*. (Note this is not the common definition for which, unfortunately, Gaussian and normal mean the same thing, i.e., $\langle X\rangle \neq 0$, in general.)

Bounded Functions. Sometimes, not all moments exist, and bounded functions of X become important. One important example (discussed further below) is the function

$$C(s) = \langle e^{isX}\rangle$$

defined for all real s which is called the *characteristic function*. Another important bounded function involves an *indicator function* defined by

$$\chi_x(y) = 1, \quad y < x\,,$$
$$\chi_x(y) = 0, \quad y > x\,.$$

Specifically, the expectation involving the indicator function

$$P(X \leq x) \equiv \lim_{\epsilon \to 0} \langle \chi_{x+\epsilon}(X) \rangle\,,$$

where ϵ goes to zero from positive values, defines the *probability that the random variable X takes on values less than or equal to the value of x.*

1.2 Probability Measures

A unified way to study probability distributions is to examine the functions

$$\mu(x) \equiv P(X \leq x)$$

defined as *probability measures*. In all cases, such functions μ have several properties in common:

$$\text{1) Normalization:} \quad \mu(-\infty) = 0, \quad \mu(\infty) = 1\,,$$
$$\text{2) Nondecreasing:} \quad \mu(x) \leq \mu(x+h), \quad h > 0\,,$$
$$\text{3) Right Continuity:} \quad \mu(x^-) \leq \mu(x) = \mu(x^+)\,,$$

where, in point 3), $\mu(x^\pm) \equiv \lim \mu(x \pm \epsilon)$ as $\epsilon \to 0$ from positive values. The properties under 1) and 2) are evident, while the last property is established as follows. On the one hand, $P(X \leq x - \epsilon) \leq P(X < x - \epsilon/2) \longrightarrow P(X < x) \leq P(X \leq x)$, while, on the other hand, $P(X \leq x + \epsilon) \longrightarrow P(X \leq x) = \mu(x)$ verifying 3).

To find all possible types of probability it is only necessary to find all types of solutions to the set of three properties that determine the probability measure μ. It is known that there are *three qualitatively different functions* that satisfy the three basic properties and that the most general μ is a normalized positive combination of the three different types.

Discrete Measures. The first type of measure is called a discrete (d) measure μ_d, and is defined as follows

$$\mu_d(x) = \sum p_j H(x - x_j)$$

where $p_j > 0$, $\sum p_j = 1$, and H is the Heaviside function

$$H(x) = 0, \quad x < 0\,,$$
$$H(x) = 1, \quad x \geq 1\,.$$

This is the measure suitable for a set of discrete outcomes for the variable X, namely the set $\{x_j\}$, like coins, dice, roulette wheels, etc. Since $H(x)$ is constant save at zero it follows that $dH(x)/dx = 0$ everywhere except at $x = 0$. From the point of view of classical analysis the (Lebesgue) integral is insensitive to the value of the integrand on a set of measure zero—and any countable collection of points on the real line is such a set—and thus we have the rather strange fact that

$$H(x) = \int_{-\infty}^{x} dH(y) \neq \int_{-\infty}^{x} \frac{dH(y)}{dy} dy \,,$$

and likewise for $\mu_d(x)$ itself. The modern (post-Dirac) physicist would argue differently asserting that $H'(y) = \delta(y)$ and consequently that $H(x) = \int_{-\infty}^{x} H'(y)dy = \int_{-\infty}^{x} \delta(y)dy$ is correct after all. This smug view, however, runs into trouble with the next category of measure.

Singular Continuous Measures. Singular continuous measures are the most difficult to visualize. We give our example as the limit of a sequence of discrete measures. Consider the sequence of discrete densities defined by

$$\mu'_N(x) = \int \delta(x - \textstyle\sum_1^N x_n) \displaystyle\prod_{n=1}^{N} \frac{1}{2}[\delta(x_n + 3^{-n}) + \delta(x_n - 3^{-n})] dx_n \,.$$

For each $N < \infty$ this example corresponds to a discrete measure with a sequence of 2^N jumps of equal height 2^{-N}. Thus $\mu_N(-\infty) = 0$ and $\mu_N(\infty) = 1$ as desired. What about the limit $N \to \infty$ and μ_∞? The result is a *continuous* (Cantor) function $\mu_\infty(x) = \hat{C}(x)$ that rises from 0 to 1, and for which $\hat{C}'(x) = 0$ for almost all x, e.g., for all irrational x. Thus, again,

$$\hat{C}(x) = \int_{-\infty}^{x} d\hat{C}(y) \neq \int_{-\infty}^{x} \frac{d\hat{C}(y)}{dy} dy \,,$$

and no formal equation involving the δ-function will help this time since the amplitude of any putative $\delta(y - y_j)$ is zero! The function $\hat{C}(x)$ is the prototype of a singular continuous measure, and more generally

$$\mu_{sc}(x) = \textstyle\sum p_j \hat{C}_j(x - x_j)$$

where $p_j > 0$, $\sum p_j = 1$, and $\hat{C}_j$ is a Cantor-type function constructed in analogy with $\hat{C}$ above.

Such measures may seem unphysical, but, in fact, they do arise in certain physical models. For instance, there is a one-body Hamiltonian with a spectrum described by such a measure [1]. In addition, it was once conjectured that Anderson localization could be understood in terms of singular continuous spectra.

Absolutely Continuous Measures. The final example, absolutely continuous measures, brings us back to familiar territory. Let $A(x)$ be a nondecreasing function for which $A(-\infty) = 0$, $A(\infty) = 1$, $A'(x) \geq 0$ almost everywhere and A' is an integrable function such that

$$A(x) = \int_{-\infty}^{x} dA(y) = \int_{-\infty}^{x} A'(y)dy \,.$$

A typical example in this case would be $A'(y) = \pi^{-\frac{1}{2}}\exp(-y^2)$ appropriate to a normal variable. Thus the most general absolutely continuous measure is given by

$$\mu_{ac}(x) = \int_{-\infty}^{x} \rho(y)dy\,,$$

where $\rho \geq 0$ and is normalized so that $\mu_{ac}(\infty) = 1$.

General Probability Measures. The above discussion describes the three basic, *pure* types of probability measures. More generally, a probability measure is given as a positive superposition of the three types, *viz.*,

$$\mu(x) = a_1\mu_d(x) + a_2\mu_{sc}(x) + a_3\mu_{ac}(x)$$

where $a_j \geq 0$ and $\sum a_j = 1$.

Remark. Of course, not all measures are probability measures, and we briefly discuss more general measures. One obvious modification deals with overall scaling such as $p(x) = a\mu(x)$, $0 < a < \infty$, for which $p(\infty) = a$ rather than unity. Another generalization leads to an overall infinite weight. For example, let $d\mu(x) = dx$ in the interval $0 \leq x < 1$. By translation we can extend this measure to $d\sigma(x) = dx$, defined for all x, $-\infty < x < \infty$.

Signed measures involve the difference of two positive measures $m(x) = p_1(x) - p_2(x)$, or $dm(x) = dp_1(x) - dp_2(x)$. To be a proper measure it is necessary that $|m|(x) \equiv p_1(x)+p_2(x)$ be finite, i.e. $|m|(\infty) < \infty$. Sometimes "signed measures" have the form $d\tau(x) = d\sigma_1(x)-d\sigma_2(x)$ such that $\tau(\infty) < \infty$ while $|\tau|(\infty) = \infty$. In this case τ is not a proper (countably additive) measure (although it does satisfy the weaker criterion of finite additivity, just as any linear functional would).

Complex measures are straightforward generalizations, namely $dq\,(x) = dm_1(x)+idm_2(x)$. If both m_1 and m_2 are signed measures then q is termed a complex measure. Even when $q(\infty) < \infty$, if either m_1 or m_2 are not proper measures then neither is q.

Countably additive measures are essential if one wants to invoke classical limiting properties such as Fatou's Lemma or Lebesgue's Dominated Convergence Theorem. For finitely additive measures counterexamples to these properties exist [2]. Hereafter, we shall use the unqualified word measure to mean a countably additive measure.

Moments in Terms of the Probability Measure. The differential probability that X lies between x and $x + dx$ is clearly

$$P(X \leq x + dx) - P(X \leq x) = \mu(x + dx) - \mu(x) \equiv d\mu(x)\,,$$

and thus the moments are given by

$$\begin{aligned}\langle X^n\rangle &= \int_{-\infty}^{\infty} x^n d\mu(x)\\ &\equiv \int x^n d\mu(x)\,;\end{aligned}$$

hereafter, all integrals from $-\infty$ to $+\infty$ will be left implicit. Likewise, the probability $P(X \leq x)$ itself is given, in the limit that $\epsilon \to 0$ from positive values, by

$$\lim \int_{-\infty}^{x+\epsilon} d\mu(x) = \lim \mu(x+\epsilon) = \mu(x^+) = P(X \leq x).$$

Another important example relates to the characteristic function.

1.3 Characteristic Function

A powerful and elegant approach to probability is afforded by the characteristic function [3]

$$C(s) \equiv \langle e^{isX} \rangle = \int e^{isx} d\mu(x)$$

which is a continuous, bounded function,

$$|C(s)| \leq \int d\mu(x) = C(0) = 1$$

for any choice of probability measure μ. Along with $C(s)$ let us also consider the averaged expression

$$C^T(t) \equiv \frac{1}{T} \int_t^{T+t} C(s) ds .$$

The three pure types of measures leave indelible imprints on the characteristic function that can be read from the asymptotic properties of C and C^T. In particular, as $|t| \to \infty$, C and/or C^T will either converge to zero or not converge at all.

For a purely discrete measure, $C_d(t)$ and $C_d^T(t)$ both fail to converge as $|t| \to \infty$; a typical example,

$$C_d(t) = \cos(t), \qquad C_d^T(t) = T^{-1}[\sin(T+t) - \sin(T)]$$

illustrates this failure to converge.

For a purely singular continuous measure $C_{sc}(t)$ fails to converge, while $C_{sc}^T(t)$ converges to zero. The example

$$C_{sc}(t) = \prod_{n=1}^{\infty} \cos(t/3^n)$$

is exactly the characteristic function for the Cantor function $\hat{C}$ given in Sec. 1.2. As t becomes large, say $t \leq 3^N$ for some N, there are many (N) factors where $\cos(t/3^n)$ is generally smaller than unity in absolute value. Thus as $t \to \infty$ and $N \to \infty$ the product defining C_{sc} goes to zero, generally speaking. Although this is not enough to ensure that C_{sc} itself converges to zero (counter example: $t_j = 2\pi 3^j$ as $j \to \infty$), it is enough to ensure that the integrated form C_{sc}^T converges to zero.

Finally, for an absolutely continuous measure the functions $C_{ac}(t)$ and $C_{ac}^T(t)$ both converge to zero. For

$$C_{ac}(t) = \int e^{itx} \rho(x) dx$$

convergence to zero is ensured by the Riemann-Lebesgue Lemma (which asserts that the Fourier transform of an L^1 function vanishes at $\pm\infty$). The convergence of C^T_{ac} follows directly from this fact (and the Lebesgue dominated convergence theorem). In the present case the function $\rho(x)$ is referred to as the ***probability density.***

1.4 Infinitely Divisible Distributions

An interesting set of distributions are those listed in the title of this section. A distribution μ is termed infinitely divisible if the m-th root of the characteristic function C for μ, i.e., $C^{1/m}$, is also a characteristic function for all m, $m = 1, 2, 3, \ldots$. From their structure it is clear that both the Poisson distribution and the Gaussian distribution are infinitely divisible distributions. For example, as follows from the moment generating function,

$$\langle e^{isX} \rangle^{1/m} = e^{(\mu/m)(e^{is}-1)}$$

leads to a new Poisson distribution with a rescaled parameter μ as indicated.

The general form for the characteristic function of an infinitely divisible distribution is easily obtained. In particular, it follows that

$$\begin{aligned}
\ln C(s) &= -\lim_{m\to\infty} m[1 - C(s)^{1/m}] \\
&= -\lim_{m\to\infty} \int [1 - e^{isu}] m d\mu_m(u) \\
&= -\lim_{m\to\infty} \left\{ \int \left(-is\frac{u}{1+u^2} + \frac{1}{2}s^2 \frac{u^2}{1+u^2} \right) m d\mu_m(u) \right. \\
&\quad \left. - \int \left(1 + is\frac{u}{1+u^2} - \frac{1}{2}s^2\frac{u^2}{1+u^2} - e^{isu} \right) m d\mu_m(u) \right\} \\
&\equiv isa - \frac{1}{2}s^2 b \\
&\quad - \int\limits_{|u|>0} \left(1 + is\frac{u}{1+u^2} - \frac{1}{2}s^2 \frac{u^2}{1+u^2} - e^{isu} \right) d\sigma(u)
\end{aligned}$$

where $a \in \mathbf{R}$, $b \geq 0$, and where $\sigma \geq 0$,

$$\int\limits_{|u|>0} \frac{u^2}{1+u^2} d\sigma(u) < \infty \,,$$

but it may well be that $\int_{|u|>0} d\sigma(u) = \infty$. As noted, the class of solutions includes the Gaussian (with $\sigma \equiv 0$) and Poisson (with $d\sigma(u) = \mu\delta(u-1)du$), as well as many others [3].

1.5 Several Variables

When several random variables are involved, say $X_1, X_2, \ldots, X_N$, $N < \infty$, then various moments arise, such as,

$$\langle X_1^{p_1} X_2^{p_2} \cdots X_N^{p_N} \rangle$$

where $p_j = 0, 1, 2, \ldots$, for each j. The characteristic function is conveniently defined in vector form as

$$C(\underset{\sim}{s}) = \langle e^{i\underset{\sim}{s}\cdot\underset{\sim}{X}} \rangle$$

where $\underset{\sim}{s} \cdot \underset{\sim}{X} = \sum_1^N s_j X_j$. Likewise the associated joint probability measure reads

$$\begin{aligned}\mu(\underset{\sim}{x}) &= P(X_j \leq x_j, j = 1, \ldots, N) \\ &= \lim_{\epsilon \to 0} \left\langle \prod_{j=1}^N \chi_{x_j+\epsilon}(X_j) \right\rangle\end{aligned}$$

where χ_x is the indicator function of Sec. 1.1.

If, for 2 random variables, X_1 and X_2,

$$\langle X_1^{p_1} X_2^{p_2} \rangle = \langle X_1^{p_1} \rangle \langle X_2^{p_2} \rangle$$

holds for all p_1 and p_2, then the two variables are said to be *independent* random variables. Alternatively, if, for all values of s_1 and s_2,

$$C(s_1, s_2) = C_1(s_1) C_2(s_2)$$

or, for all values of x_1 and x_2

$$\mu(x_1, x_2) = \mu_1(x_1) \mu_2(x_2)\,,$$

then the underlying random variables are *independent.* Such conditions can also apply to two or more groups of random variables.

Suppose $X = X_1 + X_2$, and X_1 and X_2 are independent random variables. In that case it follows that

$$\begin{aligned}C(s) &= \langle e^{is(X_1+X_2)} \rangle \\ &= \langle e^{isX_1} \rangle \langle e^{isX_2} \rangle = C_1(s) C_2(s)\end{aligned}$$

showing that the characteristic function is given by the product of the two separate characteristic functions. Furthermore, it follows that the connected moments for the variable X are just the *sum* of those for X_1 and X_2, namely that

$$\langle X^p \rangle_c = \langle X_1^p \rangle_c + \langle X_2^p \rangle_c\,, \qquad p = 1, 2, \ldots\,.$$

For a Gaussian variable X_1 the characteristic function follows directly from the moment generating function as

$$\langle e^{isX_1} \rangle = e^{is\langle X_1 \rangle - \frac{1}{2}s^2 \langle X_1^2 \rangle_c}\,.$$

Suppose X_2 is also a Gaussian variable for which a similar equation holds, i.e.,

$$\langle e^{isX_2} \rangle = e^{is\langle X_2 \rangle - \frac{1}{2}s^2 \langle X_2^2 \rangle_c}\,.$$

Given that X_1 and X_2 are separately Gaussian variables, what is the distribution of $X = X_1 + X_2$? One often hears that the sum of two Gaussian variables is also Gaussian, but in fact it is not necessarily true! Let us illustrate this point for two normal variables.

The joint distribution of X_1 and X_2 is given by a probability density $\overline{\rho}(x_1, x_2)$ that has the property, e.g., that

$$\int \overline{\rho}(x,y)dy = \rho_1(x) = e^{-x^2}/\sqrt{\pi}\,,$$

$$\int \overline{\rho}(x,y)dx = \rho_2(y) = e^{-y^2}/\sqrt{\pi}\,.$$

These relations show that X_1 and X_2 are normal variables. The probability density for $X = X_1 + X_2$ is given by

$$\rho(x) = \int \overline{\rho}(x-y,y)dy\,.$$

As an example choose

$$\overline{\rho}(x,y) = \pi^{-1}\left[e^{-(x^2+y^2)} - xy\,e^{-2(x^2+y^2)}\right],$$

in which case

$$\rho(x) = \int \overline{\rho}(x-y,y)dy = \frac{e^{-x^2/2}}{\sqrt{2\pi}} - \frac{1}{16\sqrt{\pi}}(2x^2-1)e^{-x^2}\,,$$

clearly not a Gaussian. Thus, the sum of two variables that individually are Gaussian is not necessarily Gaussian.

The proper statement in this regard refers to the sum of two variables that are *jointly* Gaussian, namely, if X_1 and X_2 are two such variables, then

$$\left\langle e^{i(s_1X_1+s_2X_2)}\right\rangle = e^{i\left(s_1\langle X_1\rangle+s_2\langle X_2\rangle\right)-\frac{1}{2}s_1^2\langle X_1^2\rangle_c-s_1s_2\langle X_1X_2\rangle_c-\frac{1}{2}s_2^2\langle X_2^2\rangle_c}\,,$$

where $\langle X_1X_2\rangle_c = \langle X_1X_2\rangle - \langle X_1\rangle\langle X_2\rangle$, holds for all s_1, s_2. It is evident that this relation equals

$$\left\langle e^{i(s_1X_1+s_2X_2)}\right\rangle = e^{i\langle s_1X_1+s_2X_2\rangle-\frac{1}{2}\langle(s_1X_1+s_2X_2)^2\rangle_c}$$

confirming, for any s_1 and s_2, that $X = s_1X_1 + s_2X_2$ is a Gaussian random variable.

1.6 Central Limit Theorem

Let $X_j, j = 1, 2, \ldots$, denote independent, identically distributed variables with mean zero, $\langle X_j\rangle = 0$. Consider the sum ($N < \infty$)

$$X_{(N)} \equiv \frac{1}{\sqrt{N}}\sum_1^N X_n$$

and the associated characteristic function

$$C_{(N)}(s) = \left\langle e^{isX_{(N)}} \right\rangle = \left\langle e^{is/\sqrt{N}\sum_1^N X_n} \right\rangle$$
$$= \left\langle e^{isX_1/\sqrt{N}} \right\rangle^N .$$

In the limit that $N \to \infty$ it follows that

$$\begin{aligned} C(s) &\equiv \lim C_{(N)}(s) \\ &= \lim \langle (1 + isX_1/\sqrt{N} - \frac{1}{2}s^2 X_1^2/N + \cdots) \rangle^N \\ &= e^{-\frac{1}{2}s^2\langle X_1^2 \rangle} , \end{aligned}$$

provided that the neglected terms $(\cdots)$ do not contribute, a condition guaranteed if $\langle |X_1|^3 \rangle < \infty$. Thus, under very general conditions, the variable

$$X = \lim_{N\to\infty} \frac{1}{\sqrt{N}} \sum_1^N X_n$$

has a *normal* distribution. With suitable rescaling nonzero means may also be accommodated leading in general to a Gaussian variable X.

1.7 Infinitely-Many Variables (Preliminaries)

An especially convenient way to discuss infinitely-many variables is with the characteristic function given as the limit of sequences having finitely-many variables. Let

$$C(s_{(N)}) \equiv \langle e^{i\sum_1^N s_n X_n} \rangle$$

denote such a characteristic function restricted to just N variables and defined for all $s_{(N)} \in \mathbf{R}^N$. As $N \to \infty$ we do not expect convergence for arbitrary sequences $s = (s_1, s_2, s_3, \ldots)$, but rather for those sequences that fall off in one way or another as $N \to \infty$. For example, assume that the X_j are independent, identically distributed normal random variables. In that case

$$C(s_{(N)}) = e^{-\frac{1}{2}A\sum_1^N s_n^2} , \quad A > 0$$

and convergence as $N \to \infty$ requires that $\sum_1^\infty s_n^2 < \infty$, or $s \in \ell^2$. Here we are using the standard convention that convergence of $\prod_{n=1}^\infty a_n$, where each $a_n \neq 0$, is equivalent to convergence of $\sum_{n=1}^\infty |a_n - 1|$. Thus, we find

$$C(s) = e^{-\frac{1}{2}A\sum_1^\infty s_n^2} , \quad s \in \ell^2 ,$$

defines the characteristic function for an infinite number of independent, identical normal variables. If the variances were *not* all identical, then

$$C(s) = e^{-\frac{1}{2}\sum_1^\infty B_n s_n^2} , \quad s \in \ell^2(B) ,$$

and the condition of acceptability would now be $\sum_1^\infty B_n s_n^2 < \infty$, which defines elements of $\ell^2(B)$. Thus the allowed space of sequences here is directly related to the characteristic function in question. This is quite generally the case.

Indeed, for the sake of completeness, we note that the maximum space of acceptable sequences, $\mathcal{V}_M$ let us say, can be taken as a complete metric space, where a suitable metric is given by [4]

$$d(s,s') = \int e^{-\lambda^2} \{1 - Re\, C(\lambda[s-s'])\}\, d\lambda\,.$$

This metric is equivalent to the natural metric in the examples given above.

Probability Distribution. As was the case for finitely-many variables the probability distribution for infinitely-many variables is conveniently related to the characteristic function through a Fourier transform. For the sequence, we first define

$$C(s_{(N)}) = \int e^{i(s,v)_N}\, d\mu_N(v)\,,$$

where $(s,v)_N = \sum_1^N s_n v_n$ and μ_N is a probability measure on $v_{(N)} \in \mathbf{R}^N$. Then, in the limit,

$$\begin{aligned} C(s) &= \lim_{N\to\infty} \int e^{i(s,v)_N}\, d\mu_N(v) \\ &\equiv \int e^{i(s,v)}\, d\mu(v)\,. \end{aligned}$$

While μ_N is supported on $\mathbf{R}^N$, the support of μ is of some interest and generally may be different than expected. To say that μ is supported on $\mathbf{R}^\infty$ is not wrong but this space is far too large. Consider the example

$$e^{-\frac{1}{2}A\sum_1^\infty s_n^2} = \int e^{i\sum_1^\infty s_n v_n}\, d\mu(v)\,.$$

It follows, for each n, that $\int v_n^2 d\mu(v) = A$, and thus for any sequence $\{\lambda_n\}_{n=1}^\infty$ with $\lambda_n > 0$ and $\sum_1^\infty \lambda_n < \infty$,

$$\int \textstyle\sum_1^\infty \lambda_n v_n^2 d\mu(v) < \infty\,.$$

Thus the support of μ is limited to sequences $\{v_n\}_{n=1}^\infty$ such that

$$\textstyle\sum_1^\infty \lambda_n v_n^2 < \infty\,, \quad \mu\text{–a.e.}\,,$$

provided that

$$\lambda_n > 0, \qquad \textstyle\sum_1^\infty \lambda_n < \infty\,.$$

Indeed, the space ℓ^2 is a set of measure *zero* in μ.

We observe that the operator T such that $(v,Tv) = \sum_1^\infty \lambda_n v_n^2$ is just one example of a *trace class operator*. The example under discussion leads to very natural generalizations for other cases involving infinitely many variables; however let us defer their discussion until the next Section which systematically deals with the case of an infinite number of random variables.

2. Stochastic Variables

2.1 Basic Concepts

An infinite sequence of real random variables, $\{X_k\}_{k=1}^{\infty}$, can be converted to a real random function of a real variable which we shall call a *stochastic variable*, say $X(t)$, with the aid of a real complete orthonormal set of functions $\{h_k(t)\}_{k=1}^{\infty}$. In particular—without discussing convergence—we have

$$X(t) = \sum_{k=1}^{\infty} X_k h_k(t)\,,$$
$$X_k = \int h_k(t)X(t)dt\,.$$

Thus properties of one ensemble are encoded in the other ensemble by means of such a transformation, and for many examples the functional form $X(t)$, $t \in \mathbf{R}$, is by far the most convenient to use. It is common to interpret t as "time" and we shall generally do so; sometimes $0 < t < \infty$, or $t_1 < t < t_2$ depending on the circumstance. To cover all these possibilities one may say that $t \in I$, an index set. In another set of applications, the variable $t \in \mathbf{R}^n$, or some open subset of $\mathbf{R}^n$, which is often to be interpreted as space time. However, we shall leave that possible generalization out of our consideration restricting attention to $n = 1$ with t interpreted as time. In addition, we shall adopt $X(t)$ as the primary stochastic variable hereafter, rather than the sequence $\{X_k\}_{k=1}^{\infty}$.

Correlation Functions. Let us assume that $X(t)$ is a well-defined random variable for each $t \in I$. Then it is common to specify the distribution by means of the *correlation functions*

$$\langle X(t_1)X(t_2)\cdots X(t_m)\rangle\,, \quad m = 1, 2, 3, \ldots\,,$$

for all values of $t_j \in I$. Observe that the correlation functions are fully symmetric in their arguments.

These correlation functions must satisfy certain coupled inequalities among themselves that are readily illustrated. For example, consider the random variable $Y = \alpha X(t_1) + \beta X(t_2)X(t_3)$. Then, since $\langle Y^2\rangle \geq 0$, it follows that

$$0 \leq \alpha^2\langle X^2(t_1)\rangle + 2\alpha\beta\langle X(t_1)X(t_2)X(t_3)\rangle + \beta^2\langle X^2(t_2)X^2(t_3)\rangle$$

holds for all real α and β and all $t_j \in I$. Similar relations relate other correlation functions as well.

Sometimes the quantity $X(t)$ is not point-wise defined as a random variable but only as a distribution-valued random variable. In that case X is referred to as a *generalized stochastic variable*, and the various correlations are properly interpreted as distributions. Let $f_m(t_1, t_2, \ldots, t_m)$ denote a test function of a suitable type, and then rather than deal with point correlation functions one considers, for all m,

$$\int f_m(t_1, t_2, \ldots, t_m)\langle X(t_1)X(t_2)\cdots X(t_m)\rangle dt_1 dt_2 \cdots dt_m$$

for a general function f_m. Clearly only symmetric functions need be considered, and such functions form a linear vector space. For many purposes it suffices to consider a total set of such functions for then any such function may be reached by linearity and suitable limits. A simple and useful set of total functions have the form

$$f_m(t_1, t_2, \ldots, t_m) = s(t_1)s(t_2)\cdots s(t_m)$$

and consequently it follows, for all m, that

$$\int s(t_1)s(t_2)\cdots s(t_m)\langle X(t_1)X(t_2)\cdots X(t_m)\rangle dt_1 dt_2 \cdots dt_m \,,$$

for all suitable s, gives just as much information as before. Observe that this correlation function may be written as

$$\langle (\textstyle\int s(t)X(t)dt)^m \rangle \,,$$

or even more concisely as

$$\langle (s, X)^m \rangle$$

in an implicit inner product notation.

These expressions, needed in the case of generalized stochastic variables, are also standard and useful even for ordinary stochastic variables. In particular, one can extract from $\langle (s,X)^m \rangle$ the desired correlation function $\langle X(t_1)\cdots X(t_m)\rangle$ just by repeated functional differentiation, i.e.,

$$(m!)^{-1}\left[\delta^m/\delta s(t_1)\delta s(t_2)\cdots \delta s(t_m)\right]\langle (s,X)^m\rangle = \langle X(t_1)X(t_2)\cdots X(t_m)\rangle \,.$$

2.2 Characteristic Functional

As a generator of correlation functions, or as a fundamental quantity in its own right, the *characteristic functional* is first among equals. The characteristic functional is defined by

$$\begin{aligned} C\{s\} &= \langle e^{i(s,X)} \rangle \\ &= \langle e^{i\int s(t)X(t)dt} \rangle \end{aligned}$$

for all $s \in \mathcal{V}$, a suitable linear vector space of test functions. A maximal such space $\mathcal{V}_M$ is given by the completion of the metric space based on the metric d defined in Sec. 1.7. We assume that for each $s \in \mathcal{V}$, a dense subset of $\mathcal{V}_M$, that $F_s(u) = C\{us\}$ is continuous in the real variable u. It then follows that $C\{s\}$ is a continuous functional with respect to d for all $s \in \mathcal{V}_M$. Of course, it is often convenient to simply restrict attention to a dense subset $\mathcal{V}$ of the maximal space $\mathcal{V}_M$ since continuity will take care of the rest.

We observe that if a given continuous functional satisfies

$$C\{0\} = 1, \qquad \textstyle\sum_{j,k=1}^{N} \alpha_j^* \alpha_k C\{s_j - s_k\} \geq 0$$

for all finite sequences ($N < \infty$) of complex numbers $\{\alpha_j\}$ and arguments $\{s_j\}$, then that function C is the characteristic functional of a unique stochastic process. On the other hand, the complete set of correlation functions may determine a stochastic process, but it may not do so uniquely in certain circumstances. The characteristic functional will always satisfy

$$C\{s\}^* = C\{-s\}\,.$$

A number of basic properties of a stochastic process may easily be defined with the aid of the characteristic functional. We briefly discuss four such properties.

Independent Stochastic Variables. Let $X(t)$ and $Y(t)$ denote two stochastic variables. The joint characteristic functional

$$\begin{aligned} C\{r,s\} &= \langle e^{i[(r,X)+(s,Y)]}\rangle \\ &= \langle e^{i\int[r(t)X(t)+s(t)Y(t)]dt}\rangle \end{aligned}$$

is easily seen to be a generating functional for all possible correlation functions, especially including joint correlation functions.

The stochastic variables X and Y are said to be *independent* if the characteristic functional factorizes, i.e., if

$$\langle e^{i\int[r(t)X(t)+s(t)Y(t)]dt}\rangle = \langle e^{i\int r(t)X(t)dt}\rangle\langle e^{i\int s(t)Y(t)dt}\rangle$$

for a dense set of arguments r and s.

Connected Correlation Functions. As was the case for random variables it is often convenient to work with an algebraic reshuffling of the correlation functions. In particular, the generating functional for connected correlation functions is defined by

$$\langle e^{i(s,X)} - 1\rangle_c = \ln\langle e^{i(s,X)}\rangle$$

or

$$\langle e^{i(s,X)}\rangle = e^{\langle e^{i(s,X)}-1\rangle_c}$$

for all s. Relations among correlation functions follow from this rule. Three examples that follow from this definition are

$$\begin{aligned} \langle X(t_1)\rangle_c &= \langle X(t_1)\rangle\,, \\ \langle X(t_1)X(t_2)\rangle_c &= \langle X(t_1)X(t_2)\rangle - \langle X(t_1)\rangle\langle X(t_2)\rangle\,, \end{aligned}$$

and, in the case that $\langle X(t)\rangle = 0$,

$$\begin{aligned} \langle X(t_1)X(t_2)X(t_3)X(t_4)\rangle &= \langle X(t_1)X(t_2)X(t_3)X(t_4)\rangle_c \\ &\quad + \langle X(t_1)X(t_2)\rangle\langle X(t_3)X(t_4)\rangle \\ &\quad + \langle X(t_1)X(t_3)\rangle\langle X(t_2)X(t_4)\rangle \\ &\quad + \langle X(t_1)X(t_4)\rangle\langle X(t_2)X(t_3)\rangle\,. \end{aligned}$$

Observe that if X and Y are independent stochastic variables, then the connected correlation functions corresponding to the stochastic variable $X+Y$ are simply the sum of the connected correlation functions of the two stochastic variables individually. In other words, if X and Y are independent, then any combination of the form

$$\langle X(t_1)\cdots X(t_p)Y(t_{p+1})\cdots Y(t_m)\rangle_c \equiv 0$$

whenever $1 \leq p < m$.

Stationary Stochastic Processes. A stationary stochastic process is one for which all the correlation functions are invariant under an arbitrary overall time translation, i.e.

$$\langle X(t_1+\tau)X(t_2+\tau)\cdots X(t_m+\tau)\rangle = \langle X(t_1)X(t_2)\cdots X(t_m)\rangle$$

holds (as distributions if necessary) for all $m \geq 1$, all t_j and all τ. Stationary processes are defined for all time ($I = \mathbf{R}$). If X depends on additional arguments, such as spacial coordinates, then the concept of stationarity extends in a natural way to these variables as well.

Let $s_\tau(t) \equiv s(t-\tau)$, then stationarity of a stochastic process is encoded in the characteristic functional by the equation

$$C\{s_\tau\} = C\{s\}$$

which should hold for all $\tau \in \mathbf{R}$ and a dense set of s.

Observe that these equalities do *not* mean that $X(t+\tau)$ *equals* $X(t)$ for all τ, but rather that these two variables are *stochastically equivalent*, i.e. their *statistical properties* are equal.

Ergodicity. Ergodicity refers to the equivalence of ensemble averages and the limit of long-time averages of stationary stochastic processes. In other words temporal averages equal ensemble averages (for stationary ensembles). The key idea involves the independence of two sets of random variables as their temporal separation diverges, namely,

$$\begin{aligned}\lim_{|\tau|\to\infty} &\langle X(t_1)\cdots X(t_p)X(t_{p+1}+\tau)\cdots X(t_m+\tau)\rangle \\ &= \langle X(t_1)\cdots X(t_p)\rangle\langle X(t_{p+1})\cdots X(t_m)\rangle\end{aligned}$$

holds for all $t_j \in \mathbf{R}$ and $1 \leq p < m$. In terms of the characteristic function it follows that

$$\lim_{|\tau|\to\infty} C\{r+s_\tau\} = C\{r\}C\{s\}$$

for a dense set of test functions r and s. A stochastic process for which such a factorization occurs may be said to satisfy *clustering*.

Let us exploit clustering to examine long-time averaging. Let

$$W(T,s) \equiv \frac{1}{2T}\int_{-T}^{T} d\tau e^{i\int s(t+\tau)X(t)dt},$$

and let us examine the convergence of $W(T,s)$ as $T \to \infty$. It follows that

$$\langle |W(T,s) - C\{s\}|^2\rangle = \frac{1}{4T^2}\int_{-T}^{T}\int_{-T}^{T} d\tau_1 d\tau_2 \left[C\{s_{\tau_1} - s_{\tau_2}\} - |C\{s\}|^2\right].$$

Since

$$\lim_{T\to\infty}\frac{1}{4T^2}\int_{-T}^{T}\int_{-T}^{T}d\tau_1 d\tau_2 C\{s_{\tau_1}-s_{\tau_2}\} = \lim_{T\to\infty}\frac{1}{4T^2}\int_{-T}^{T}\int_{-T}^{T}d\tau_1 d\tau_2 C\{s_{\tau_1}\}C\{-s_{\tau_2}\} = |C\{s\}|^2,$$

it follows that

$$\lim_{T\to\infty} W(T,s) = C\{s\}$$

in the sense that

$$\lim_{T\to\infty}\langle |W(T,s)-C\{s\}|^2\rangle = 0\,.$$

It must be emphasized that convergence is not absolutely assured but occurs only with probability one, i.e., not everywhere but almost everywhere (apart from a set of paths of measure zero). Nonetheless, for ergodic ensembles the condition $W(T,s)\to C\{s\}$ provides the strictest support conditions on the functional measure, as we shall see.

2.3. Examples of Stochastic Processes

Gaussian Stochastic Processes. Gaussian stochastic processes provide a rich source of important examples, and these, in turn, are defined by the first two connected correlation functions with all higher-order connected correlation functions vanishing. In particular, the characteristic functional for a Gaussian process has the general form

$$C\{s\} = e^{i\int s(t)\langle X(t)\rangle dt - \frac{1}{2}\int s(t_1)s(t_2)\langle X(t_1)X(t_2)\rangle_c dt_1 dt_2}$$

and is completed determined by the two functions

$$m(t) \equiv \langle X(t)\rangle\,,$$
$$c(t_1;t_2) \equiv \langle X(t_1)X(t_2)\rangle_c\,,$$

which are assumed to be distributions. A *normal* stochastic process is one for which $\langle X(t)\rangle = 0$ for all t, and in that case

$$\langle X(t_1)X(t_2)\rangle_c = \langle X(t_1)X(t_2)\rangle\,.$$

A *stationary* Gaussian process is one for which

$$m(t) = \langle X(t)\rangle = \langle X(0)\rangle \equiv m$$
$$c(t_1;t_2) = \langle X(t_1)X(t_2)\rangle_c = \langle X(t_1-t_2)X(0)\rangle_c \equiv c(t_1-t_2)\,.$$

In as much as

$$\int s(t_1)s(t_2)c(t_1;t_2)dt_1 dt_2 \geq 0$$

for arbitrary real functions s, it follows for a stationary process that, as a distribution,

$$c(t) = \int e^{i\omega t} d\rho(\omega)$$

where ρ is a positive measure, possibly infinite. Note this relation holds true for any stochastic process the two-point correlation function of which is time translation invariant.

The nature of the allowed test functions s is determined by the measure ρ (since $|m| < \infty$ it does not limit the test functions). In particular, the bound

$$\int s(t_1)s(t_2)c(t_1 - t_2)dt_1 dt_2 = \int |\tilde{s}(\omega)|^2 d\rho(\omega) < \infty$$

defines acceptable test functions. If $\int d\rho(\omega) < \infty$ then $\tilde{s}(\omega) \equiv be^{-i\omega t_0}/\sqrt{2\pi}$, or $s(t) = b\delta(t - t_0)$ are acceptable functions for any b and t_0, in which case the Gaussian process is pointwise defined. However, if $\int d\rho(\omega) = \infty$, then such functions are *not* acceptable and the process is a generalized stochastic process.

Although the process is stationary it may or may not be ergodic as well. Since the measure ρ refers to the connected two-point function, ergodicity requires that

$$\begin{aligned}\lim_{\tau\to\pm\infty} &\int r(t)s_\tau(u)c(t-u)dtdu \\ &= \lim_{\tau\to\pm\infty} \int \tilde{r}^*(\omega)\tilde{s}(\omega)e^{i\tau\omega}d\rho(\omega) \\ &= 0\end{aligned}$$

for a dense set of test functions r and s. If the measure ρ has a discrete component, as described in Sec. 1.2, then the desired limit does not hold and the process is clearly *not* ergodic [take $d\rho(\omega) = \delta(\omega - \omega_0)d\omega$, for example]. The same conclusion also holds if the process has a singular continuous component; for example, let $\rho(\omega) = \hat{C}(\omega)$, the Cantor function defined in Sec. 1.2. Then a dense set of $\tilde{r}$ and $\tilde{s}$ may well include those that are, e.g., constant in the interval $0 \le \omega \le 1$, and so the present discussion reduces exactly to that of the previous Section.

However, suppose the process is purely absolutely continuous. In that case $d\rho(\omega) = \rho'(\omega)d\omega$ where $\rho'(\omega) \ge 0$ and is locally integrable. Since acceptable test functions satisfy $\int |\tilde{s}(\omega)|^2\rho'(\omega)d\omega < \infty$, it follows that $|\tilde{s}(\omega)|^2\rho'(\omega) \in L^1(\mathbf{R})$. Consequently,

$$\lim_{\tau\to\pm\infty} \int \tilde{r}^*(\omega)\tilde{s}(\omega)e^{i\omega\tau}\rho'(\omega)d\omega = 0$$

as a direct consequence of the Riemann-Lebesgue lemma.

In summary, then, we observe that ergodicity holds for a stationary Gaussian distribution if and only if the spectral measure ρ of the two-point function is strictly absolutely continuous. Observe that this conclusion is independent of whether the process is point-wise defined [$\int d\rho(\omega) < \infty$] or not point-wise defined [$\int d\rho(\omega) = \infty$].

White Noise. A normal stochastic process of fundamental importance is *white noise*. White noise is a mean-zero Gaussian with a δ-correlated two-point function, i.e.,

$$\begin{aligned}\langle N(t)\rangle &= 0\,, \\ \langle N(t_1)N(t_2)\rangle &= c\delta(t_1 - t_2)\end{aligned}$$

where we have introduced N as the stochastic variable to stand for white noise. If the positive constant $c = 1$, then one arrives at *standard white noise*. For white noise, $d\rho(\omega) = c\,d\omega/2\pi$ which implies that N is an ergodic generalized (normal) stochastic process. The all important characteristic functional for white noise is then given by

$$C_{WN}\{s\} = e^{-\frac{1}{2}c\int s^2(t)dt}$$

Acceptable test functions s in this case are such that $s \in L^2(\mathbf{R})$.

All correlation functions follow from the characteristic functional and in particular we observe that

$$\begin{aligned}&\langle N(t_1)N(t_2)N(t_3)N(t_4)\rangle \\ &= c^2[\delta(t_1-t_2)\delta(t_3-t_4)+\delta(t_1-t_3)\delta(t_2-t_4)+\delta(t_1-t_4)\delta(t_2-t_3)]\,,\end{aligned}$$

and moreover that

$$\langle N(t_1)N(t_2)N(t_3)N(t_4)\rangle_c = 0$$

as do all connected correlation functions except the two-point function. One says that N is a *δ-correlated stochastic process.*

The white noise characteristic functional admits a functional Fourier representation based on a white-noise functional measure,

$$e^{-\frac{1}{2}c\int s^2(t)dt} = \int e^{i(s,\Lambda)}d\mu(\Lambda)$$

and many properties of μ may be deduced from this relation. As is implicit in Sec. 1.7, the variables Λ are distributions and the support of μ is contained in the space of Λ for which $(\Lambda, T\Lambda) < \infty$, where T is an arbitrary positive trace class operator. In particular, if $\{h_n(t)\}_{n=1}^{\infty}$ denotes a complete real orthonormal set of functions on $\mathbf{R}$, and $\{p_n\}_{n=1}^{\infty}$ is a set of real numbers such that $p_n > 0$ for all n and $\Sigma p_n < \infty$, then the support of μ is concentrated on those Λ for which

$$\textstyle\sum_{n=1}^{\infty} p_n(h_n,\Lambda)^2 < \infty\,.$$

There are other sharper support conditions for μ which are discussed below.

2.4. Measures for Stochastic Processes

Every characteristic functional is given, in turn, as the functional Fourier transform of some nonnegative normalized measure on the space of generalized functions, namely by

$$C\{s\} = \int e^{i(s,\Lambda)}d\mu(\Lambda)\,.$$

In many cases an indirect approach to the measure μ is useful. In this method attention is focused on a finite dimensional subspace of test functions, e.g.,

$$s_{(N)}(t) = \sum_{n=1}^{N} h_n(t)(h_n,s)\,,\quad N<\infty\,,$$

where $\{h_n\}$ are a set of real orthonormal functions and $s_n = (h_n, s)$ corresponding real variables. The result is a characteristic function for N variables $C(s_{(N)})$ which admits a Fourier representation

$$C(s_{(N)}) = \int e^{i\Sigma_1^N s_n\Lambda_n}d\mu_N(\Lambda_N)$$

where $\Lambda_n \equiv (h_n, \Lambda)$. By construction

$$\lim_{N\to\infty} C(s_{(N)}) = C\{s\},$$

and, in turn, the measures μ_N converge weakly to the measure μ. As measures on finite-dimensional spaces the μ_N are standard measures having, in general, discrete, singular continuous, and absolutely continuous components. The measures μ_N are conventional and easy to deal with, at least in principle. On the other hand, the measures μ are far less easy to appreciate due to their dependence on an infinite number of variables.

One is tempted to set

$$d\mu(\Lambda) = w(\Lambda)\mathcal{D}\Lambda$$

in analogy with the absolutely continuous case and in which $\mathcal{D}\Lambda$ is the infamous flat "measure" on function space. Such relations are not in principle inappropriate and typically take the form

$$d\mu(\Lambda) = \rho(\Lambda)d\sigma(\Lambda)$$

where the measure μ assigns a zero measure to all sets for which the measure σ assigns a zero measure. In that case $\rho(\Lambda)$ is denoted by $d\mu(\Lambda)/d\sigma(\Lambda)$ and is referred to as the Radon-Nikodym derivative. One says that μ is *absolutely continuous with respect to* σ. If additionally σ is absolutely continuous with respect to μ then the two measures are said to be *quasi-equivalent.* On the other hand, if neither μ nor σ is absolutely continuous with respect to the other, then the two measures are *mutually singular*, a common and important case, for which, in essence, the measures "live" on (partially) disjoint sets.

With the foregoing as background, let us return to

$$d\mu(\Lambda) = w(\Lambda)\mathcal{D}\Lambda$$

and examine why this relation is strictly speaking inappropriate but nonetheless is heuristically of considerable utility. The issue centers on whether or not $\mathcal{D}\Lambda$ is a measure. It makes little difference in this question whether the range of Λ is finite or infinite and we will illustrate both cases. We formally interpret $\mathcal{D}\Lambda = \prod_{n=1}^{\infty} d\Lambda_n$, an infinite product of Lebesgue measures, where each Λ_n lies in the interval $-\infty \leq a < \Lambda_n < b \leq \infty$, $a < b$. Observe that such a "measure" leads naturally to a ***nonseparable*** space of square integrable functions, as may be illustrated by giving a continuous set of mutually orthogonal normalized square integrable "functions". One set of such "functions" is labeled by the continuous variable ϵ, $0 \leq \epsilon < b$, where $(a = 0,\ 0 < b < \infty)$,

$$\begin{aligned} f_\epsilon\{\Lambda\} &= \textstyle\prod_{n=1}^{\infty}(b-\epsilon)^{-\frac{1}{2}}, \quad \textit{if}\ \ 0 \leq \Lambda_n \leq b-\epsilon,\ \textit{for all}\ n \\ &= 0, \quad \textit{otherwise}. \end{aligned}$$

A clear indication of the formal nature here is the vanishing or diverging product defining f_ϵ (save for $b - \epsilon = 1$); these concepts make sense as direct products rather than ordinary products. To pick out a separable subspace one must choose a fixed, nonnegative, normalized integrable element as a "base" from which finitely many changes are made, and which is then completed by including Cauchy sequences. But this very "base" along with the flat "measure" is sufficient to define a measure $d\sigma(\Lambda)$!

For example, one common "base" is the uniform weight $(b-a)^{-1}$ on each Λ_n variable, assuming still that $b-a<\infty$. Choosing this example, let us consider sets of the kind

$$B(\beta,\alpha;P)=\{a\leq\alpha_p\leq\Lambda_p\leq\beta_p\leq b\,,p=1\,,\ldots,P<\infty\}$$

with Λ_n, $n>P$, unrestricted. To this particular set one assigns the measure

$$m(\beta,\alpha;P)=\textstyle\prod_{p=1}^{P}(\beta_p-\alpha_p)/(b-a)\,,$$

which may be extended naturally to other sets. In this sense one may give rigorous meaning to an infinite product of normalized Lebesgue measures. Since $b-a<\infty$ the resultant measure is not translation invariant, although even that invariance may be partially restored if each Λ_n space has the topology of the circle and the functions to be integrated have the required periodicity.

For an infinite interval the "base" element cannot be uniform. A standard example for the real line ($a=-\infty$, $b=\infty$) is to use a formal "base" element

$$\textstyle\prod_{n=1}^{\infty}(2\pi\sigma^2)^{-1/2}\exp(-\Lambda_n^2/2\sigma^2)\,.$$

Observe, if $2\pi\sigma^2>1$, this product "vanishes" for all Λ values. Nevertheless, the essence of this example is contained in the following correct characterization. In the present case, consider sets of the kind

$$B(\beta,\alpha;P)=\{-\infty\leq\alpha_p<\Lambda_p<\beta_p\leq\infty\,,p=1\,,\ldots,P<\infty\}$$

with Λ_n, $n>P$, unrestricted. To this particular set assign the measure

$$m(\beta,\alpha;P)=\prod_{p=1}^{P}\left\{\int_{\alpha_p}^{\beta_p}e^{-y^2/2\sigma^2}\,dy/\sqrt{2\pi\sigma^2}\right\}$$

which may be extended naturally to other sets. In this sense rigorous meaning is given to an infinite product of normalized normal measures.

Although $\mathcal{D}\Lambda$ is not a countably additive measure and the associated multiplier $w(\Lambda)$ may "vanish" or "diverge" for sets of relevant Λ, there is nevertheless some merit in the use of such formal expressions. In many ways, $w(\Lambda)\mathcal{D}\Lambda$ may expose the dependence of $d\rho(\Lambda)$ on certain parameters more clearly than otherwise would be the case. Moreover, the measure $d\rho$ can be approached as the (weak) limit of a sequence of measures $w_N(\Lambda_N)\prod_{n=1}^{N}d\Lambda_n$. One danger of using $w(\Lambda)\mathcal{D}\Lambda$ in place of $d\rho(\Lambda)$ is the possible misconception that $w(\Lambda)$, or some function of $w(\Lambda)$, for example, may be really used separately. If we firmly swear to keep the symbols $w(\Lambda)$ and $\mathcal{D}\Lambda$ always linked together—as in the form $w(\Lambda)\mathcal{D}\Lambda$—then this "single" combined symbol is nothing but an alternative notation for $d\rho(\Lambda)$; nothing more, nothing less. Of course, that doesn't prevent us from *discussing*—but not *using*—either $w(\Lambda)$ or $\mathcal{D}\Lambda$ separately.

Having argued in favor of the notation $w(\Lambda)\mathcal{D}\Lambda$ as an alternative to $d\rho(\Lambda)$ we must strongly stress that there always must be a well defined measure ρ *first*. One cannot use $w(\Lambda)\mathcal{D}\Lambda$ to *define* ρ; it must always be the other way round!

Measures on function spaces often have properties different than their finite-dimensional analogs. For example, the measures represented, for $N < \infty$, by

$$\prod_{n=1}^{N} e^{-x_n^2/2\sigma^2} dx_n/\sqrt{2\pi\sigma^2} \,,$$

are quasi-equivalent for any two different values of σ. On the other hand, the measures represented by the same formal expression, but for $N = \infty$, are mutually singular for any two different values of σ. This fact does not invalidate the notation, but merely serves to suggest caution at interpreting the meaning of parameters in such expressions in the $N = \infty$ case in the same way they are interpreted when $N < \infty$.

The support properties of the preceding example have already been discussed in Sec. 1.7. In particular, in the case $N = \infty$, the measure is concentrated on those sequences for which $\sum_1^\infty p_n x_n^2 < \infty$ for any sequence where $p_n > 0$ and $\sum_1^\infty p_n < \infty$. This condition holds for any value of σ. Tighter support conditions also hold because of the ergodic nature of the underlying ensemble. Specifically, it follows that

$$\begin{aligned}\lim_{M\to\infty} \frac{1}{M} \sum_{m=1}^{M} e^{i\sum_1^\infty s_n x_{n+m-1}} &= C\{s\} \\ &= e^{-\frac{\sigma^2}{2}\sum_1^\infty s_n^2}\end{aligned}$$

holds with probability one. Consider carefully what this relation means. The left hand side is constructed out of almost any sequence $\{x_n\}_{n=1}^\infty$ supported by the appropriate measure, and the limit in question has the value of the right hand side, which is *not* stochastic. Moreover, in the present example almost every sequence $\{x_n\}$ is able to generate its own value of σ! Thus, of necessity, the measures for two different σ values have disjoint support, i.e., are mutually singular. The preceding relation has many consequences. One standard one follows if we let $s_n = s\delta_{n1}$ and expand both sides to order s^2. We then learn that

$$\lim_{M\to\infty} \frac{1}{M} \sum_{m=1}^{M} x_m^2 = \sigma^2$$

holds with probability one, signifying, roughly speaking, that the support of μ lies on an infinite-dimensional sphere whose average (radius)^2 is σ^2. From this fact alone it follows that the space $\ell^2 = \left\{\{x_n\} : \sum_1^\infty x_n^2 < \infty\right\}$ is of measure zero in these examples.

The consequences derived above for the normal distributions hold, then, in the case of white noise expressed in sequence form, and show that for two unequal values of the parameter c the associated white-noise measures are mutually singular.

If one deals with $\Lambda(t)$ rather than Λ_n then much of the preceding discussion carries over to the new case. We consider directly the most general case. For a point-wise defined process introduce sets of the kind

$$B_P(\beta,\alpha;t) = \{\alpha_p \le \Lambda(t_p) \le \beta_p \,, p = 1,\ldots,P < \infty,\ t_p \in \mathbf{R},\ t_p \ne t_{p'},\ if\ p \ne p'\}$$

and to this particular set assign the measure

$$m_P(\beta,\alpha;t) = \left\{\int_\alpha^\beta dm_P(y;t)\,, y\in \mathbf{R}^P\,, P<\infty\right\}$$

for some compatible set of finite and suitably normalized measures $m_P, P = 1,2,3,\ldots,$ depending on the time set $t = (t_1,\ldots,t_P)$ as well.

A useful expression to represent $m_P(\beta,\alpha;t)$ is given by

$$m_P(\beta,\alpha;t) = \int_\alpha^\beta \langle \textstyle\prod_{p=1}^P \delta(X(t_p)-y_p)\,\rangle\, d^P y\,,$$

an expression that shows the manifestly nonnegative character of the probability density and implicitly contains the requisite compatibility conditions. In case the process is not point-wise defined then the preceding formula should be generalized to read

$$m_P(\beta,\alpha;f_P) = \int_\alpha^\beta \langle \int \textstyle\prod_{p=1}^P \delta(X(t_p)-y_p) f_P(t) d^P t\,\rangle\, d^P y\,.$$

The support properties of the underlying functional measure are easily characterized when the process is ergodic. In particular, it follows from Sec. 2.2 that

$$\lim_{T\to\infty}\frac{1}{2T}\int_{-T}^{T} d\tau e^{i\int s(t+\tau)X(t)dt} = C\{s\}$$

holds with probability one. In other words, support of the functional measure μ is concentrated on those stochastic variables $X(\cdot)$ that satisfy the preceding equation. Observe, then, that any two processes for which the characteristic functionals are not identical necessarily have mutually singular functional measures. If two processes have identical characteristic functionals they are stochastically equivalent, and evidently they have equal functional measures. The preceding remarks are the strongest statements about the support of measures for ergodic processes that are possible.

2.5 Brownian Motion and Related Processes

Definition. Without doubt the most important stochastic process is that associated with Brownian motion, which also is generally referred to as a Wiener process. In brief, a *Wiener process* $W^c_{x_0;t_0}(t), t\geq t_0$, may be defined as the time integral of a standard white noise $N(\tau)$ times a factor c and shifted by x_0, i.e.,

$$W^c_{x_0;t_0}(t) = c\int_{t_0}^t N(\tau)d\tau + x_0\,.$$

Observe that N is a generalized stochastic process with test functions, as noted previously, which are arbitrary square-integrable functions. Since the indicator function $\chi_{t;t_0}(\tau) = 1$ if $t_0\leq\tau<t$, and zero otherwise is such a function, it follows that $W^c_{x_0;t_0}$ is a well-defined stochastic variable. As the linear superposition of jointly Gaussian variables, $W^c_{x_0;t_0}$ is likewise a Gaussian variable. In general, it suffices to consider a *standard Wiener process* for which $c=1$ and $t_0=x_0=0$, denoted simply by W, and defined by

$$W(t) = \int_0^t N(\tau)d\tau ,$$

which in fact is a normal stochastic process. Hereafter, we generally confine our attention to standard Wiener variables.

As a Gaussian stochastic process, W is characterized by the fact that

$$W(0) = 0 , \qquad \langle W(t) \rangle = 0 ,$$

$$\langle W(t)W(u) \rangle = \min(t, u) ,$$

the latter following from the relation

$$\begin{aligned} \langle W(t)W(u) \rangle &= \int_0^t \int_0^u \langle N(t')N(u') \rangle dt'du' \\ &= \int_0^t \int_0^u \delta(t' - u')dt'du' \\ &= \int_0^{\min(t,u)} dt' . \end{aligned}$$

Although N is a stationary stochastic process, W is not stationary. Observe that $W(0) = 0$, a fixed value, and one says that the standard Wiener process is *pinned* at $t = 0$.

A great many important properties about W follow from the form of the covariance function and the fact that W is a normal variable. It follows that

$$\begin{aligned} \langle [W(t) - W(u)]^2 \rangle &= \langle W^2(t) \rangle + \langle W^2(u) \rangle - 2\langle W(t)W(u) \rangle \\ &= t + u - 2 \min(t, u) \\ &= |t - u| , \end{aligned}$$

and since the connected four-point correlation function vanishes it follows that

$$\langle [W(t) - W(u)]^4 \rangle = 3|t - u|^2 ,$$

and

$$\begin{aligned} &\langle [W(t_4) - W(t_3)]^2 [W(t_2) - W(t_1)]^2 \rangle \\ &= \langle [W(t_4) - W(t_3)]^2 \rangle \langle [W(t_2) - W(t_1)]^2 \rangle \\ &= |t_4 - t_3||t_2 - t_1| \end{aligned}$$

provided that $t_4 \geq t_3 \geq t_2 \geq t_1$. The latter result evinces an aspect of independence of disjoint increments, a fact that is already obvious from the relations

$$W(t_4) - W(t_3) = \int_{t_3}^{t_4} N(\tau)d\tau \equiv W_{43} ,$$

$$W(t_2) - W(t_1) = \int_{t_1}^{t_2} N(\tau)d\tau \equiv W_{21} ,$$

since $N(\tau)$ is δ-correlated and therefore stochastically independent for each τ. Based on the characteristic functional for white noise it follows that

$$\langle e^{i(rW_{43}+sW_{21})}\rangle = e^{-\frac{1}{2}(r^2|t_4-t_3|+s^2|t_2-t_1|)}$$

establishing the independence of these two increment pairs. It is clear from this construction than any number of disjoint increments are also independent. Indeed Brownian motion is just one of a number of *independent increment processes.*

Continuity Constraints. Let us take up the question of the nature of the Brownian paths. For any normal variable X it follows that

$$\begin{aligned}\langle e^{-\frac{1}{2}X^2}\rangle &= (2\pi)^{-\frac{1}{2}}\int e^{-\frac{1}{2}s^2}\langle e^{isX}\rangle ds\\ &= (2\pi)^{-\frac{1}{2}}\int e^{-\frac{1}{2}s^2}e^{-\frac{1}{2}s^2\langle X^2\rangle}ds\\ &= \left[1+\langle X^2\rangle\right]^{-\frac{1}{2}}.\end{aligned}$$

Thus if $\langle X^2\rangle = \infty$, it follows that $\langle e^{-\frac{1}{2}X^2}\rangle = 0$, i.e., $X^2 = \infty$, a.e. Likewise if $\langle X^2\rangle = 0$ then clearly $X = 0$, a.e. Let us apply these facts to the increments of a Wiener process.

Let $X = |t-u|^{-\beta}[W(t)-W(u)]$ and observe that $\langle X^2\rangle = |t-u|^{1-2\beta}$. If $\beta < \frac{1}{2}$, then $X^2 \to 0$, a.e., as $u \to t$, while if $\beta > \frac{1}{2}$, then $X^2 \to \infty$, a.e., as $u \to t$. Stated otherwise it follows, for $\epsilon > 0$ and $|t-u| \leq 1$, that

$$|W(t)-W(u)| \leq C_\epsilon |t-u|^{\frac{1}{2}-\epsilon}, \quad |t-u| \leq 1,$$

holds for some stochastic variable C_ϵ with probability *one*, while

$$|W(t)-W(u)| \leq D_\epsilon |t-u|^{\frac{1}{2}+\epsilon}, \quad |t-u| \leq 1,$$

holds for some stochastic variable D_ϵ with probability *zero.*

These relations establish, with probability one, that the Wiener paths are *everywhere continuous* but are *nowhere differentiable.*

Invariance Transformations of Brownian Path Ensembles. Certain transformations of Wiener processes yield Wiener processes again. These amount to nothing more than a reshuffling of the sample paths that make up the ensemble of Brownian motion paths. Recall that the criteria for a Gaussian process to be a standard Wiener process are simply $W(0) = 0$, $\langle W(t)\rangle = 0$, and $\langle W(t)W(u)\rangle = \min(t,u)$. One elementary example is to let $W(t) \to \overline{W}(t) = -W(t)$. Clearly $\overline{W}$ satisfies the requisite conditions to be a standard Wiener process. One may say that $W(t) \to -W(t)$ is an *invariance* of the Wiener process ensemble.

Another invariance relates to a time translation and a reset to zero. Let $\tau > 0$ and consider $W(t) \to \tilde{W}(t) = W(t+\tau) - W(\tau)$. Then clearly $\tilde{W}(0) = 0$, $\langle\tilde{W}(t)\rangle = 0$, and

$$\begin{aligned}\langle\tilde{W}(t)\tilde{W}(u)\rangle &= \langle W(t+\tau)W(u+\tau)\rangle - \langle W(t+\tau)W(\tau)\rangle\\ &\quad - \langle W(\tau)W(u+\tau)\rangle + \langle W(\tau)W(\tau)\rangle\\ &= \min(t+\tau, u+\tau) - \tau\\ &= \min(t,u).\end{aligned}$$

The implication of this invariance is that Brownian motion which at any future time τ has its clock restarted and is reset to a zero value will again reproduce Brownian motion.

Consider the previous process run backwards, namely $W(t) \to \breve{W}(t) = W(\tau - t) - W(\tau)$, in the interval $0 \le t \le \tau$. In that interval $\breve{W}(0) = 0$, $\langle \breve{W}(t) \rangle = 0$, and

$$\begin{aligned}\langle \breve{W}(t)\breve{W}(u) \rangle &= \min(\tau - t, \tau - u) - \tau + t - \tau + u + \tau \\ &= \min(t, u)\,,\end{aligned}$$

so (within the indicated time span) Brownian motion run backwards in time is again Brownian motion.

Yet another invariance is given by

$$W(t) \to W_k(t) = kW(t/k^2)\,, \quad k \neq 0\,.$$

Since $W_k(0) = 0$, $\langle W_k(t) \rangle = 0$, and $\langle W_k(t)W_k(u) \rangle = k^2 \min(t/k^2, u/k^2) = \min(t, u)$ it follows that $W(t) \to kW(t/k^2)$ is also an invariance of the ensemble. This invariance relates to the appearance of similar fluctuation behavior on all time scales.

Lastly we illustrate the transformation

$$W(t) \to \hat{W}(t) = tW(t^{-1})$$

for which $\langle \hat{W}(t)^2 \rangle = t$ implies $\hat{W}(0) = 0$ with probability one, $\langle \hat{W}(t) \rangle = 0$, and lastly

$$\begin{aligned}\langle \hat{W}(t)\hat{W}(u) \rangle &= tu \min(t^{-1}, u^{-1}) \\ &= \min(t, u)\,.\end{aligned}$$

Hence the ensemble also exhibits invariance under $W(t) \to tW(t^{-1})$.

Probability Distribution of Brownian Paths. The characteristic function for $W(t)$ follows immediately from the relation

$$\begin{aligned}\left\langle e^{isW(t)} \right\rangle &= \left\langle e^{is \int_0^t N(\tau)d\tau} \right\rangle = e^{-\frac{1}{2}s^2 t}\,, \quad t \ge 0 \\ &= \int e^{isx} p(x, t)dx\,,\end{aligned}$$

which implies that

$$\begin{aligned}p(x, t) &= \frac{1}{2\pi} \int e^{-isx - \frac{1}{2}s^2 t} ds \\ &= \frac{1}{\sqrt{2\pi t}} e^{-x^2/2t}\,.\end{aligned}$$

This is a distribution that starts out as $\delta(x)$ at $t = 0$ and continually spreads out with a normal distribution (as a normal variable, it could hardly be otherwise!) with a variance given by t.

In like manner it follows for $t > u > 0$ that

$$\begin{aligned}\left\langle e^{is[W(t) - W(u)]} \right\rangle &= \left\langle e^{is \int_u^t N(\tau)d\tau} \right\rangle = e^{-\frac{1}{2}s^2(t-u)} \\ &= \int e^{isx} p(x, t-u)dx\,,\end{aligned}$$

where, as before,

$$p(x,t-u)=\frac{1}{\sqrt{2\pi(t-u)}}e^{-x^2/2(t-u)}\,.$$

This result may be restated as

$$p(x,t-u)=\langle\delta(W(t)-W(u)-x)\rangle\,,$$

which leads to the value already obtained.

Joint distributions are no harder to find thanks to the independence of disjoint increments. Define the multiple *joint probability distribution* as (using $x_0=t_0=0$)

$$\begin{aligned}
&p_K(x_K,t_K;\cdots;x_2,t_2;x_1,t_1)\\
&=\langle\textstyle\prod_{k=1}^{K}\delta(W(t_k)-x_k)\rangle\\
&=\langle\textstyle\prod_{k=1}^{K}\delta(W(t_k)-W(t_{k-1})-x_k+x_{k-1})\rangle\\
&=\textstyle\prod_{k=1}^{K}\langle\delta(W(t_k)-W(t_{k-1})-x_k+x_{k-1})\rangle\\
&=\textstyle\prod_{k=1}^{K}[2\pi(t_k-t_{k-1})]^{-\frac{1}{2}}\exp\left[-\dfrac{(x_k-x_{k-1})^2}{2(t_k-t_{k-1})}\right]\,,
\end{aligned}$$

where we have introduced $x_0\equiv 0$ and $t_0\equiv 0$, corresponding to the starting position and time of a standard Wiener process. For a nonstandard Wiener process (but with $c=1$) this same formula holds for any value of x_0 and t_0.

Wiener Measure. The joint probability distribution derived above leads to a unique functional measure on Brownian motion processes. This measure is not unrelated to that for standard white noise which we describe here in a formal fashion. We set

$$d\mu_{WN}(\Lambda)=\mathcal{N}'e^{-\frac{1}{2}\int\Lambda^2(t)dt}\prod d\Lambda(t)$$

where the integral and the product run over all t values, and $\mathcal{N}'$ is a formal normalization constant. In as much as $\dot{W}(t)=N(t)$ it is entirely natural that the formal expression for a standard Wiener measure is given by

$$d\mu_W(\Lambda)=\mathcal{N}e^{-\frac{1}{2}\int\dot{\Lambda}^2(t)dt}\prod d\Lambda(t)$$

where $\mathcal{N}$ is a new formal normalization constant absorbing the Jacobian of $\Lambda\to\dot{\Lambda}$, which is a Λ-independent formal constant. Here it is understood that $t\geq 0$ and $\Lambda(0)=0$. Observe that our formal prescription is fully in accord with the characteristic functional for Brownian motion given by

$$\begin{aligned}
C\{s\}&=e^{-\frac{1}{2}\int s(t)s(u)\min(t,u)dtdu}\\
&\equiv\int e^{i\int s(t)\Lambda(t)dt}d\mu_W(\Lambda)\,.
\end{aligned}$$

Moreover, the formal expression when given a Riemann-sum-like approximation to the integral in the exponent closely resembles the formula for the joint probability distribution. Clearly, in a formal sense at least, a limit involving the joint probability distribution as $\max|t_{k+1} - t_k| \to 0$ and $K \to \infty$ leads to the formal expression for $d\mu_W(\Lambda)$ in a natural way.

Brownian Bridge. All Wiener processes begin at some value at some initial time, e.g., $x_0 = t_0 = 0$. A subset of all the Wiener paths pass through a fixed x-window at a future time T, and by rescaling the measure appropriately a limit may be taken to extract those paths that also have a fixed value at time T. These are called *pinned Wiener processes* or, more picturesquely, *Brownian bridges*. The simplest and standard such bridge is a normal process that satisfies $B(0) = 0$ and $B(T) = 0$ and is a linear function of a standard Wiener process. Let us choose

$$B(t) = W(t) - (t/T)W(T), \quad 0 \le t \le T.$$

This has the desired features $B(0) = B(T) = 0$ and $\langle B(t)\rangle = 0$. Of course B cannot have the variance of W and still be pinned at time T. The variance of our expression reads

$$\begin{aligned}\langle B^2(t)\rangle &= t - 2t^2/T + t^2/T \\ &= t(1 - t/T),\end{aligned}$$

Of course other candidates such as $W(t) - (t/T)^2W(T)$ would satisfy the basic criteria but would not have a time reversal symmetry that B has. In particular, define

$$\tilde{B}(t) = W(T - t) - [(T - t)/T]W(T), \quad 0 \le t \le T$$

which runs the previous bridge process backwards in time. Observe that $\tilde{B}(0) = 0 = \tilde{B}(T)$, $\langle \tilde{B}(t)\rangle = 0$, and, for $t \le u$, that

$$\begin{aligned}\langle \tilde{B}(t)\tilde{B}(u)\rangle &= (T - u)[1 - (T - t)/T] \\ &= t(1 - u/T)\end{aligned}$$

as required for the appropriate time reversal symmetry.

To make a Brownian bridge that starts at x_0 and ends at x_T it suffices to consider

$$B_{x_T, x_0}(t) \equiv (1 - t/T)x_0 + (t/T)x_T + B(t),$$

a process that also exhibits the appropriate time reversal symmetry. As $T \to \infty$ it follows that $\langle B(t)B(u)\rangle \to \min(t, u)$ and so $B(t) \to W(t)$. This is also evident because the additional term, $(t/T)W(T)$, satisfies $\langle[(t/T)W(T)]^2\rangle = t^2/T \to 0$ implying that as $T \to \infty$ the additional term vanishes almost everywhere. In the same limit $B_{x_T, x_0}(t) \to W_{x_0}(t)$.

If the paths of a Brownian bridge become Brownian as $T \to \infty$ one should also be able to see their Brownian character within the bridge itself if one looks on a fine enough time scale. To do so let $0 < \tau < T$ and $\tau \ge t$, $\tau \ge u$, specifically choosing t and u very close to τ, i.e., $(u + t - 2\tau) \ll T$. In that case it follows that

$$\langle [B(t) - B(\tau)][B(u) - B(\tau)]\rangle = \min(t - \tau, u - \tau) + (t - \tau)(u - \tau)/T,$$

which illustrates Brownian behavior on a fine scale when the second term can be neglected.

Lastly, we illustrate a Brownian bridge that holds for times t such that $T \leq t \leq U$, where $0 < T < U$. In that case

$$\overline{B}(t) = W(t) - [(U-t)/(U-T)]W(T) - [(t-T)/(U-T)]W(U).$$

It is not difficult to show, for $T \leq t \leq u \leq U$, that

$$\langle \overline{B}(t)\overline{B}(u)\rangle = (t-T)[1-(u-T)/(U-T)].$$

If one chooses $\tilde{B}(t) = B(t)$ for $0 \leq t \leq T$ and $\tilde{B}(t) = \overline{B}(t)$ for $T \leq t \leq U$, then one has a process $\tilde{B}$ for $0 \leq t \leq U$ that is pinned at $t = 0$, $t = T$, and $t = U$. Evidently this procedure may be continued for any number of pinnings that may be desired.

Other Processes Constructed from Brownian Motion. Some transformations of Brownian motion paths result in a different ensemble and important examples can often be constructed in this fashion. Two trivial examples serve to illustrate the idea. For $t \geq 0$ let

$$W_{x_0}(t) = W(t) + x_0\,, \quad x_0 \neq 0\,,$$
$$W^c(t) = cW(t), \quad c^2 \neq 1\,.$$

These transformations take a standard Wiener process into nonstandard ones that are completely disjoint from the standard ensembles.

More interesting is the transformation

$$U(t) = e^{-t}W(e^{2t})\,, \quad -\infty < t < \infty\,,$$

which defines a process for all t. It follows that

$$\langle U(t)\rangle = 0\,,$$

and

$$\begin{aligned}\langle U(t)U(s)\rangle &= e^{-(t+s)}\langle W(e^{2t})W(e^{2s})\rangle \\ &= e^{-(t+s)+2\min(t,s)} \\ &= e^{-|t-s|}\,.\end{aligned}$$

Observe that the U-process is Gaussian and (unlike W) it is *stationary*. Moreover, since

$$e^{-|t|} = \frac{1}{\pi}\int \frac{e^{-it\omega}}{1+\omega^2}d\omega$$

it follows that the U-process is *ergodic*. The U-process is an important one and it is known as the *Ornstein-Uhlenbeck process* [16].

Non-Gaussian processes may also be constructed from W as, for example, the process

$$Y(t) = W^3(t)\,,$$

etc. A systematic way to relate two processes, one of which is Brownian, is with the aid of stochastic differential equations, the subject of the next Section.

3. Stochastic Differential Equations, a.k.a. Langevin Equations

3.1 Stochastic Calculus

Elementary Differentials and Integrals. Let $W(t)$, $t \geq 0$, denote a standard Wiener process, which although composed of continuous paths, with probability one, these paths are nondifferentiable. Moreover each path is of unbounded variation so it cannot be used as a (signed) measure in the standard sense. Nevertheless appropriate *stochastic* integrals can be defined. Introduce

$$dW(t) = W(t + dt) - W(t)$$

as a differential. It is clear, however, that this differential is not $O(dt)$ as would hold if the paths were differentiable. Instead we have

$$\begin{aligned} \langle dW(t)^2 \rangle &= \langle [W(t+dt) - W(t)]^2 \rangle \\ &= dt \end{aligned}$$

which shows, roughly speaking, that $dW(t)$ is $O\left(\sqrt{dt}\right)$. This fact will be important in the sequel.

Let us take a lattice of time steps $t_k = k\epsilon, k = 0, 1, 2, \ldots, \epsilon > 0$. Then clearly $(K > J)$

$$W(K\epsilon) - W(J\epsilon) = \sum_{k=J}^{K-1} [W((k+1)\epsilon) - W(k\epsilon)]$$

In the limit that $\epsilon \to 0$ and $K\epsilon = t$, $J\epsilon = u$ are fixed, then

$$W(t) - W(u) = \lim_{\epsilon \to 0} \sum [W((k+1)\epsilon) - W(k\epsilon)] \equiv \int_u^t dW(\tau).$$

If $f(t)$ is a continuous, differentiable function on a finite time interval, then we can define

$$\int f(t) dW(t) \equiv f(t)W(t) \Big| - \int W(t) df(t)$$

namely by integration by parts. Alternatively, we have

$$\int f(t) dW(t) \equiv \lim_{\epsilon \to 0} \int f(k\epsilon) \Big[W((k+1)\epsilon) - W(k\epsilon)\Big].$$

For the example in question the same result arises whether we use $f(k\epsilon)$ or $f\big((k+1)\epsilon\big)$ or an average of the two.

These latter integrals can be viewed alternatively if we introduce $N(t) = dW(t)/dt$ as a generalized stochastic process for which f is a suitable test function. In other words, $\dot{W}(t)$ does exist in the sense of distributions. We shall employ both points of view in this Section where appropriate.

Stochastic Differentials and Integrals. Due to the fluctuating character of Brownian motion paths, differentials and integrals of functions of W are somewhat delicate, a fact already implied by the relation

$$\langle dW^2(t)\rangle = dt\,.$$

Indeed, one can say much more. Consider first the expression

$$\langle [W(t+\Delta)-W(t)]^2 W^2(\tau)\rangle\,,\qquad \Delta>0\,,$$

which is given by

$$\begin{aligned}&= \langle [W(t+\Delta)-W(t)]^2\rangle\langle W^2(\tau)\rangle\\ &\quad + 2\langle [W(t+\Delta)-W(t)]\,W(\tau)\rangle^2\\ &= \Delta\tau + 2\big[\min(t+\Delta,\tau)-\min(t,\tau)\big]^2\\ &= \Delta\tau + 2\Delta^2 b\,,\end{aligned}$$

where $0\le b\le 1$. This division into an $O(\Delta)$ and an $O(\Delta^2)$ contribution persists and, in general, leads to

$$\langle [W(t+\Delta)-W(t)]^2 e^{\int s(\tau)W(\tau)d\tau}\rangle = \Delta\langle e^{\int s(\tau)W(\tau)d\tau}\rangle + O(\Delta^2)\,.$$

It follows, for example, that

$$\begin{aligned}&\lim_{\epsilon\to 0}\sum f(k\epsilon)\langle [W((k+1)\epsilon)-W(k\epsilon)]^2 e^{\int s(\tau)W(\tau)d\tau}\rangle\\ &\qquad = \int f(t)dt\langle e^{\int s(\tau)W(\tau)d\tau}\rangle\end{aligned}$$

from which one concludes, with probability one, that

$$dW(t)^2 = dt$$

as a deterministic fact without the need of an ensemble average. This is one of the celebrated rules of the Itô calculus [6].

Next let us consider the differential of a function of W, $df(W(t))$, as the limit of

$$\Delta f(W(t)) \equiv f(W(t)+\Delta W(t)) - f(W(t))\,,$$

where $\Delta W(t)\equiv W(t+\Delta)-W(t)$, $\Delta>0$. By Taylor's theorem it follows that

$$\Delta f(W(t)) = f'(W(t))\Delta W(t) + \frac{1}{2}f''(W(t))\Delta W(t)^2 + \cdots\,,$$

which in the limit reads

$$df(W(t)) = f'(W(t))dW(t) + \frac{1}{2}f''(W(t))dt\,,$$

another of the celebrated Itô rules.

Although this derivation may seem inevitable one may approach this limit in a different way be taking the average of a Taylor series of both terms. Thus

$$\Delta f(W(t)) = \frac{1}{2}\left[f'(W(t))\Delta W(t) + \frac{1}{2}f''(W(t))\Delta W(t)^2 + \cdots\right]$$
$$+\frac{1}{2}\left[f'(W(t) + \Delta W(t))\Delta W(t) - \right.$$
$$\left. -\frac{1}{2}f''(W(t) + \Delta W(t))\Delta W(t)^2 + \cdots\right].$$

In the limit we have

$$df(W(t)) = f'(W(t)) \circ dW(t)$$

with the second-order terms dropping out. We have added the symbol $\circ$ to denote a different rule, due to ***Stratonovich*** [7], by which differentials are calculated. With the abbreviation $f'(W(k\epsilon)) = f'_k$ and $W(k\epsilon) = W_k$, one writes the two rules as

$$(I) \qquad f'(W(t))dW(t) = \lim_{\epsilon \to 0} f'_k(W_{k+1} - W_k),$$

$$(S) \quad f'(W(t)) \circ dW(t) = \lim_{\epsilon \to 0} \frac{1}{2}\left(f'_{k+1} + f'_k\right)(W_{k+1} - W_k).$$

For functions of continuous *and* differentiable paths these two expressions would be the same; for Brownian paths this is not the case, in general.

The primary advantage of the Stratonovich prescription—also called the *midpoint rule*—is that the ordinary rules of calculus apply. The Itô prescription has some advantages when dealing with integrals. One can always pass from one form to the other form by the relation

$$f'(W) \circ dW = f'(W)dW + \frac{1}{2}f''(W)dt.$$

Perhaps it is worth adding why such ambiguities in the product arise from the point of view of distribution theory. The issue is to define the product, i.e., $f'(W) \cdot \dot{W}$, of two distributions, and this may generally be done in different ways, if it can be done at all. As an elementary example of such an ambiguity consider the product $\theta(t)\delta(t)$ where $\theta(t) = 0$ if $t < 0$ and $\theta(t) = 1$ if $t > 0$. If, for example, we define this product as

$$\theta(t)\delta(t) \equiv \lim_{\epsilon \to 0}\left[\lambda\theta(t)\delta(t-\epsilon) + (1-\lambda)\theta(t)\delta(t+\epsilon)\right]$$

as $\epsilon \to 0$ from positive values, the result is $\theta(t)\delta(t) = \lambda\delta(t)$ which depends sensitively on our chosen definition.

Stochastic Integrals. In elementary calculus the relation

$$f(y(t)) - f(y(u)) = \int_u^t df(y(\tau)) = \int_u^t f'(y(\tau))dy(\tau)$$

holds when f and y are both C^1 functions. For $f \in C^1$ and Wiener process paths W it follows that

$$\int_u^t f'(W(\tau)) \circ dW(\tau) = \int_u^t df(W(\tau)) = f(W(t)) - f(W(u))$$

using the Stratonovich rule. On the other hand the Itô rule implies that

$$\int_u^t f'(W(\tau))dW(\tau) = \int_u^t [df(W(\tau)) - \frac{1}{2}f''(W(\tau))d\tau]$$
$$= f(W(t)) - f(W(u)) - \frac{1}{2}\int_u^t f''(W(\tau))d\tau\,.$$

An important fact about Itô integrals relates to their expectation

$$\begin{aligned}\langle \textstyle\int_u^t f'(W(\tau))dW(\tau)\rangle &= \lim_{\epsilon\to 0}\sum\langle f_k'(W_{k+1}-W_k)\rangle \\ &= \lim_{\epsilon\to 0}\sum\langle W_{k+1}-W_k\rangle\langle f_k'\rangle \\ &= 0\end{aligned}$$

thanks to the fact of independent increments. Indeed such an integral will always have a vanishing expectation whenever f depends on values $W(\sigma)$ for $\sigma \le \tau$. Such functions are called *nonanticipating*. Consider the following example: Let $f'(x) = 2x$ so that

$$2\int_u^t W(\tau)dW(\tau) = [W^2(t) - W^2(u)] - (t-u)$$

Since the expectation of the left side vanishes it follows that

$$\langle W^2(t) - W^2(u)\rangle = t-u\,,$$

which is in fact correct. This identity can be rewritten in a different form if we introduce

$$:W^2(t) :\equiv W^2(t) - t$$

in which case it follows that

$$\langle : W^2(t) :\rangle = \langle : W^2(u) :\rangle$$

which means its value is *independent of time* and can therefore be evaluated at any convenient time such as $t = 0$ where $\langle : W^2(0) :\rangle$ clearly vanishes [since $: W^2(0) := 0$].

If in addition to W the function f has an explicit dependence on time then the preceding formulas are readily generalized. Let $\dot f = \partial f/\partial t$ and $f' = \partial f/\partial x$. Then it follows that

$$\begin{aligned}df(W(t),t) = f'(W(t),t)dW(t) + \frac{1}{2}f''(W(t),t)dt \\ + \dot f(W(t),t)dt\end{aligned}$$

or

$$df(W(t),t) = f'(W(t),t)\circ dW(t) + \dot f(W(t),t)dt\,.$$

Based on the first formula it follows that

$$\begin{aligned}\int_u^t f'(W(\tau),\tau)dW(\tau) &= f(W(t),t) - f(W(u),u) \\ &\quad -\frac{1}{2}\int_u^t f''(W(\tau),\tau)d\tau - \int_u^t \dot f(W(\tau),\tau)d\tau\,.\end{aligned}$$

As an example let $f(x,t) = \exp(x - t/2)$. In that case

$$\int_u^t e^{W(\tau)-\tau/2} dW(\tau) = e^{W(t)-t/2} - e^{W(u)-u/2}$$
$$- \frac{1}{2}\int_u^t e^{W(\tau)-\tau/2} d\tau + \frac{1}{2}\int_u^t e^{W(\tau)-\tau/2} d\tau$$
$$= e^{W(t)-t/2} - e^{W(u)-u/2}\,.$$

Hence, with

$$: e^{W(t)} :\equiv e^{W(t)-t/2}$$

it follows that

$$\langle : e^{W(t)} : \rangle = \langle : e^{W(u)} : \rangle$$

independent of time, and since $: e^{W(0)} := 1$,

$$\langle : e^{W(t)} : \rangle = 1\,.$$

The validity of this result is easily seen when it is recognized that

$$: e^{W(t)} := \frac{e^{W(t)}}{\langle e^{W(t)} \rangle}\,.$$

The notation is no accident. If these relations remind the reader of properties of normal ordered operators in quantum mechanics it is because, in fact, there is a connection.

We have seen two stochastic variables with the property that their mean values are independent of time. Any such stochastic variable $M(t)$ has the property that $\langle M(t) \rangle = \langle M(0) \rangle$ for all $t \geq 0$. This property is one of the important ones associated with stochastic variables called *martingales* [8].

As a final example let $f'(x) = e^x$ which leads to the stochastic integral

$$\int_u^t e^{W(\tau)} dW(\tau) = e^{W(t)} - e^{W(u)} - \frac{1}{2}\int_u^t e^{W(\tau)} d\tau\,.$$

We leave it as an exercise for the reader to verify that the mean of the right hand side vanishes. Note that, unlike the two previous examples, the right hand side does not factor into the form $F(t) - F(u)$ suitable for a martingale.

3.2 Stochastic Differential Equations

As we shall use them stochastic differential equations provide a map from a Wiener process W to another stochastic process X. The connection between the two processes is embodied in a simple-looking equation conventionally written as

$$dX(t) = a\big(X(t)\big)dt + b\big(X(t)\big)dW(t)\,.$$

This relation implicitly determines X as a functional of W. As written the product $b\ dW$ is interpreted in the sense of Itô and b corresponds to a nonanticipating functional so that $\langle dX(t) \rangle = \langle a\big(X(t)\big) \rangle dt$. The term a is referred to as the *drift* since it governs the average change of $dX(t)$.

The proper meaning of this expression is obtained upon integration which leads to

$$X(t) - X(u) = \int_u^t a(X(\tau))d\tau + \int_u^t b(X(\tau))dW(\tau).$$

This result may be arrived at just as well if we introduce the generalized stochastic processes $\dot{W}(t) = dW(t)/dt$ and $\dot{X}(t) = dX(t)/dt$, and the distributional equation

$$\dot{X}(t) = a(X(t)) + b(X(t))\dot{W}(t).$$

These equations are known as *Langevin equations* as he was the first to study such equations (for a special choice of a and b). If the Langevin equation is tested with an indicator function then the well-defined integral equation again emerges.

The equation

$$\dot{X}(t) = a(X(t))$$

is a deterministic one with a unique solution depending on only the initial value $X(u)$. The addition of the term $b(X(t))\dot{W}(t)$ adds a component of *noise* to the equation and thereby to the solution X that disturbs, loosely speaking, the solution of the deterministic equation. Indeed, the term $b\dot{W}$ is referred to as the *noise term.* It is conventional to divide the noise term into two classes: If $b =$ constant (typically unity), then one speaks of *additive noise*; if $b \neq$ constant, then one speaks of *multiplicative noise.*

Under coordinate transformations the form of the stochastic differential equation (or the Langevin equation) will generally change. To see how this comes about let

$$Y(t) = f(X(t)).$$

In that case the Itô calculus leads to

$$dY(t) = f'(X(t))dX(t) + \frac{1}{2}f''(X(t))[dX(t)]^2.$$

Now

$$\begin{aligned}[dX(t)]^2 &= \left[a(X(t))dt + b(X(t))dW(t)\right]^2 \\ &= b^2(X(t))dt\end{aligned}$$

with all other terms vanishing as is evident if they were used in a time integration. These facts, i.e., $dW(t)dt = 0 = ([dW(t)]^3)$ and $(dt)^2 = 0 = ([dW(t)]^4)$, are also part of the Itô calculus rules. With $[dX(t)]^2$ identified it follows that

$$\begin{aligned}dY(t) &= f'(X(t))\left[a(X(t))dt + b(X(t))dW(t)\right] \\ &\quad + \frac{1}{2}f''(X(t))b^2(X(t))dt \\ &\equiv \overline{a}(Y(t))dt + \overline{b}(Y(t))dW(t),\end{aligned}$$

where, with $X(t) = g(Y(t))$,

$$\begin{aligned}\overline{a}(y) &= f'(g(y))a(g(y)) + \frac{1}{2}f''(g(y))b^2(g(y)), \\ \overline{b}(y) &= f'(g(y))b(g(y)).\end{aligned}$$

Observe, if $[b(x)]^{-1}$ exists, that one may always choose f in such a way that $\overline{b} = 1$, at the price, in general, of complicating the drift term.

Let us illustrate this kind of change of variables with an example. Assume that

$$dX(t) = -X(t)dt + dW(t),$$

an important and simple case, which up to constants, was the original example studied by Langevin. While this example is so simple that it is soluble (see below) our purpose here is to discuss its form under a coordinate change. Let $Y(t) = X^3(t)/3$, for which $f'(x) = x^2$ and $f''(x) = 2x$. In this case

$$\begin{aligned} dY(t) &= X^2(t)dX(t) + X(t)dt \\ &= X^2(t)\big[-X(t)dt + dW(t)\big] + X(t)dt \\ &= \big[X(t) - X^3(t)\big]dt + X^2(t)dW(t) \end{aligned}$$

or

$$dY(t) = \big\{[3Y(t)]^{\frac{1}{3}} - 3Y(t)\big\}dt + [3Y(t)]^{\frac{2}{3}}dW(t).$$

Clearly, if performed in reverse, this particular coordinate transformation would be more impressive!

Solution of Elementary Langevin Equations. Let us first discuss the solution of the basic equation $dX = -X\,dt + dW$. As a linear equation it follows that

$$X(t) = e^{-t}X(0) + e^{-t}\int_0^t e^u dW(u),$$

which illustrates an explicit example of X as a nonanticipating functional of W. In this solution $X(0)$ denotes a potentially independent random variable that may be chosen in any desired fashion. One particularly interesting choice is to set

$$X(0) = \int_{-\infty}^0 e^u dW(u)$$

where we have extended $W(u)$ to negative times in the fashion indicated earlier. Conversely, one could also note that $\dot{W}(u) = N(u)$, a standard white noise, which is a stationary process and therefore defined for all time, $-\infty < t < \infty$. With this choice it follows that

$$X(t) = e^{-t}\int_{-\infty}^t e^u dW(u).$$

Clearly X is a *normal process* with a two-point function given by

$$\begin{aligned} \langle X(t)X(s)\rangle &= e^{-t-s}\int_{-\infty}^t\int_{-\infty}^s e^{u+v}\langle dW(u)dW(v)\rangle \\ &= e^{-t-s}\int_{-\infty}^{\min(t,s)} e^{2u}du \\ &= \frac{1}{2}e^{-t-s+2\min(t,s)} \\ &= \frac{1}{2}e^{-|t-s|}, \end{aligned}$$

which is exactly of the form appropriate to the (stationary, ergodic) Ornstein-Uhlenbeck process discussed above; in particular, $X(t)$ is stochastically equivalent to $U(t)/\sqrt{2}$.

Another soluble example of a Langevin equation is given by

$$dr(t) = \frac{(p-1)dt}{2r(t)} + dW(t)$$

where $p = 2, 3, \ldots$, and $r(0) = r_0$, where r_0 is a fixed positive constant. It is clear that if the noise reduces r and makes it small, then the tendency of the drift term is to increase r. The solution to this equation is given by

$$r(t) = \left[\sum_{j=1}^{p} W_j^2(t)\right]^{\frac{1}{2}}$$

where the W_j are independent Wiener processes for which the Itô calculus rule $dW_j(t)dW_k(t) = \delta_{jk}dt$ applies. We choose $W_1(t)$ nonstandard in that $W_1(0) \equiv r_0$; all other $W_j(t)$, $j \geq 2$, are standard. Let us verify this answer:

$$dr(t) = \frac{\sum W_k(t)dW_k(t) + \frac{1}{2}\sum dW_k^2(t)}{[\sum W_\ell^2(t)]^{\frac{1}{2}}} - \frac{1}{2}\frac{\sum W_j(t)W_k(t)dW_j(t)dW_k(t)}{[\sum W_\ell^2(t)]^{\frac{3}{2}}}$$

i.e.,

$$dr(t) = \frac{(p-1)}{2}\frac{dt}{r(t)} + dZ(t)$$

where

$$dZ(t) \equiv \frac{\sum W_k(t)dW_k(t)}{[\sum W_\ell^2(t)]^{\frac{1}{2}}} .$$

It follows that $\langle dZ(t) \rangle = 0$ and

$$[dZ(t)]^{?} = \frac{\sum W_j W_k dW_j dW_k}{\sum W_\ell^2} = dt$$

while $[dZ(t)]^p = 0$ for $p \geq 3$. Thus Z is just another form of a standard Wiener process confirming the stated solution. This particular process is called the *p-dimensional Bessel process* since it deals with a radial process in p-dimensions.

Iteration Solutions. Consider two stochastic differential equations given by

$$dX(t) = a\big(X(t)\big)dt + dW(t),$$
$$dY(t) = A\big(Y(t)\big)dt + dW(t).$$

Suppose we know the solution for X and wish to find the solution for Y in terms of X rather than W. This approach would be useful if the difference $A - a$ was small in some sense. Let $\Delta(t)$ denote the difference stochastic process,

$$Y(t) = X(t) + \Delta(t),$$

then it follows, by subtracting the two basic equations, that

$$d\Delta(t) = \big[A\big(X(t) + \Delta(t)\big) - a\big(X(t)\big)\big]\,dt\,,$$

or in integral form

$$\Delta(t) = \Delta(u) + \int_u^t \big[A\big(X(\tau) + \Delta(\tau)\big) - a\big(X(\tau)\big)\big]\,d\tau\,.$$

In turn this equation may be studied by iteration, and with suitable Lipshitz conditions on A and a an iteration may be shown to converge. Beyond that, and especially for examples where $t - u$ and $A - a$ are suitably small, this technique of solution has been shown to work well numerically [9]. In principle, this approach is also suitable for a singular stochastic differential equation, such as the Bessel process equation, and perturbations about that equation.

3.3 The Fokker-Planck Equation

The Fokker-Planck equation is a differential equation that governs the single-time distribution function

$$p(y,t) \equiv \big\langle \delta\big(X(t) - y\big)\big\rangle\,,$$

and there is an important relation between the Fokker-Planck equation and the stochastic differential equation that is straightforward to derive. Consider a general function F of the stochastic variable X and its stochastic differential equation, namely

$$dF\big(X(t)\big) = F'\big(X(t)\big)dX(t) + \frac{1}{2}F''\big(X(t)\big)\big[dX(t)\big]^2\,.$$

Let us adopt the stochastic differential equation

$$dX(t) = -\frac{1}{2}S'\big(X(t)\big)dt + dW(t)$$

in which the drift has been defined as $-S'/2$. It follows that $\big[dX(t)\big]^2 = \big[dW(t)\big]^2 = dt$, and that

$$dF\big(X(t)\big) = \frac{1}{2}\big[-F'\big(X(t)\big)S'\big(X(t)\big) + F''\big(X(t)\big)\big]dt + F'\big(X(t)\big)dW(t)\,.$$

The mean of this equation leads to

$$\big\langle dF\big(X(t)\big)\big\rangle = \frac{1}{2}\big\langle\big[F''\big(X(t)\big) - F'\big(X(t)\big)S'\big(X(t)\big)\big]\big\rangle dt$$

since $\big\langle F'\big(X(t)\big)dW(t)\big\rangle = \big\langle F'\big(X(t)\big)\big\rangle\big\langle dW(t)\big\rangle = 0$.

These expressions may be alternatively represented in terms of the single-time distribution function using the fact that

$$\begin{aligned}\big\langle F\big(X(t)\big)\big\rangle &= \int F(y)\big\langle\delta\big(X(t) - y\big)\big\rangle dy\\ &= \int F(y)p(y,t)dy\end{aligned}$$

for a general function F assuming that p falls off sufficiently fast for large $|y|$. A temporal differential of this expression leads to

$$\langle dF(X(t))\rangle = \int F(y)[\frac{\partial p(y,t)}{\partial t}dt]\,dy\,.$$

In turn, it follows that

$$\begin{aligned}\int F(y)\frac{\partial p(y,t)}{\partial t}dy&\\ &= \frac{1}{2}\int [F''(y) - F'(y)S'(y)]p(y,t)dy\\ &= \frac{1}{2}\int F(y)[\frac{\partial^2}{\partial y^2} + \frac{\partial}{\partial y}S'(y)]p(y,t)dy\end{aligned}$$

where we have dropped the surface terms assuming that p falls off at $y = \pm\infty$. Since this relation holds for a general function F we learn that the single-time distribution function must satisfy

$$\frac{\partial p(y,t)}{\partial t} = \frac{1}{2}\frac{\partial}{\partial y}[\frac{\partial}{\partial y} + S'(y)]p(y,t)\,,$$

which is the celebrated *Fokker-Planck equation* [10].

An immediate consequence of this equation is the conservation of probability. From its definition $p(y,t) \geq 0$ and satisfies $\int p(y,t)dy = 1$ for all t. This normalization is also consistent with the Fokker-Planck equation as is clear from the relation

$$\begin{aligned}\frac{\partial}{\partial t}\int p(y,t)dy &= \frac{1}{2}\int \frac{\partial}{\partial y}[\frac{\partial}{\partial y} + S'(y)]p(y,t)dy\\ &= \frac{1}{2}[\frac{\partial}{\partial y} + S'(y)]p(y,t)\,\Big|_{-\infty}^{\infty}\\ &= 0\end{aligned}$$

assuming p falls off sufficiently fast.

In order to analyze solutions of the Fokker-Planck equation it is useful to introduce $f(y,t)$ defined through

$$p(y,t) = e^{-S(y)/2}f(y,t)\,.$$

In this case it follows that

$$\begin{aligned}\frac{\partial}{\partial t}e^{-S(y)/2}f(y,t) &= \frac{1}{2}\frac{\partial}{\partial y}[\frac{\partial}{\partial y} + S'(y)]e^{-S(y)/2}f(y,t)\\ &= e^{-S(y)/2}\frac{1}{2}[\frac{\partial}{\partial y} - \frac{1}{2}S'(y)]\,[\frac{\partial}{\partial y} + \frac{1}{2}S'(y)]\,f(y,t)]\,,\end{aligned}$$

or, stated otherwise,

$$\frac{\partial}{\partial t}f(y,t) = [\frac{1}{2}\frac{\partial^2}{\partial y^2} - V(y)]\,f(y,t)\,,$$

where

$$V(y) = \frac{1}{8}[S'(y)]^2 - \frac{1}{4}S''(y).$$

The differential operator

$$H = -\frac{1}{2}\frac{\partial^2}{\partial y^2} + V(y)$$

is referred to as the ***Fokker-Planck Hamiltonian.*** Analysis of this Hamiltonian proceeds much as it does in quantum mechanics. Assume for simplicity that H has a purely discrete spectrum, $Hu_n(y) = \lambda_n u_n(y)$, $n = 1, 2, 3, \ldots$, where the u_n are real, and $\int u_n(y)u_m(y)dy = \delta_{nm}$. Such an assumption holds true whenever $V(y) \to \infty$ as $y \to \pm\infty$. Granting that case we see that the solution for f may be expressed as

$$f(y,t) = \sum_{n=1}^{\infty} f_n e^{-\lambda_n t} u_n(y).$$

In fact one eigenfunction and eigenvalue of this kind is known. From its definition it follows that

$$He^{-\frac{1}{2}S(y)} = 0.$$

By construction $u_1(y) = \exp[-\frac{1}{2}S(y)]$ has the form of a ground state (no nodes) and so $\lambda_n > \lambda_1 = 0$, $n \geq 2$. Stated otherwise, it follows that

$$f(y,t) = f_1 e^{-\frac{1}{2}S(y)} + \sum_{n=2}^{\infty} f_n e^{-\lambda_n t} u_n(y).$$

Let us assume that p is properly normalized, $\int p(y,t)dy = 1$, or that

$$\int u_1(y) f(y,t) dy = \sum_{n=1}^{\infty} e^{-\lambda_n t} f_n \int u_1(y) u_n(y) dy = 1.$$

The orthonormality of the real eigenfunctions implies that only u_1 contributes and that $f_1 = 1$. Furthermore, whatever the initial state $p(y, 0)$, it follows, as $t \to \infty$, that

$$\lim_{t\to\infty} p(y,t) = e^{-S(y)},$$

which then represents the *equilibrium distribution* to which a general initial distribution relaxes. Note that, as defined, S is such that $\exp(-S)$ is normalized; if this is not the case then

$$\lim_{t\to\infty} p(y,t) = e^{-S(y)} \Big/ \int e^{-S(x)} dx.$$

These relations establish a connection between the Langevin equation,

$$\dot{X}(t) = -\frac{1}{2}S'(X(t)) + \dot{W}(t),$$

the Fokker-Planck equation,

$$\frac{\partial p(y,t)}{\partial t} = \frac{1}{2}\frac{\partial}{\partial y}\left[\frac{\partial}{\partial y} + S'(y)\right]p(y,t),$$

as well as the equilibrium distribution,

$$p(y) = e^{-S(y)} \Big/ \int e^{-S(x)} dx.$$

For a multidimensional stochastic variable X_k, $k = 1, \dots, K$, a similar set of relations exist that connect the Langevin equation,

$$\dot{X}_k(t) = -\frac{1}{2}\nabla_k S(X(t)) + \dot{W}_k(t),$$

the Fokker-Planck equation,

$$\frac{\partial p(y,t)}{\partial t} = \frac{1}{2}\sum_{k=1}^{K}\frac{\partial}{\partial y_k}\left[\frac{\partial}{\partial y_k} + \frac{\partial S(y)}{\partial y_k}\right]p(y,t)$$

and the equilibrium distribution,

$$p(y) = e^{-S(y)} \Big/ \int e^{-S(x)} d^K x\,.$$

The conditions that ensure that $\exp(-S/2)$ is a unique ground state of the Fokker-Planck Hamiltonian also ensure that the stationary stochastic process under consideration is ergodic. Consequently, it follows that

$$\lim_{T\to\infty}\frac{1}{T}\int_0^T A(X(\tau))d\tau = \lim_{t\to\infty}\langle A(X(t))\rangle$$

holds with probability one. Observe that

$$\lim_{t\to\infty}\langle A(X(t))\rangle = \int A(y)p(y)d^K y\,.$$

Thus we conclude that

$$\lim_{T\to\infty}\frac{1}{T}\int_0^T A(X(\tau))d\tau = \frac{\int A(y)e^{-S(y)}d^K y}{\int e^{-S(y)}d^K y}\,.$$

This equation holds for a general A.

This latter relation lies behind the quantization scheme due to Parisi and Wu [11]. In this scheme the right hand side above is regarded as an approximate Euclidean-time path integral expectation value the evaluation of which is sought. The equality with the left hand side implies that this expectation value may be calculated by first solving the relevant Langevin equation and subsequently taking a suitable long-term average involving that solution.

Such an approach provides an alternative to standard (Metropolis) Monte-Carlo methods that have been applied in a variety of numerical studies. When S is *complex* the standard Monte-Carlo approach is inappropriate; nevertheless the route through the Langevin equation may well still be applicable. When S is complex the Fokker-Planck Hamiltonian involves a complex potential, and thus the eigenfunctions and eigenvalues are generally complex. Still $u_1(y) = \exp[-S(y)/2]$ and $\lambda_1 = 0$. The form of the general solution still holds, and provided that $Re\lambda_n > 0$, $n \geq 2$, the limit as $t \to \infty$ leads to the proper expression for the complex weight function. In such a case the complex Langevin method provides an effective way to estimate the appropriate statistical average [12].

3.4 The Path Integral Connection

There is a close connection between the Feynman-Kac [13] formula and Langevin equations that does not pass directly through a partial differential equation, although, of course, all three concepts are very closely related. Consider the stochastic differential equation

$$dU(t) = A\big(W(t),t\big)U(t)dW(t)\,.$$

The solution to this equation is given by the Prokhorov formula [14],

$$U(t) = e^{\int_0^t A\left(W(\tau),\tau\right)dW(\tau) - \frac{1}{2}\int_0^t A^2\left(W(\tau),\tau\right)d\tau}$$

subject to the initial condition $U(0) = 1$. To verify this solution we expand $U(t + dt) - U(t)$ out to the proper order involving dW, which leads to

$$\begin{aligned} dU(t) = \big\{&A\big(W(t),t\big)dW(t) + \frac{1}{2}A^2\big(W(t),t\big)[dW(t)]^2 \\ &- \frac{1}{2}A^2\big(W(t),t\big)dt\big\}U(t) \end{aligned}$$

as required. Since $\langle dU(t)\rangle = 0$ it follows that $\langle U(t)\rangle$ is independent of time and thus $\langle U(t)\rangle = \langle U(0)\rangle = 1$. In fact, $U(t)$ is a martingale.

Let B be a primitive integral of A, namely,

$$B(y,t) = \int^y A(x,t)dx$$

and therefore $\partial B(y,t)/\partial y \equiv B'(y,t) = A(y,t)$. If we let $\dot{B}(y,t) = \partial B(y,t)/\partial t$, then

$$dB\big(W(t),t\big) = B'\big(W(t),t\big)dW(t) + \frac{1}{2}B''\big(W(t),t\big)dt + \dot{B}\big(W(t),t\big)dt$$

which we can use to reexpress $U(t)$ in the form

$$U(t) = \exp\big\{B\big(W(t),t\big) - B(0,0) - \int_0^t \big[\dot{B} + \frac{1}{2}(B'' + B'^2)\big]d\tau\big\}$$

We can relate this expression to the Feynman-Kac formula as follows [15]. First let

$$V(x,\tau) = \frac{1}{2}\big[B''(x,\tau) + B'(x,\tau)^2\big] + \dot{B}(x,\tau)\,,$$

or if we introduce

$$C(x,\tau) \equiv e^{B(x,\tau)}\,,$$

then it follows that

$$\frac{\partial C(x,\tau)}{\partial \tau} = -\frac{1}{2}\frac{\partial^2 C(x,\tau)}{\partial x^2} + V(x,\tau)C(x,\tau)\,,$$

which holds for $0 \le \tau \le t$. The desired solution of this equation is one for which $C(x,t) > 0$ is given, and it may be shown that there is a unique solution such that $C(x,\tau) > 0$ in the required domain. The quantity

$$e^{-B(0,0)} = N = \frac{1}{C(0,0)}$$

denotes a normalization factor to ensure that $\langle U(t)\rangle = 1$, or stated otherwise so that

$$\begin{aligned} C(0,0) &= \langle C\big(W(t),t\big)e^{-\int_0^t V\left(W(\tau),\tau\right)d\tau}\rangle \\ &= \int C\big(\Lambda(t),t\big)e^{-\int_0^t V\left(\Lambda(\tau),\tau\right)d\tau}d\mu_W(\Lambda)\,. \end{aligned}$$

Here we see, for the special case of $x = \tau = 0$, the desired solution of the partial differential equation given by means of a functional integral.

It is straightforward to modify this result to give the form of the solution more generally, namely, for any x and σ, $0 \le \sigma \le t$,

$$C(x,\sigma) = \int C\big(\Lambda(t-\sigma)+x,t\big)e^{-\int_0^{t-\sigma} V\left(\Lambda(\tau)+x,\tau+\sigma\right)d\tau}d\mu_W(\Lambda)\,.$$

This version of the Feynman-Kac formula is due to Cameron [16].

Further interesting relations beyond the few we have indicated may be found in the published literature, e.g.,[17].

References

[1] Pearson, D.B.: *Comm. Math. Phys.* **60** (1978) 13

[2] Klauder, J.R.: in "Progress in Quantum Field Theory," Eds. Ezawa, H. and Kamefuchi, S. (Elsevier, B.V., 1986), p. 31

[3] Lukacs, E.: "Characteristic Functions," 2nd Ed. (Hafner, New York, 1970)

[4] Hegerfeldt, G., Klauder, J.R.: *Comm. Math. Phys.* **16** (1970) 329

[5] Uhlenbeck, G.E., Ornstein, L.S.: *Phys. Rev.* **36** (1930) 823; Wang, M.C. Uhlenbeck, G.E.: *Rev. Mod. Phys.* **17**(1945) 323

[6] Itô, K.: *Proc. Imp. Acad. Tokyo* **20** (1944) 519; *Mem. Am. Math. Soc.*, No. 4 (1951)

[7] Stratonovich, R.L.: "Topics in the Theory of Random Noise," (Gordon and Breach, New York, 1963)

[8] Doob, J.L.: "Stochastic Processes" (John Wiley and Sons, New York, 1953)

[9] Klauder, J.R.: in "Colloquia Mathematica Societatis Janos Bolyai 27, Random Fields," (1979) 617

[10] van Kampen, N.: "Stochastic Processes in Physics and Chemistry," (North-Holland, Amsterdam, 1987)

[11] Parisi, G., Wu, Y.-S.: *Sci. Sinica* **24** (1981) 483

[12] Gausterer, H., Klauder, J.R.: *Phys. Rev.* **D33** (1986) 3678; Klauder J.R.: *Acta Physica Austriaca, Suppl.* XXV, (1983) 251

[13] Kac, M.: in "Proceedings Second Berkeley Symposium on Mathematical Statistics and Probability," (Univ. of California Press, Berkeley, 1951), p. 18

[14] Prokhorov, Yu. V.: *Theor. Prob. Appl.* **1** (1956) 157

[15] Benes, V., Shepp, L.A.: *Theor. Prob. Appl.* **13** (1968) 475

[16] Cameron, R.H.: *Ann. of Math.* **59** (1954) 434

[17] Ezawa, H., Klauder, J.R., Shepp, L.A.: *Ann. of Physics* **88** (1974) 588

This article was processed by the author using the TEX macro package from Springer-Verlag.

Computer Assisted Proofs

Oscar E. Lanford III

Mathematics Department, ETH-Zürich, 8092 Zürich, Switzerland

1 Introduction

The subject of these lectures is the use of computers to establish mathematically rigorous quantitative estimates. Such estimates have turned out to have considerable usefulness in the proof of qualitative mathematical assertions which have been resistant to proof by other means. It may be useful to begin by describing schematically one application; we will return later to give a more systematic treatment.

The application is the proof of the validity of the *Feigenbaum renormalization group picture* for period-doubling accumulation. The general setup is as follows: We consider mappings f of an interval to itself with a single turning-point. We can evidently take the interval to be $[-1,1]$ and the turning point to be a maximum at $x=0$. Although this is less evident, it turns out that there is no loss of significant generality in restricting our attention to even mappings sending 0 to 1. Thus, we consider f's defined on $[-1,1]$, with $f(-x)=f(x)$ and $f(0)=1$, strictly decreasing on $(0,1]$. The standard examples are the quadratic mappings

$$f_\mu(x) = 1 - \mu x^2,$$

with μ a parameter between 0 and 2.

Some interval mappings of the general type considered above have the property that they exchange two disjoint intervals, one – which we can take to be symmetric – containing the turning point 0, and the other containing the image 1 of 0. If f has this property, then $f \circ f$ maps the central interval into itself with a single turning point, again at 0, but this time a minimum. By a linear rescaling, with a negative rescaling factor, we can reduce $f \circ f$ on the central interval to a new mapping of the general form we are considering. We denote this new mapping by $\mathcal{R}f$, the *renormalized* f. Explicitly:

$$(\mathcal{R}f)(x) := \frac{1}{\lambda_f} f(f(\lambda_f x)) \quad \text{where } \lambda_f := f^2(0) = f(1) < 0.$$

The *renormalization group* picture is as follows:

1. $\mathcal{R}$ has a fixed point g. (This fixed point is analytic on a neighborhood of $[-1,1]$, even, and is given to two decimal places by $g(x) \approx 1 - 1.528x^2 + 0.105x^4 + 0.027x^6$.)

2. The derivative (or linearization) $D\mathcal{R}(g)$ of $\mathcal{R}$ at the fixed point g has a single simple expanding eigenvalue; the rest of its spectrum is strictly contracting.

Starting from this picture, the standard renormalization group analysis leads to results about "universality of characteristic exponents" which, in the case at hand, translate into assertions of "universality" of the rate of period-doubling accumulation. We will not give this analysis here; see, for example, Collet and Eckmann (1980), Chap. III.3. Our interest is rather in the proof of the two basic assertions 1. and 2. above. These are precise, qualitative mathematical statements, and ought to be accessible to proof by the standard conceptual methods of functional analysis. In fact, such a proof can now be given, but it is surprisingly long and difficult. The first full proof, which preceded the conceptual proof by nearly ten years, used instead detailed numerical estimates proved by computer.

The general strategy of the "computer-assisted" approach to this problem is to find a good approximate fixed point and to show that a simplified variant on Newton's method, starting from this approximate fixed point, converges to an exact fixed point. The heart of the analysis is to get good quantitative control over $D\mathcal{R}(f)$ for f in a small ball in an appropriate Banach space. The computations use floating point arithmetic with strict control of round-off error.

2 Floating Point Computation and Round-off Error

The essential feature of floating point computation is that one works with a fixed number of digits of precision, and that most arithmetic operations, when performed on n-digit operands, produce results with more than n digits. (Multiplication typically produces results with about $2n$ digits; division infinitely many digits; and addition/subtraction produces results with a number of digits depending on the relative magnitudes of the operands.) What the computing environment returns, as the result of a requested operation, is an n-digit *approximation* to the exact result[1].

Digression. It may be useful to formulate explicitly here a point of view: We view floating point numbers as *exact rational numbers* and not as some kind of "fuzzy" quantity. In this regard, floating point numbers are quite different from, e.g., the results of physical measurement. In a given floating point environment, there are only finitely many numbers, which we will refer to as the *representable numbers* in that environment.

Floating point environments available in practice differ considerably in their degree of "cleanliness." One desideratum, often but not always achieved, is that, when the exact result of a requested operation is representable, the exact result is indeed returned. Exactly representable results, in any event, represent the exception and not the typical occurrence. In the general case, some realistic specifications (in order of increasing quality) are:

1. the returned result differs by no more than one unit in the last place from the exact result.

[1] We are simplifying here in ignoring overflow and underflow. In practice, these have of course to be dealt with.

2. the returned result is the exact result truncated to n digits.
3. the returned result is the representable number ***nearest to*** the exact result, with a rule for breaking ties if the exact result happens to lie exactly half-way between its two nearest representable neighbors.

The last of these possibilities is often referred to as *correct rounding.* It might appear that correct rounding in all cases would be difficult and expensive to realize. In fact, with the standard hardware organization for floating point arithmetic, it requires very little additional hardware to provide correct rounding[2] (as opposed to, say, providing correct truncation). Floating point environments conforming to the IEEE standard 754 – in particular, those provided by the 80x87 and 68881 numerical coprocessors – do provide correct rounding[3] . Working in the IEEE-754 environment has the further advantage that floating point computations should be exactly reproducible on different machines.

The above is concerned with the error in individual operations. There are several possible approaches to estimating the error in the results of complex computations. The most straightforward approach is to start from some information about the error in individual operations – for example, that each operation produces a result with error no larger than one unit in the last place – and then to apply elementary formulas for error propagation. The complexity of the algebra required in this approach grows rapidly and is in practice manageable only for very simple computations. A second approach is that of *backward error analysis.* The idea here is to show that the result of the floating point computation is the exact solution to a slightly perturbed problem and then to apply standard kinds of perturbation theory to determine how much a small change in the problem can affect the solution. This approach can be very effective, but seems to be applicable mostly to computations with nice mathematical structure, as, for example, the solution of systems of linear equations. We will discuss here an alternative method, known as *interval arithmetic,* in which the computer is programmed to generate strict upper and lower bounds on all quantities arising in the computation. That is: Each quantity encountered is described, not by a single representable number giving an approximation to the exact quantity, but by an *interval* with representable endpoints which is guaranteed to contain the corresponding exact quantity. To convert a standard floating-point computation to one giving guaranteed bounds, it is only necessary to replace the standard elementary floating point operations $+$, $-$, $\times$, $\div$ by procedures for "doing arithmetic on intervals." This method is quite general and can easily be extended to control the effects of truncation error and of uncertainty in the input. As we will see shortly, however, it does have limitations. Interval arithmetic was originated by R. E. Moore. The classic exposition is Moore (1966); for a more recent treatment, see Moore (1979).

To explain concretely what we mean by procedures for doing arithmetic on intervals, let us look at the example of multiplication. What we need is a procedure which, given two intervals $[a_1, b_1]$ and $[a_2, b_2]$, with representable endpoints, produces a third

[2] We stress that we are speaking here of elementary arithmetic operations – addition, subtraction, multiplication, division, (and square root) – and not of logarithms, exponentials, and trigonometric functions.

[3] As the default. They also provided *directed rounding,* to be discussed below.

$[a_3, b_3]$ such that

$$x_1 \in [a_1, b_1] \quad \text{and} \quad x_2 \in [a_2, b_2] \Longrightarrow x_1 \times x_2 \in [a_3, b_3],$$

where x_1 and x_2 may be general real numbers – i.e., not necessarily representable – and $x_1 \times x_2$ means the mathematically exact product. One way to find a_3, b_3 is as follows:

1. Compute the four exact products $a_1 \times a_2$, $a_1 \times b_2$, $b_1 \times a_2$, $b_1 \times b_2$. If a_1, a_2, b_1, b_2 have at most n digits, the products have each at most $2n$ digits.
2. Find the smallest of these products and round down to n digits: The results is a_3.
3. Find the largest of the products and round up to n digits: The result is b_3.

The above procedure has the advantage of conceptual simplicity; in practice, it is possible to improve on it.

1. Usually, one needs only two of the four products; which product is largest and which smallest can be determined just from sign information. For example, if $b_1 \leq 0$ (so that the first interval is contained in the negative axis), then the smallest product is $a_1 \times b_2$ if $b_2 \geq 0$ and $b_1 \times b_2$ otherwise. A similar case-by-case analysis works as long as 0 is not in the interior of both $[a_1, b_1]$ and $[a_2, b_2]$.
2. It is not really necessary to have all $2n$ digits of the products. An n-digit approximation and an error bound suffice to construct a correct – if perhaps less than optimal – $[a_3, b_3]$.
3. In a floating point environment conforming to the IEEE floating point standard, *directed rounding* is provided. That is: the hardware can be instructed to produce, as "the" product $a_1 \times b_2$, the largest representable number not greater than the exact product, i.e., the exact result rounded toward $-\infty$. With the appropriate hardware, this computation takes essentially no longer than that of the ordinary (correctly rounded) product, and, with this facility, it becomes both easy and efficient to find the optimal a_3 and b_3.

In any event: Given a set of procedures for doing the four basic arithmetic operations on intervals, it is a routine matter to convert a large class of programs doing "naive" floating point computations into ones producing strict error bounds: One simply replaces all floating point variables by interval variables and floating point arithmetic operations by invocations of the corresponding interval arithmetic procedures. This process can even be automated with the aid, for example, of programming languages like C^{++} which provide "operator overloading."

3 Problems with Interval Arithmetic

3.1 Conditional Branches

One major limitation of this approach needs to be mentioned at once: It works only for program segments which do not contain branches conditional on the result of comparing two floating point numbers. A little reflection indicates that the most useful meaning to give to

$$[a_1, b_1] < [a_2, b_2]$$

is that *every* element of the first interval lies to the left of *every* element of the second, i.e., that $b_1 < a_2$. There is a similarly natural interpretation for $[a_1, b_1] \geq [a_2, b_2]$. The difficulty is that trichotomy fails, that is, it can happen that *neither* $[a_1, b_1] < [a_2, b_2]$ nor $[a_1, b_1] \geq [a_2, b_2]$. Thus, code which does one thing if $x < y$ and something else if $x \geq y$ cannot be translated in a simple mechanical way into code generating strict error bounds.

3.2 An Example: Newton's Method

Aside from this problem, the mechanical approach is often not very effective. Newton's method for root-finding in one dimension provides an illuminating example. In its standard form, the iteration for finding a root of $f(x)$ is

$$x_{n+1} = x_n - f(x_n)/f'(x_n).$$

We will use the following notation. If J denotes a (representable) interval, we write $f(J)$ for an interval guaranteed to contain $f(x)$ for all $x \in J$, and similarly $f'(J)$ for an interval guaranteed to contain $f'(x)$ for all $x \in J$. Suppose we are given an interval J_0 which contains a root of f, and we want to shrink it as much as possible preserving root-inclusion. It is *not* a good idea to iterate via

$$J_{n+1} = J_n - f(J_n)/f'(J_n),$$

where $-$ and $/$ denote the interval operations. The reason is that – as is nearly trivial to see – the length of the interval sum of two intervals is at least equal to the length of the longer summand, and so the above sequence of intervals J_n has non-decreasing length.

On the other hand, we can start from the following simple observation: Let $\hat{x}$ be a root of f in some interval J_0, and let x_0 be any point of J_0. By the mean value theorem, there is an x' in J_0 such that

$$f'(x') = \frac{f(x_0) - f(\hat{x})}{x_0 - \hat{x}} = \frac{f(x_0)}{x_0 - \hat{x}},$$

from which it follows that

$$\hat{x} = x_0 - f(x_0)/f'(x').$$

Again, this is an *exact* formula (but we do not know x'.) If we now let $[x_0]$ denote the degenerate interval containing only x_0, it then follows that

$$\hat{x} \in [x_0] - f([x_0])/f'(J_0),$$

where the expression on the right is to be evaluated using interval arithmetic. Thus, if we take any x_0 in J_0, the expression on the right of the above expression is another interval guaranteed to contain the root $\hat{x}$ of f. This suggest the following algorithm (interval Newton's method): Given an interval J_n known to contain a root of f, we choose any x_n in J_n (for example, the midpoint of J_n), and we put

$$J_{n+1} = J_n \cap \big([x_n] - f([x_n])/f'(J_n)\big).$$

Taking the intersection with J_n evidently does not spoil the property of containing $\hat{x}$ and has the advantage of producing a *non-increasing* sequence of intervals. One sensible termination criterion is simply to continue until $J_{n+1} = J_n$. (Note that, in any event, this method will only work if f' does not vanish on the initial interval J_0 or, more precisely, if $f'([J_0])$ does not contain 0.) It can also easily be shown that this algorithm shares the very rapid convergence of Newton's method, i.e., that the length of J_{n+1} is approximately a constant times the square of the length of J_n, as long as, roughly speaking, the length of J_n stays large compared with the round-off error in evaluating $f([x_n])$.

The lesson I would like to draw from this example is that the straightforward translation of the standard floating-point algorithm into interval arithmetic is useless, whereas a simple but not-quite-obvious variant gives an effective root-finding method.

3.3 Pessimistic Error Estimates

Perhaps the most serious drawback with interval arithmetic is that, although it gives *correct* error estimates, it often gives *outrageously pessimistic* ones. The standard – and important – example is numerical linear algebra. There is a particularly successful theoretical error analysis for the solution of systems of simultaneous linear equations, and it was learned very early that blind application of interval arithmetic often gives much worse estimates than the theoretical analysis. We will not discuss this example further here, since it involves some delicate technical analyses. Instead, we look at a simple "toy" example which gives some idea of what can go wrong.

The example is the computation of the sequence

$$x_{n+1} = x_n - h \times x_n,$$

that is, solving the differential equation $x' = -x$ by Euler's method. For simplicity, let us take h to be a degenerate interval whose single element is > 0, and let x_0 be an interval, degenerate or not, but not $[0, 0]$. A simple theoretical analysis gives the mathematically rigorous estimate

$$|X_{n+1}| \geq (1 + h)|X_n|,$$

where the X_n's are the computed intervals and $|X_n|$ denotes the length of x_n. (This is to be understood as a standard mathematical formula; the addition and multiplication are mathematically exact, not some finite-precision or interval version.) From this estimate – and a simple argument showing that, even if $|X_n|$ is initially zero, it becomes non-zero after a few iterations – shows that $|X_n|$ goes to infinity with n (and that the rate of divergence is realistically something like e^{nh} times the roundoff unit.)

On the other hand: It is also not difficult to show, under almost any reasonable assumptions about the error in individual floating point operations, that the result of applying the above iteration naively in floating point arithmetic will not differ much from the mathematically correct result $(1 - h)^n x_0$ as long as this result stays above the *underflow* threshold.

The moral is: *Naive computation gives results of good accuracy up to quite large values of n, whereas the information provided by interval arithmetic soon becomes* *worthless.* There is a further lesson to be learned from this example. If we rewrite the recurrence in the mathematically equivalent form

$$x_{n+1} = (1 - h)x_n,$$

and then apply interval arithmetic, the bounds produced become much more realistic. This is an instance of a general phenomenon: Interval arithmetic is *even more pathologically sensitive to mathematical reworking of formulas* than is ordinary floating point arithmetic. In this simple example, it is not difficult to determine which formula will give good results, but, in more realistic cases, this is usually not so evident.

We mention a few other concrete causes of pessimistic error estimates: The first of them, which might be called the x^2 problem, it that, if J is an interval with 0 in its interior, then the interval product $J \times J$ contains negative as well a positive numbers, and so is not a very good approximation to the minimal interval containing x^2 for all $x \in J$.

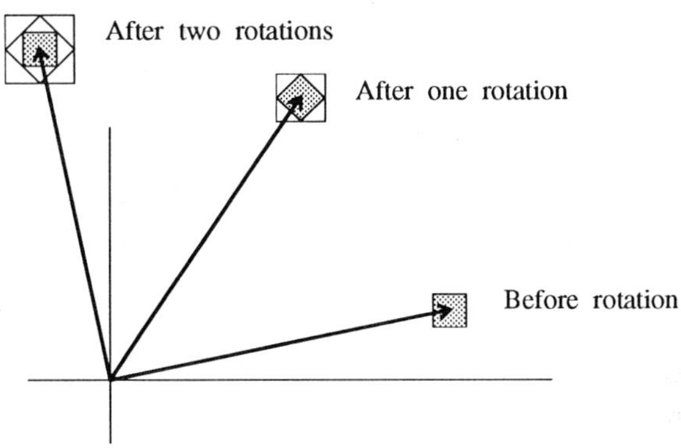

Fig. 1. The rotation problem

The second cause of pessimistic error estimates might be called the rotation problem. For the sake of concreteness, consider a two-vector quantity (x_1, x_2), represented as a pair of intervals. Again for the sake of concreteness, suppose that these two intervals have equal length. Then the "uncertainty region" is a square with sides parallel to the coordinate axes. See Fig. 1. We now rotate this vector by $\pi/4$. Ignoring any roundoff error involved in the rotation, the uncertainty region becomes a square turned $\pi/4$ with respect to the coordinate axes. Because interval arithmetic permits us to consider only uncertainty regions of the form of rectangles with sides parallel to the axes, it cannot do better than to enclose this rotated square in a new square with edges parallel to the axes but with linear dimensions larger by a factor of $2^{1/2}$ than those of the original square. Although we have describe this occurrence only for a single rotation by $\pi/4$, it is clear that the effect is about the same if we make many small successive rotation with total rotation angle $\pi/4$. Thus: Rotating through 2π in small steps multiplies error bounds by about 16. Because of this phenomenon, interval arithmetic gives very bad results in following, for example, a stable oscillation of a differential equation over many periods. Again in this case: simple theoretical considerations show that standard Runge-Kutta methods applied in the straightforward way give quite reasonable accuracy.

3.4 Summary

To sum up the lessons of the above: Although interval arithmetic is potentially a very general and powerful method, it is not a panacea. There are many circumstances in which it works less well than might have been hoped. There are also circumstances in which it does work well, but only if applied in a thoughtful and non-mechanical way. By good fortune, it has turned out to work quite well in a number of applications to computer assisted proofs.

4 The Feigenbaum Problem

We return to the problem described briefly at the beginning: the computer assisted proof of the validity of the Feigenbaum picture. We recall that the renormalization operator is defined (on an appropriate domain) by

$$(\mathcal{R}f)(x) = \frac{1}{\lambda_f} f(f(\lambda_f x)) \quad \text{where } \lambda_f \text{ denotes } f(1).$$

The objective is to prove that $\mathcal{R}$ has a fixed point g and that the linearization of $\mathcal{R}$ at this fixed point has all its spectrum inside the unit disk except for a single simple expanding eigenvalue. Before taking up the specifically computer-based part of the proof, we need some preliminaries.

4.1 A Simplified Newton Iteration

The first remark is that, since we know at the outset that there is an expanding direction at the fixed point, we cannot hope to find the fixed point simply by iterating the renormalization operator: $\mathcal{R}$ is not contractive in a neighborhood of the fixed point. Newton's method and its relatives provide a convenient and effective way of converting the original fixed point problem to a contracting one. In our context – we are looking for a fixed point of $\mathcal{R}$ or, equivalently, a zero of $\mathcal{R}f - f$ – the Newton iteration has the form

$$f_{n+1} = f_n - \left(D\mathcal{R}(f_n) - \mathbb{1}\right)^{-1}\left(\mathcal{R}f_n - f_n\right).$$

A fixed point g for $\mathcal{R}$ is a very strongly contractive fixed point for the operator on the right, provided that 1 is not in the spectrum of $D\mathcal{R}(g)$. Although we could have used Newton's method itself for our proof, it has the disadvantage that a convergence proof requires estimates on the second derivative of $\mathcal{R}$, and these are inconveniently complicated. What works better for our purposes is to choose at the outset an explicit linear operator Γ which is supposed to be a reasonably good approximation to $D\mathcal{R}(g)$ and which is such that $\Gamma - \mathbb{1}$ has a bounded inverse. We then replace $D\mathcal{R}(f_n)$ by Γ in Newton's iteration, that is, we consider the iteration

$$f_{n+1} = \Phi f_n \quad \text{where} \quad \Phi f := f - (\Gamma - \mathbb{1})^{-1} (\mathcal{R}f - f).$$

It is evident that Φ and $\mathcal{R}$ have the same fixed points. The essential observation here is that

$$\begin{aligned} D\Phi(f) &= \mathbb{1} - (\Gamma - \mathbb{1})^{-1} (D\mathcal{R}(f) - \mathbb{1}) \\ &= (\Gamma - \mathbb{1})^{-1} ((\Gamma - \mathbb{1}) - (D\mathcal{R}(f) - \mathbb{1}) = -(\Gamma - \mathbb{1})^{-1} (D\mathcal{R}(f) - \Gamma). \end{aligned}$$

Thus, intuitively, Φ will be contractive in a neighborhood of g provided Γ is a good enough approximation to $D\mathcal{R}(g)$. It is easy to convert this observation into a simple but very useful theorem:

Proposition 1 (Kantorovich(?)). *Let $\mathcal{R}$ be a continuously differentiable operator on the closed ball of radius ρ about g_0. Suppose that there exist Γ (operator) and $\kappa < 1$ (contractivity constant) such that*

1. $\left\|(\Gamma - \mathbb{1})^{-1} (D\mathcal{R}(f) - \Gamma)\right\| \le \kappa$ *for* $\|f - g_0\| \le \rho$.
2. $\left\|(\Gamma - \mathbb{1})^{-1} (\mathcal{R}g_0 - g_0)\right\| \le (1 - \kappa)\rho$.

Then $\mathcal{R}$ has one and only one fixed point in $\{f : \|f - g_0\| \le \rho\}$.

The proof is very simple: One verifies that Φ as defined above maps the ρ-ball about g_0 contractively into itself.

We will apply this result with:

- g_0 an accurate approximate fixed point (produced by careful but uncontrolled numerical computation; error of the order of 10^{-5}.)
- Γ a crude – but not too crude – approximation to the numerically-computed $D\mathcal{R}(g_0)$. We make, roughly,

$$\left\|(\Gamma - \mathbb{1})^{-1} (D\mathcal{R}(g_0) - \Gamma)\right\| \approx 1/2.$$

- ρ small ($\approx 2\,\|(\Gamma - \mathbb{1})^{-1}(\mathcal{R}g_0 - g_0)\|$.)

We will concentrate here on describing the proof of estimate 1. of Proposition 1. Estimate 2. is proved using similar methods, and the proposition then gives the existence of the fixed point. A straightforward extension of the proof of the proposition, combined with the fact that our approximation Γ to $D\mathcal{R}(g)$ has all but one eigenvalue inside the unit circle, gives the assertion about the spectrum of $D\mathcal{R}(g)$.

4.2 Function Space and Norm

The analysis of the preceding subsection is abstract and general. We next need to discuss exactly what function space we are going to apply it in. Our choice here is guided by a non-rigorous but careful computation of the fixed point and exploration of its properties – notably its domain of analyticity. The fixed point we are looking for is analytic (not entire, but with a largish analyticity domain) and even. We can thus write

$$g(x) = G(x^2);$$

by numerical computation, $G(z)$ is analytic in a disk of radius ≈ 10.5 about the origin. It is absolutely crucial for the analysis that the fixed point g has a property of *analyticity improvingness* on appropriate domains. This means the following: The fixed point satisfies the functional equation

$$g(x) = \frac{1}{\lambda} g(g(\lambda x)).$$

There exist domains Ω, for example

$$\Omega = \left\{ x : |x^2 - 1| < 2.5 \right\},$$

on which g is analytic and such that

$$x \longmapsto g(\lambda x)$$

maps a neighborhood of the closure $\overline{\Omega}$ into Ω. Thus, the functional equation gives an analytic continuation of g from Ω to a larger domain. Heuristically, this means that the choice of domain is not critical, provided that it is taken to have this analyticity improvingness property. We will use the explicit Ω defined above. On a more concrete technical level, analyticity improvingness implies that $D\mathcal{R}(g)$ is a compact operator on the space of functions analytic on Ω.

The f's on which $\mathcal{R}$ acts are subject to constraints: They are even and normalized by $f(0) = 1$. It is convenient to make a "change of variable in function space" to eliminate these constraints. We write, accordingly,

$$f(x) = 1 + x^2 h((x^2 - 1)/2.5);$$

the "constraint-free coordinate" h is a function analytic on the unit disk. We can think of $\mathcal{R}$ as acting on h rather than f, and we will continue to denote the operator by $\mathcal{R}$.

Next we choose a norm. Since our functions h are analytic on the unit disk, we can represent them by Taylor series:

$$h(z) = \sum_{n=0}^{\infty} h_n z^n.$$

We define

$$\|h\| := \sum_{n=0}^{\infty} |h_n|,$$

and we work in space $\mathcal{A}$ of functions such that $\|h\| < \infty$. Saying that h is in $\mathcal{A}$ says a little more than that it is analytic in the open unit disk, but a little less than that it is analytic on a neighborhood of the closed unit disk. Evidently, $\mathcal{A}$ is isomorphic to ℓ^1; the isomorphism is realized by expansion in the "natural basis" $e_n(z) = z^n$.

Working in the ℓ^1 norm has the great practical advantage that it is easy to compute norms of linear operators: A bounded operator T on ℓ^1 is in a natural way identified with an infinite matrix (t_{ij}), and it is easy to check that

$$\|T\| = \sup_j \sum_i |t_{ij}|.$$

Translated into the $\mathcal{A}$-language, the assertion is that, for any bounded operator T,

$$\|T\| = \sup_j \|Te_j\|,$$

where (e_j) denotes the natural basis introduced above. The norm has the additional convenience that, for u, v in $\mathcal{A}$,

$$\|u\,v\| \leq \|u\| \, \|v\|,$$

where the product $u\,v$ on the left denotes the pointwise product.

We can now describe how Γ is chosen: We take it to be an $m \times m$ upper-left submatrix of a numerically computed $D\mathcal{R}(g)$, with $m \approx 10$.

4.3 Algebra

Recall what our objective is: We want to show that

$$\left\|(\Gamma - \mathbb{1})^{-1}(D\mathcal{R}(h) - \Gamma)\right\| \leq \kappa \approx 1/2$$

on a smallish ball about the approximate fixed point (which we will continue to denote by g_0.) We will estimate this norm in a simple-minded way as

$$\|(\Gamma - \mathbb{1})^{-1}\| \, \|D\mathcal{R}(h) - \Gamma\|,$$

and calculation shows that $\|(\Gamma - \mathbb{1})^{-1}\|$ is no more than about 10. Thus, what we really want to arrange is

$$\|D\mathcal{R}(h) - \Gamma\| \leq \hat{\kappa} \approx 1/20.$$

By the above discussion of how to compute norms of operators with respect the the ℓ^1-like norm on $\mathcal{A}$, this reduces to showing

$$\|D\mathcal{R}(h)e_n - \gamma_n\| \leq \hat{\kappa} \quad \text{for all } n,$$

where γ_n denotes the n-th column of (the matrix representation of) Γ.

Our next task, therefore, is the derivation of an explicit formula for $D\mathcal{R}(h)e_n$. Recall that

1. In the f language,

$$(\mathcal{R}f)(x) = \frac{1}{\lambda_f} f(f(\lambda_f x)), \quad \lambda_f := f(1).$$

2. Everything is to be expressed in the h language via
$$f(x) = 1 + x^2 h((x^2 - 1)/2.5).$$

3. $D\mathcal{R}(h)\delta h$ is a function analytic on the unit disk; it can be defined informally as the coefficient of ϵ in
$$\mathcal{R}(h + \epsilon\delta h) - \mathcal{R}(h).$$

These remarks specify in principle how to compute $D\mathcal{R}(h)e_n$. The full computation is moderately complicated and not too important for our purposes; we will go through only a part of the computation – schematically – to try to give some feeling for how it works. Loosely speaking f appears four times in the expression for $\mathcal{R}f$, and there are correspondingly four terms in the expression for the derivative. We will look at just one of these, the one arising from the inner f in the composition. In the f language, this term contributes
$$\frac{1}{\lambda} f'(f(\lambda x))\, \delta f(\lambda x)$$
(we have suppressed the subscript on λ_f), and
$$\delta f(x) = x^2 \delta h(z) \quad \text{where } z = (x^2 - 1)/2.5.$$
In this expression, we want to insert
$$\delta h(z) = e_n(z) = z^n,$$
that is,
$$\delta f(\lambda x) = \lambda^2 x^2 \left(\frac{\lambda^2 x^2 - 1}{2.5} \right)^n .$$
The contribution to $D\mathcal{R}(h)e_n$ is therefore – in a mixed notation –
$$\delta \bar{h}(z) = \frac{1}{x^2} \left[\frac{1}{\lambda} f'(f(\lambda x))\, \lambda^2\, x^2 \left(\frac{\lambda^2 x^2 - 1}{2.5} \right)^n \right],$$
where x is to be eliminated throughout in favor of z via $x^2 = 2.5z + 1$. The factor of $1/x^2$ can now be explicitly cancelled, and we can then rewrite the expression as
$$\delta \bar{h}(z) = \underbrace{\lambda f'(f(\lambda x))}_{b(z)} \underbrace{\left(\lambda^2 z + \left(\frac{\lambda^2 - 1}{2.5} \right) \right)^n}_{(v(z))^n} .$$

The other terms can be handled similarly; the final result can be put into the general form
$$D\mathcal{R}(h)e_n(z) = a(z)(u(z))^n + b(z)(v(z))^n + \delta_{n,0}c(z),$$
where a, b, c, u, v are explicit expressions in h, for example,
$$v(z) = \lambda^2 z + \frac{\lambda^2 - 1}{2.5} \quad \text{with } \lambda = h(0) + 1.$$
An important feature of the above expression for $D\mathcal{R}(h)$ is that
$$\|u\| < 1 \quad \text{and} \quad \|v\| < 1,$$
uniformly in h on a small neighborhood of g_0. This means that crude estimates involving norms only will suffice for all sufficiently large n (greater than perhaps 20, in fact) leaving only finitely many n's for which more detailed estimates are required.

4.4 Rectangles in Function Space

The above has been purely mathematical analysis, and shows is that the problem we want to solve can be reduced to a finite computation. We now turn to the question of how this computation can be carried out. The general organizational scheme is inspired by, and built on top of, interval arithmetic. What we do is:

1. Devise a data structure giving a finite parametrization of a sufficiently rich class of regions in the function space $\mathcal{A}$, just as intervals with representable endpoints parametrize regions in $\mathbb{R}$. We will refer to such a region as a *rectangle in function space.*
2. Construct procedures for "doing elementary operations" on these rectangles. The list of elementary operations includes things like pointwise addition, pointwise multiplication, composition, and differentiation.
3. Transcribe the prescriptions for constructing a, b, c, u, v out of h into a sequence of calls to these procedures.
4. Run the resulting program with the input h taken to be a rectangle containing the ρ-ball about the approximate fixed point. From the result, verify
   (a) $\|u\| < 1$ and $\|v\| < 1$ so $\|au^n + bv^n\| \leq \hat{\kappa}$ for $n > n_0$, where, for simplicity, we take n_0 large enough so that γ_n vanishes for $n > n_0$.
   (b) $\|au^n + bv^n + \delta_{n,0} - \gamma_n\| \leq \hat{\kappa}$ for $n = 0, 1, \ldots, n_0$.

In a first approximation, a rectangle in function space is specified by N (representable) intervals $J_0, \ldots, J_{N-1}$ and a non-negative representable number ϵ. The corresponding region $\mathcal{U}(J_0, \ldots, J_{N-1}, \epsilon)$ is the set of all

$$u = \sum_j u_j z^j \in \mathcal{A}$$

such that

$$u_j \in J_j \quad \text{for} \quad 0 \leq j < N \quad \textit{and} \quad \sum_{j=N}^{\infty} |u_j| \leq \epsilon.$$

The number N of terms estimated in detail is adjusted to fit the problem being treated; in the Feigenbaum problem, $N = 10$ is adequate.

To see what is involved in performing elementary operations on such rectangles, we consider the example of pointwise multiplication. The problem is the following: We are given $J_0^{(1)}, \ldots, J_{N-1}^{(1)}, \epsilon^{(1)}$ and $J_0^{(2)}, \ldots, J_{N-1}^{(2)}, \epsilon^{(2)}$, and we want to find $J_0^{(3)}, \ldots, J_{N-1}^{(3)}, \epsilon^{(3)}$ so that

$$\left.\begin{array}{l} u_1 \in \mathcal{U}(J_0^{(1)}, \ldots, J_{N-1}^{(1)}, \epsilon^{(1)}) \\ u_2 \in \mathcal{U}(J_0^{(2)}, \ldots, J_{N-1}^{(2)}, \epsilon^{(2)}) \end{array}\right\} \Longrightarrow u_1 \cdot u_2 \in \mathcal{U}(J_0^{(3)}, \ldots, J_{N-1}^{(3)}, \epsilon^{(3)}).$$

The estimates on the low-order terms in the product are easy to obtain: We simply multiply the polynomials in interval arithmetic, that is,

$$J_k^{(3)} = \sum_{j=0}^{k} J_j^{(1)} \times J_{k-j}^{(2)},$$

with, as indicated, interval $+$ and $\times$. High order terms in $u_1 \cdot u_2$ can come from

– (high order in u_1)×(high order in u_2)
– (low order in u_1)×(high order in u_2)
– (high order in u_1)×(low order in u_2)
– (low order in u_1)×(low order in u_2), with total order $\geq N$.

Thus, we can take

$$\begin{aligned}\epsilon^{(3)} = \epsilon^{(1)} \times \epsilon^{(2)} &+ \Big(\sum_j \operatorname{abs}(J_j^{(1)}\Big)\epsilon^{(2)} \\ &+\epsilon^{(1)}\Big(\sum_j \operatorname{abs}(J_j^{(2)})\Big) \\ &+ \sum_{k=N}^{2N-2} \operatorname{abs}\Big(\sum_{k-N<j<N} J_j^{(1)} \times J_{k-j}^{(2)}\Big),\end{aligned}$$

where

$$\operatorname{abs}(J) := \sup\{|x| : x \in J\} = \max\{-a, b\} \quad \text{if } J = [a, b].$$

All arithmetic in the above expression for $\epsilon^{(3)}$ is to be protected against roundoff error by the methods of interval arithmetic.

With the above as an illustration of the general method, we now want to mention two complications. The first arises in estimating the composition of two functions: We are given bounds (of the indicated form) on u and v, and we want to obtain corresponding bounds on $u \circ v$. In order to be able to do this at all, we need to know that v maps the unit disk into itself, and an easily verifiable sufficient condition for this is that $\|v^j\| < 1$ for some j. For simplicity, we will assume here $\|v\| < 1$. We write

$$u(z) = p(z) + h(z),$$

with $p(z)$ a polynomial of degree $< N$ and $h(z)$ the terms of higher order in the Taylor expansion. If

$$p(z) = u_0 + u_1 z + \cdots + u_{N-1} z^{N-1},$$

then

$$p \circ v = u_0 + u_1 v + \cdots + u_{N-1} v^{N-1},$$

i.e., is built up out of the more elementary primitives multiplication and addition.

The term $h \circ v$ will normally be small, because

$$h \circ v = \sum_{j=N}^{\infty} u_j v^j,$$

so

$$\begin{aligned}\|h \circ v\| &\leq \sum_{j=N}^{\infty} |u_j|\, \|v^j\| \\ &\leq \|v\|^N \sum_{j=N}^{\infty} |u_j| \leq \|v\|^N \epsilon^{(u)}.\end{aligned}$$

The problem here is that, although this term is small, it need not be high-order. The obvious way to take it into account in estimating $u \circ v$ is therefore to widen *all* the intervals (on both ends) by a bound on $\|v\|^N \epsilon^{(u)}$ (and to add this bound to the bound on the high-order terms.) Doing this gives crude estimates: Roughly speaking, it multiplies the uncertainty by an unnecessary factor of $N+1$. This loss in sharpness can be avoided by introducing a slightly more complicated rectangle structure. The idea is that we allow for an "indeterminate error", that is, a general element u of a rectangle $\mathcal{U}$ has the form

$$u(z) = p(z) + h(z) + r(z),$$

with $p(z)$ and $h(z)$ as above and subject to the sorts of bounds explained previously but with $r(z)$ unspecified except for a bound on its norm. Formally, it is only necessary to modify the prescription for rectangles by replacing the single positive number ϵ by two such, ϵ_h and ϵ_r, bounding respectively $\|h\|$ and $\|r\|$. In composition, the estimate for $\|h \circ v\|$ is added to ϵ_r; in most other operations, only terms arising from the r component of one of the operands contribute to the output ϵ_r. In well-behaved applications of this scheme, we put $\epsilon_r = 0$ in the input rectangles, and ϵ_r then remains much smaller than other uncertainties.

A second complication concerns differentiation. The formulas for $D\mathcal{R}(h)$ really do involve derivatives of h, so some sort of differentiation primitive is needed. The problem is that differentiation is not a continuous operation on $\mathcal{A}$, so one cannot estimate u' in terms of imprecise information about u: On fundamental grounds it is impossible to construct a procedure for differentiation. We can work around this problem by observing that, whenever derivatives occur in the expressions we want to evaluate, they always occur in expressions of the general form $u' \circ v$ with v mapping the unit disk into itself. This is not an accident: It reflects the analyticity-improving property of the functional equation. It *is* possible to bound $u' \circ v$ in terms of imprecise information about u and v provided that, for example, $\|v\| < 1$. Thus, we have only to treat

$$(u, v) \longmapsto u' \circ v \quad (\|v\| < 1)$$

rather than differentiation by itself as an elementary operation.

5 Other Applications

In the interests of simplicity and concreteness, we have organized the preceding exposition around a particular application of estimates proved by computer. Although the details vary from problem to problem, quite similar ideas have been applied successfully to a number of different kinds of problems. Among them are:

1. Existence and properties of solutions of other functional equations arising from renormalization group analyses in dynamical systems theory: Eckmann et. al. (1984), Eckmann and Wittwer (1985), Mestel (1985).
2. Existence of a non-Gaussian fixed point for a hierarchical lattice field theory: Koch and Wittwer (1986).
3. Realistic estimates on the ground state energies of many-particle hamiltonians: Fefferman and de la Llave (1986), Seco (1991).

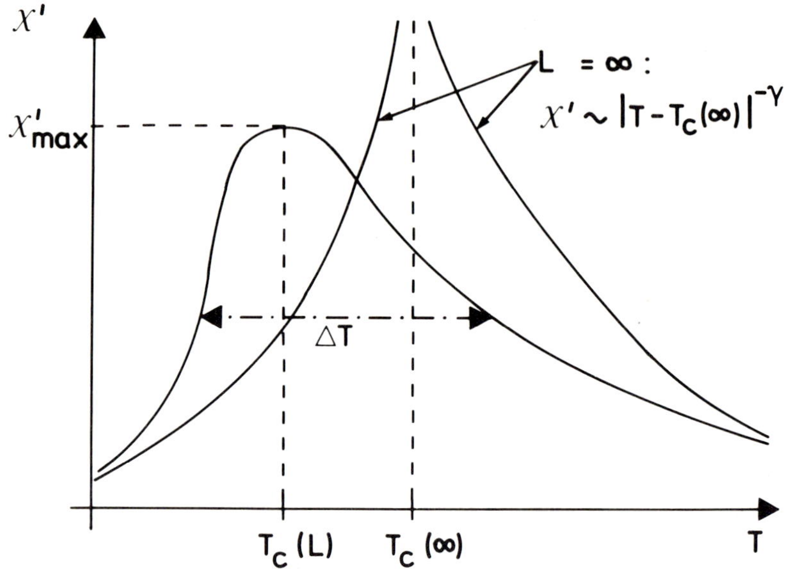

Fig. 1. Qualitative sketch of the susceptibility χ' (3) of a system undergoing a phase transition from a ferromagnetic phase to a paramagnetic phase, as the temperature is raised. Both the behavior in the thermodynamic limit and of the finite system is shown. From K. Binder, in Ref.[6], p.173.

For instance, let us consider a magnetic spin system (for example, an Ising model, where each lattice site i carries a spin $S_i = \pm 1$) as function of temperature. For $L \to \infty$ a second-order phase transition (for example, from a ferromagnetic to a paramagnetic state, the order parameter M then simply is the magnetization per spin, $\langle M \rangle = \sum_i \langle S_i \rangle / L^d$) is assumed to occur at a critical temperature $T_c(\infty)$. For finite L it is "smeared out" (or "rounded") over a temperature region $\Delta T_{\text{rounding}}$,

$$\Delta T_{\text{rounding}} \propto L^{-\theta_T} . \tag{1}$$

The center of this region over which the rounding occurs, $T_c(L)$, may also be shifted away from $T_c(\infty)$,

$$T_c(L) - T_c(\infty) \propto L^{-\lambda_T} . \tag{2}$$

In (1),(2), the quantities θ_T and λ_T are phenomenological exponents to describe this rounding and shifting. Further such exponents are needed to describe the quantities themselves that would be singular at the phase transition temperature $T_c(\infty)$ as $L \to \infty$ but are nonsingular for finite L. Quantities that have a critical divergence at a second order transition, for example, the susceptibility χ' [7,8]

$$\chi' = (L^d / k_B T)(\langle M^2 \rangle_T - \langle |M| \rangle_T^2) \propto |T - T_c(\infty)|^{-\gamma}, \quad L \to \infty, \tag{3}$$

or the specific heat C_H ($\mathcal{H}$ is the Hamiltonian describing the potential energy of the system)

$$C_H = (L^d/k_BT^2)(\langle\mathcal{H}^2\rangle_T - \langle\mathcal{H}\rangle_T^2) \propto |T - T_c(\infty)|^{-\alpha}, \quad L \to \infty, \tag{4}$$

reach a maximum of finite height only:

$$\chi'_{\max} \propto \chi'(L, T = T_c(L)) \propto L^{\gamma_c}, \tag{5}$$

$$C_H^{\max} \propto C_H(L, T = T_c(L)) \propto L^{\alpha_c}. \tag{6}$$

Note that it is not implied that the temperatures $T_{\chi'_{\max}}$ of the susceptibility maximum and $T_{C_H^{\max}}$ of the specific heat maximum agree with each other - thus taking $T_c(L) = T_{\chi'_{\max}}$ as done in Fig.1 is a definition with some arbitrariness, of course.

Similarly, the order parameter $\langle|M|\rangle_{T,L}$ for finite L does not vanish at $T_c(\infty)$ as it does for $L \to \infty$,

$$\langle|M|\rangle_{T,L\to\infty} \propto [1 - T/T_c(\infty)]^{\beta}, \tag{7}$$

rather one has

$$\langle|M|\rangle_{T_c,L} \propto L^{-\beta_c}. \tag{8}$$

Of course, it also is possible to study the phase transition not for $H = 0$ as function of T, but choose $T = T_c(\infty)$ and vary H: In general, we expect a rounding over a region of fields

$$\Delta H_{\text{rounding}} \propto L^{-\theta_H} \tag{9}$$

and shifting to

$$H_c(L) \propto L^{-\lambda_H} \tag{10}$$

as well.

These critical exponents θ_T, θ_H, λ_T, λ_H, α_c, β_c, γ_c, ... characterize the size dependence for the quantities of interest at second-order phase transitions. These exponents, associated prefactors ("critical amplitudes") and associated *scaling functions* that describe the crossover from the bulk singular behavior to the rounded nonsingular behavior of the finite system are the main topic of the present lecture notes. Actually we shall not restrict ourselves to second order phase transitions but ask analogous questions about rounding and shifting of first order phase transitions as well.

The literature on this subject is longstanding and rich [6,9-13] and hence only a small selection of material will be presented. In order to introduce some of the key ideas in a simple and plausible way, we shall first discuss finite size effects for the percolation problem (Sect.2). The percolation transition [14] is the simplest case of a purely geometrical problem exhibiting critical phenomena and one is not yet hampered by the problem how a broken symmetry appears. Sect. 3 then deals with the problem alluded to in Fig.1, namely finite size effects at thermal second order transitions. We then will pay attention to the question what symmetry breaking means for finite lattices in Sect.4, where we deal with the order parameter probability distribution. The following parts then deal with somewhat more specialized topics: the extension of finite size scaling theory to critical slowing down (Sect.5), to critical phenomena for which the hyperscaling relation does not hold (Sect.6), to first-order transitions (Sect.7), to anisotropic geometries and anisotropic critical behavior (Sect.8), and to crossover phenomena from one "universality class" to another (Sect.9). Finally Sect.10 contains some concluding remarks.

2 Finite Size Effects and the Percolation Problem

Let us consider an infinite lattice where each site is randomly occupied with probability p and empty with probability $1-p$ ("site percolation") [14]. Neighboring occupied sites are said to form "clusters" (Fig.2). There exists a critical probability p_c such that for $p < p_c$ only clusters of finite size ℓ exist on the lattice ($\ell = 1,2,3,\ldots$, see Fig.2), while for $p \geq p_c$ an infinite cluster has formed that extends from one boundary of the system throughout the lattice to the opposite boundary. As an "order parameter" of this transition, one takes the percolation probability $P_\infty(p)$, that is the probability that an occupied site is part of the percolating cluster. Defining also $n_\ell(p)$ as the average number of clusters per lattice site that contain exactly ℓ occupied sites, one can introduce a percolation susceptibility as follows,

$$\chi(p) = \sum_{\ell=1}^{\infty}{}' \ell^2 n_\ell(p)/p \quad ; \tag{11}$$

here the prime means that the largest cluster (for $p > p_c$ this is the percolating infinite cluster) is omitted from the summation. $P_\infty(p)$ and $\chi(p)$ show singularities analogous to (3),(7), namely (considering always the limit $|p - p_c| \to 0$)

$$P_\infty(p) = \hat{B}_p \left(\frac{p}{p_c} - 1\right)^{\beta_p}, \quad p > p_c, \tag{12}$$

while $P_\infty(p) \equiv 0$ for $p < p_c$, and

$$\chi(p) = \begin{cases} \hat{\Gamma}_p^+ (1 - \frac{p}{p_c})^{-\gamma_p}, \, p < p_c, \\ \hat{\Gamma}_p^- (\frac{p}{p_c} - 1)^{-\gamma_p}, \, p > p_c. \end{cases} \tag{13}$$

Here β_p, γ_p are the critical exponents (distinguished by an index p from those of a thermal transition) and $\hat{B}_p$, $\hat{\Gamma}_p^-$, $\hat{\Gamma}_p^+$ are the associated critical amplitudes.

In a finite lattice, $\chi(p)$ cannot diverge but reaches a maximum of finite height only: (11) then is a finite sum over clusters of finite "mass" ℓ, infinitely large clusters would not fit on a finite lattice. Similarly, the percolation probability $P_\infty(p)$ does not vanish at any $p > 0$, but must attain small nonzero values as soon as $p > 0$ (even the probability that all sites are occupied is nonzero in a finite system as soon as $p > 0$, namely it is $p^{L^d} = \exp(L^d \ln p)$ in a d-dimensional system; percolation only requires a string of L occupied sites running through the system, which occurs with probability $p^L = \exp(L \ln p)$ as $p \to 0$.) Thus in a finite lattice the singularities described by (12),(13) are smoothed out: this rounding of the transition is easily understood geometrically.

For a quantitative description of this rounding, we have to study the behavior of the percolation clusters near the percolation threshold p_c in more detail [14,15]. For the infinite lattice, the cluster mass distribution $n_\ell(p)$ behaves as (calling ℓ as the "mass" of a cluster in the following)

$$n_\ell(p) = \ell^{-\tau} \tilde{n} \left\{ \ell^\sigma \left(1 - \frac{p}{p_c}\right) \right\}, \quad p \to p_c, \quad \ell \to \infty, \tag{14}$$

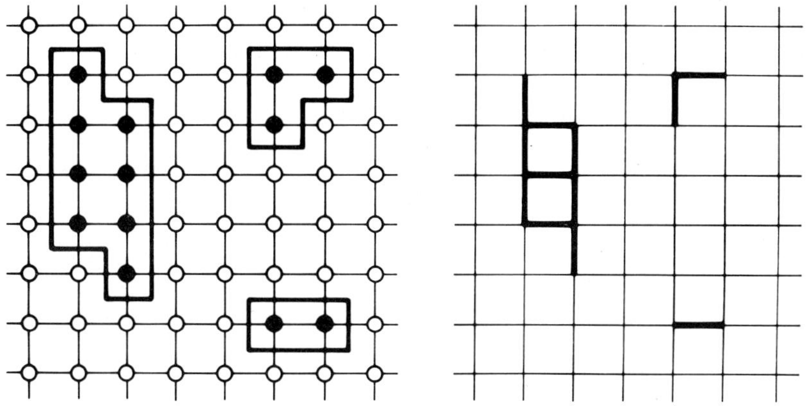

Fig. 2. (a) Clusters on the square lattice for the site percolation problem. Sites are occupied with probability p (dots) or empty with probability $1-p$ (circles). Neighboring occupied sites (encircled) form clusters. The 8×8 lattice shown here hence contains clusters with $\ell = 2$, $\ell = 3$ and $\ell = 8$ occupied sites.
(b) Clusters on the square lattice for the bond percolation problem. Bonds are occupied with probability p (thick lines) or empty with probability $1-p$ (thin lines). Occupied bonds which are connected to each other form clusters. The lattice shown contains clusters with $\ell = 1$, $\ell = 2$ and $\ell = 9$ occupied bonds. From Binder and Heermann [1].

where $\tilde{n}$ is a scaling function, and the exponents τ, σ are related to the exponents β_p, γ_p defined in (12),(13) as

$$\tau = 2 + \beta_p/(\gamma_p + \beta_p), \qquad \sigma = 1/(\gamma_p + \beta_p). \tag{15}$$

The large clusters that appear near the percolation threshold p_c are "fractal objects", that is their "mass" ℓ and radius r_ℓ are not related via the dimensionality of space d but a smaller "fractal dimensionality" d_f,

$$r_\ell = \hat{r}\ell^{1/d_f}, \ell \to \infty. \tag{16}$$

First we establish a geometrical relation between d_f and τ, by noting that in a finite subsystem of linear dimension L a percolating cluster has $r_\ell = L$, and the probability $P_s^{(L)}(p_c)$ that such a cluster spanning the subsystem occurs is of order unity,

$$P_S^{(L)}(p_c) \approx L^d \int_{r_\ell = L}^{\infty} n_\ell(p_c) d\ell \approx L^d \tilde{n}(0) \int_{(L/\hat{r})^{d_f}}^{\infty} \ell^{-\tau} d\ell \approx 1. \tag{17}$$

In a finite subsystem, percolation can occur via any cluster of size $r_\ell = L$ or larger, but these possibilities are mutually exclusive in this subsystem, and thus their probabilities simply add up. For large ℓ, we can replace summation over ℓ by an integration, and apply (14),(16). Now working out the integral in (17) yields

$$\tilde{n}(0)L^{d+d_f(1-\tau)}\hat{r}^{-d_f(1-\tau)}/(\tau-1) = 1, \tag{18}$$

which must hold for any large L which is possible only if the exponent of L vanishes, that is

$$d_f = d/(\tau-1) = d/(1+\beta_p/(\gamma_p+\beta_p)) = d(\beta_p+\gamma_p)/(2\beta_p+\gamma_p) = d-\beta_p/\nu_p, \tag{19}$$

where in the last step the hyperscaling relation between critical exponents is used,

$$d\nu_p = \gamma_p + 2\beta_p, \tag{20}$$

and ν_p is the critical exponent of the correlation length,

$$\xi_p \propto |p-p_c|^{-\nu_p}. \tag{21}$$

From (19),(16) we conclude that the number of sites ℓ in a cluster that spans the distance L (that is $r_\ell = L$) simply is

$$\ell_L = (L/\hat{r})^{d_f} \propto L^{d-\beta_p/\nu_p}. \tag{22}$$

Since a subsystem of linear dimension L contains L^d sites, the fraction of its sites that belong to such a spanning cluster must be of the order of

$$P_\infty^{(L)}(p_c) = L^{-d}\ell_L \propto L^{-\beta_p/\nu_p}. \tag{23}$$

We readily obtain the desired results on finite size effects in truly finite systems, if we assume that the above reasoning for finite subsystems of an infinite system can be essentially carried over to finite systems. Thus we imply that relations such as (16) are essentially true in finite systems as long as $r_\ell < L$: all what the finite system size essentially does, is to provide a cutoff to the cluster size distribution $n_\ell(p)$ (14) at the value of ℓ corresponding to $r_\ell \approx L$, that is at $\ell_L = (L/\hat{r})^{d_f}$.

From this argument we can immediately find the rounding of the divergence of the percolation susceptibility, (13), by noting that the summation in (11) cannot extend to infinity in the finite system, but must be cutoff at ℓ_L:

$$\chi^{(L)}(p) \approx (1/p)\sum_{\ell=1}^{\ell_L} \ell^2 n_\ell(p), \tag{24}$$

which yields for $p = p_c$, using (14)

$$\begin{aligned}\chi^{(L)}(p_c) &\approx [\tilde{n}(0)/p_c]\sum_{\ell=1}^{\ell_L} \ell^{2-\tau} \approx [\tilde{n}(0)/p_c]\int_0^{\ell_L} d\ell \quad \ell^{2-\tau} \\ &= [\tilde{n}(0)/p_c]\ell_L^{3-\tau}/(3-\tau).\end{aligned} \tag{25}$$

Using again (22) this yields with the help of (19),(20)

$$\chi^{(L)}(p_c) \propto L^{d_f(3-\tau)} = L^{d[(\beta_p+\gamma_p)/(2\beta_p+\gamma_p)][\gamma_p/(\beta_p+\gamma_p)]} = L^{\gamma_p/\nu_p}. \tag{26}$$

Equations (23),(26) yield the exponents β_c, γ_c (5),(7) for the percolation problem as $\beta_c = \beta_p/\nu_p$ and $\gamma_c = \gamma_p/\nu_p$.

Next we show that (21) merely is a consequence of the scaling assumption for the cluster mass distribution (14) and the relation between cluster mass and cluster radius, (16):

$$\xi_p^2 \equiv \frac{\sum_{\ell=1}^{\infty}{}' r_\ell^2 \ell^2 n_\ell(p)}{\sum_{\ell=1}^{\infty}{}' \ell^2 n_\ell(p)} \approx \frac{\hat{r}^2 \int_0^\infty \ell^{(2/d_f+2-\tau)} \tilde{n}\{\ell^\sigma(1-p/p_c)\} d\ell}{\int_0^\infty \ell^{(2-\tau)} \tilde{n}\{\ell^\sigma(1-p/p_c)\} d\ell}. \tag{27}$$

Choosing $\ell(1-p/p_c)^{1/\sigma} = x$ as new integration variable, we readily obtain

$$\xi_p^2 = (1-p/p_c)^{-2/(d_f\sigma)} \frac{\int_0^\infty x^{2/d_f+2-\tau} \tilde{n}(x^\sigma) dx}{\int_0^\infty x^{2-\tau} \tilde{n}(x^\sigma) dx} \propto (1-p/p_c)^{-2\nu_p}. \tag{28}$$

This shows the selfconsistency of our scaling description of percolation.

In a spirit similar to the above treatment, we can also obtain a prediction for the finite-size behavior of $\chi^{(L)}(p)$ for $p \neq p_c$. Using again (24),(14),(22)

$$\chi^{(L)}(p) \approx (1/p) \sum_{\ell=1}^{\ell_L} \ell^{2-\tau} \tilde{n}\{\ell^\sigma(1-p/p_c)\} \approx (1/p) \int_0^{(L/\hat{r})^{d-\beta_p/\nu_p}} \ell^{2-\tau} \tilde{n}\{\ell^\sigma(1-\frac{p}{p_c})\} d\ell. \tag{29}$$

Transforming integration variables as in (27),(28) we further get

$$\chi^{(L)}(p) \propto \left(1-\frac{p}{p_c}\right)^{-(3-\tau)/\sigma} \int_0^{(L/\hat{r})^{d-\beta_p/\nu_p}(1-p/p_c)^{1/\sigma}} x^{2-\tau} \tilde{n}(x^\sigma) dx = \left(1-\frac{p}{p_c}\right)^{-\gamma_p} \int_0^{(L/\hat{r})^{(\beta_p+\gamma_p)/\nu_p}(1-p/p_c)^{(\beta_p+\gamma_p)}} x^{2-\tau} \tilde{n}(x^\sigma) dx, \tag{30}$$

where in the last step scaling relations (15),(19) were used. We see that apart from the power law $(1-p/p_c)^{-\gamma_p}$, which is the behavior in the thermodynamic limit, (13), $\chi^{(L)}(p)$ depends on L, p only via the scaled combination $L^{1/\nu_p}(1-p/p_c)$, or $L(1-p/p_c)^{+\nu_p}$ or L/ξ_p, respectively. Thus we can write

$$\chi^{(L)}(p) = \left(1-\frac{p}{p_c}\right)^{-\gamma_p} \tilde{F}\left\{L(1-p/p_c)^{+\nu_p}\right\} \propto \xi_p^{\gamma_p/\nu_p} \tilde{F}(L/\xi_p). \tag{31}$$

Splitting off a factor $(L/\xi_p)^{\gamma_p/\nu_p}$ from the scaling function $\tilde{F}$ we can also write the percolation susceptibility in the form

$$\chi^{(L)}(p) = L^{\gamma_p/\nu_p}\tilde{\chi}(L/\xi_p), \tilde{\chi}(\mathcal{Z}) \propto \mathcal{Z}^{-\gamma_p/\nu_p}, \mathcal{Z} \to \infty. \tag{32}$$

Thus (30)-(32) describe a smooth crossover from the behavior of the infinite system (13) to the rounded behavior of the finite system (26). Clearly the variable which controls the rounding is L/ξ_p, the ratio of the linear dimension L and correlation length ξ_p. The width Δ_p around p_c where the deviations from (13) are significant, clearly is given as $\Delta_p \propto L^{-1/\nu_p}$, that is the rounding exponent $\theta_p = 1/\nu_p$.

In principle, the scaling function $\tilde{\chi}(\mathcal{Z})$ in (32) could be calculated explicitly from (30) if the scaling function $\tilde{n}(x^\sigma)$ of the cluster distribution function is known. But the use of a sharp cutoff at $r_\ell = L$ for the integration clearly is an approximation, and while we consider the derivation that has been presented here as a valid justification of the finite size scaling structure, (32), it is not an accurate method for an actual calculation of the scaling function. Thus the actual location of the maximum $\mathcal{Z}_m$ of $\tilde{\chi}(\mathcal{Z})$ cannot be estimated. But from the existence of the maximum and the scaling structure it already follows that the position p_m corresponding to $\mathcal{Z}_m$ scales with L as $p_m - p_c \propto L^{-1/\nu_p}$ as well, that is the shift exponent is $\lambda_p = 1/\nu_p$. This treatment can be extended to the percolation probability $P_\infty^{(L)}(p)$, one finds

$$P_\infty^{(L)}(p) = L^{-\beta_p/\nu_p}\tilde{P}(L/\xi_p), L \to \infty, p \to p_c \tag{33}$$

with

$$\tilde{P}(\mathcal{Z} \gg 1) \propto \mathcal{Z}^{\beta_p/\nu_p} \propto L^{\beta_p/\nu_p}(p - p_c)^{\beta_p}. \tag{34}$$

Again a smooth interpolation between the behavior in the thermodynamic limit (12) and the rounded behavior in the finite system (23) is achieved. Finally, we again consider the probability $P_s^{(L)}(p)$ that a cluster spanning a distance L occurs: we have $P_s^\infty(p < p_c) = 0$ and $P_s^\infty(p > p_c) = 1$ in the thermodynamic limit. In a finite system, again a scaling behavior with the ratio L/ξ_p is found,

$$P_s^{(L)}(p) = \tilde{P}_s(L/\xi_p), L \to \infty, p \to p_c. \tag{35}$$

From (35) one sees that for $p = p_c$ where $\xi_p \to \infty$ one expects that for all L $P_s^{(L)}(p_c)$ should take an L-independent universal value, $\tilde{P}_s(0)$. This fact makes the spanning probability a very useful quantity for the numerical location of p_c from simulation data: Plotting $P_s^{(L)}(p)$ as function of p for various choices of L all curves should intersect at a unique point, and the abscissa of this common intersection point yields p_c. This intersection method works indeed well in practice [16]. Also the scaling predictions (32) and (33) are nicely verified by numerical data [17], see Figs. 3,4.

It must be noted, however, that rather large values of L have been used for this "data collapsing". This is necessary because all these finite size scaling relations hold only asymptotically in the limit $L \to \infty$: for finite L, there are systematic corrections to finite size scaling which lead to systematic deviations from perfect data collapsing. Such corrections are plausible for various reasons: there are already corrections to our basic cluster distribution scaling assumption, (14), which also holds for very large clusters only, and also corrections result when sums are transformed to integrals (25),(29), and so on.

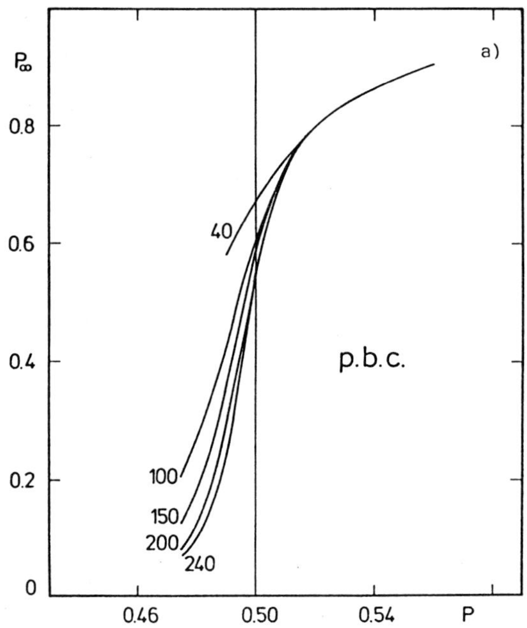

Fig. 3. (a) Percolation probability $P_\infty^{(L)}(p)$ plotted versus p, for the bond percolation problem on the square lattice, using various choices of L (indicated on the curves) and periodic boundary conditions. All lengths are measured in units of the lattice spacing. Note that $p_c = 1/2$ is known from duality exactly [14].
(b) Percolation susceptibility $\chi^{(L)}(p)$ plotted vs. p, for the bond percolation problem on a 100×100 square lattice, comparing periodic boundary conditions (p.b.c.) and free edge boundary conditions (f.e.).
(c) Same as (b) but for a 240 lattice. From Heermann and Stauffer [17].

The physical interpretation of finite size scaling for percolation phenomena in any case is pretty obvious: the correlation length ξ_p is the typical radius r_ℓ of clusters near p_c (27),(28). When ξ_p is comparable to L, clusters no longer fit easily onto the lattice, and hence can no longer contribute towards the singular behavior of $P_\infty(p)$ and $\chi(p)$. For $L \gg \xi_p$, however, clusters of typical size are still unaffected by the finite extent of the lattice. Thus it is clear that a comparison of the two lengths L and ξ_p controls the finite-size effects.

3 Finite Size Effects at Thermal Second Order Phase Transitions

We now discuss ordinary, thermally driven phase transitions, where the state of the system changes from the disordered phase at high temperatures to a spontaneously

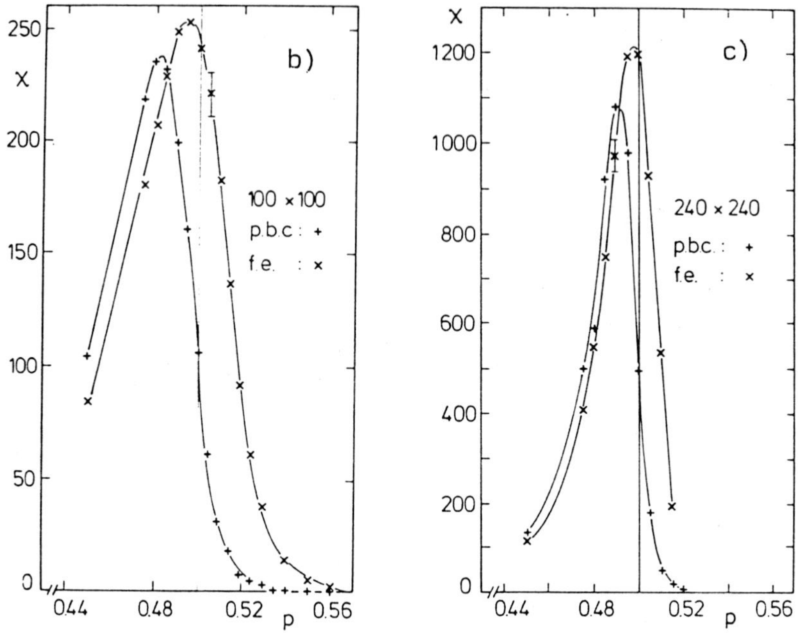

Fig. 3. Cont.

ordered state at temperatures T below the critical temperature T_c of a second-order transition. As a prototype example, we discuss the Ising ferromagnet. The low temperature phase then is a state with nonzero spontaneous magnetization, $\pm|M|$ for zero applied field H. Of course, this ordering implies a spontaneous symmetry breaking, since the Ising model Hamiltonian

$$\mathcal{H} = -J \sum_{\langle i,j \rangle} S_i S_j - H \sum_i S_i \tag{36}$$

for $H = 0$ is invariant against a change of sign of all the spins. This symmetry no longer exists if there is a spontaneous magnetization. Such a symmetry breaking can occur in the thermodynamic limit only: for a finite system, there is always a nonzero probability that the system may pass from a state near $+|M_{sp}|$ to a state near $-|M_{sp}|$, as well as in the opposite direction, see Fig.5 [18]. Therefore, the magnetization M at zero field vanishes for all nonzero temperatures,

$$M(T, H = 0) = \frac{1}{N} \sum_{i=1}^{N} \langle S_i \rangle_{T,H=0} = 0, \tag{37}$$

irrespective of the value of $N = L^d$. Thus the first principles definition of the order parameter M_{sp} considers $M(T, H)$, first taking the thermodynamic limit $N \to \infty$,

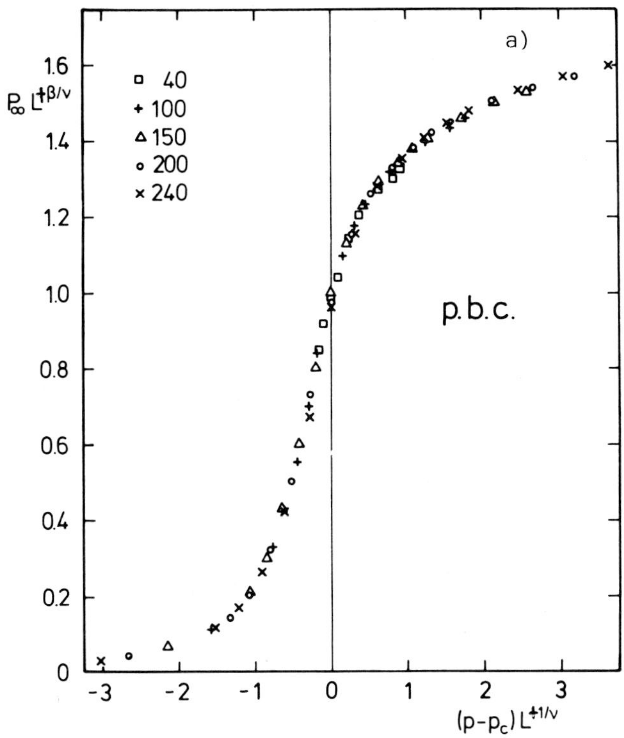

Fig. 4. (a) Finite size scaling plot of the data shown in Fig.3a for the percolation probability $P_\infty^{(L)}(p)$. $L^{\beta_p/\nu_p} P_\infty^{(L)}(p)$ is plotted versus $(p-p_c)L^{+1/\nu_p}$. Various symbols denote different choices of L as indicated. $\beta_p = 0.139$ and $\nu_p = 4/3$ is used.
(b) Finite size scaling plot of Monte Carlo data for the percolation susceptibility $\chi^{(L)}(p)$. $L^{-\gamma_p/\nu_p}\chi^{(L)}(p)$ is plotted versus $(p-p_c)L^{-1/\nu_p}$. $\gamma_p = 2.41$ and $\nu_p = 4/3$ is used. All data refer to periodic boundary conditions. From Heermann and Stauffer [17].

and afterwards letting H tend to zero

$$M_{\text{sp}} = \lim_{H\to O}\lim_{N\to\infty} M(T,H). \tag{38}$$

This double limiting procedure is inconvenient for numerical simulations, and thus one tries to avoid it. In Ising ferromagnets below T_c (but not too close to T_c) one finds a magnetization M (or $-M$, respectively, depending on the initial condition) being metastable over a sufficiently long observation time. Hence reasonably accurate estimates for M_{sp} can be obtained although one does not sample the full equilibrium ensemble [19]. However, even above T_c a small magnetization will typically be found, due to fluctuations which have not completely averaged out in a finite system observed over a finite time. One will find a value $\pm\delta M$, where δM depends both on the size

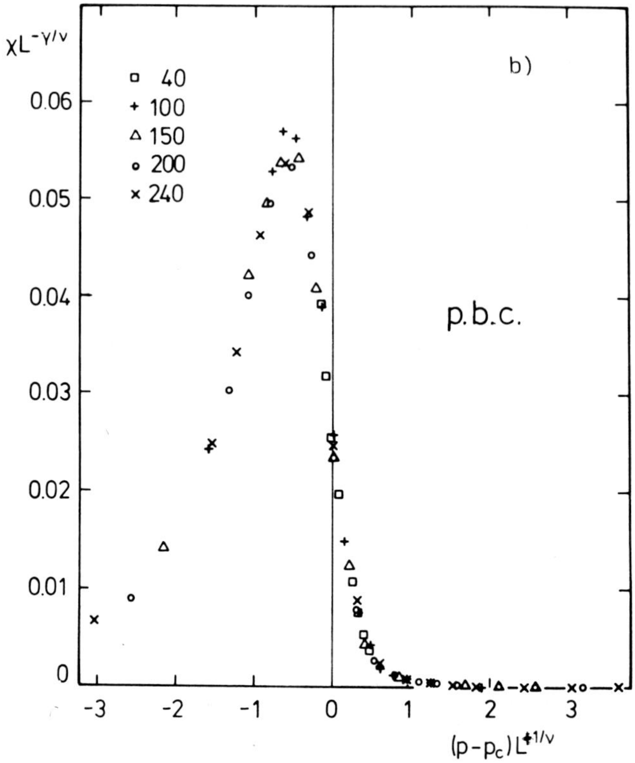

Fig. 4. Cont.

of the system and the observation time t_{obs}. Similarly, below T_c the magnetization fluctuates in the range $M \pm \delta M$ or $-M \mp \delta M$, and one cannot make δM arbitrarily small by making t_{obs} larger and larger: If t_{obs} becomes of the order of the ergodic time t_e, which is the time needed to observe transitions from $+M$ to $-M$ or vice versa, one would start averaging the magnetization to zero. This is what is actually observed (Fig.6) [20]. This situation becomes particularly cumbersome near T_c, where M itself strongly decreases, while δM increases until the fluctuations become comparable with the order parameter itself.

To avoid these problems, the standard recipe is to record the root mean square order parameter [21]

$$M_{\text{rms}}(T) = \sqrt{\langle M^2 \rangle_T} = \left\langle \left(\sum_{i=1}^{N} S_i/N \right)^2 \right\rangle_T^{1/2} = \frac{1}{N} \left(\sum_{i,j=1}^{N} \langle S_i S_j \rangle \right)^{1/2} . \tag{39}$$

In particular, (39) must be used for isotropic spin systems, such as XY or Heisenberg

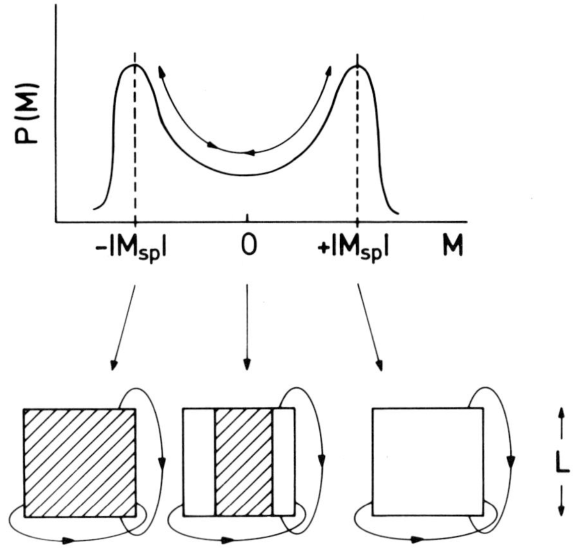

Fig. 5. Probability distribution $P(M)$ for a finite Ising ferromagnet at a temperature T less than the critical temperature T_c where in the thermodynamic limit a spontaneous magnetization $\pm|M_{\text{sp}}|$ appears. While $P(M)$ has a peak near $M = \pm|M_{\text{sp}}|$, where the finite lattice has a more or less uniformly negative magnetization (shaded), and a second peak where it has a more or less uniformly positive magnetization (unshaded), in between $P(M)$ has a minimum which is nonzero even for $M = 0$. This state corresponds to a nonuniform distribution of magnetization in the finite lattice [18]: making use of the periodic boundary conditions, two domain walls run through the lattice and lead to coexistence of two domains of opposite magnetization. In a finite lattice, the system spontaneously makes excursions from states with uniformly negative magnetization through this intermediate mixed-phase state to states with uniformly positive magnetization, and vice versa. From Binder and Heermann [1].

models

$$\mathcal{H}_{XY} = -J\sum_{\langle i,j\rangle}(S_i^x S_j^x + S_i^y S_j^y) - H_x \sum_i S_i^x, (S_i^x)^2 + (S_i^y)^2 = 1, \tag{40}$$

$$\mathcal{H}_{\text{Heisenberg}} = -J\sum_{\langle i,j\rangle}(\mathbf{S}_i \cdot \mathbf{S}_j) - H_z \sum_i S_i^z, (S_i^x)^2 + (S_i^y)^2 + (S_i^z)^2 = 1, \tag{41}$$

where one has a vector order parameter $\mathbf{M}_{\text{sp}}$, whose orientation is not even metastable but rather unstable: one observes a sort of "rotational diffusion" of the unit vector along $\mathbf{M}_{\text{sp}}$ on the unit sphere [21,22]. Hence each component of $\langle \mathbf{M}_{\text{sp}} \rangle$ steadily decreases with increasing observation time.

In a finite system, the order parameter M_{rms} is nonzero at all temperatures. Even at infinite temperature, where $\langle S_i S_j \rangle = \delta_{ij}$, one still obtains $M_{\text{rms}} = 1/\sqrt{N}$ [21]. At temperatures slightly above T_c, where the correlation function

$$G(\mathbf{r}_{ij}) \equiv \langle S_i S_j \rangle_T, \qquad \mathbf{r}_{ij} = \mathbf{r}_i - \mathbf{r}_j \tag{42}$$

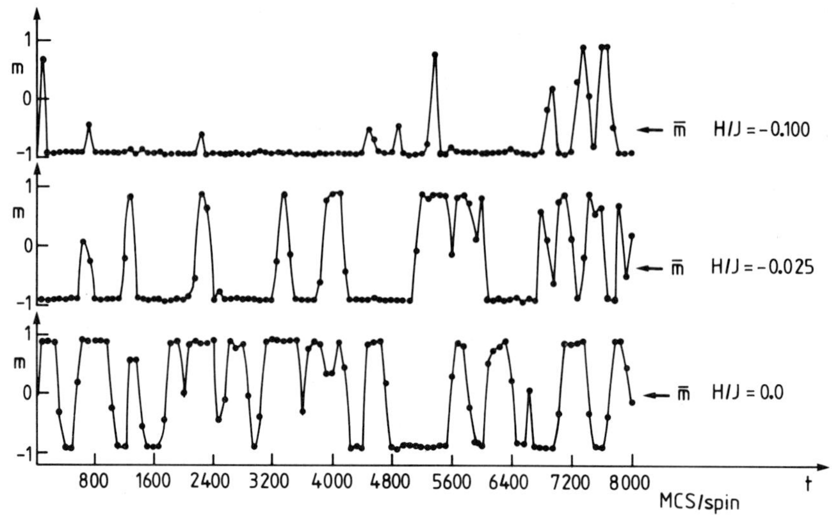

Fig. 6. Evolution of the magnetization $m(t)$ as a function of observation time t [measured in Monte Carlo steps (MCS)/spin] in an Ising nearest-neighbor square lattice of linear dimension $L = 6$ with periodic boundary conditions at a temperature $k_B T/J = 2.1$ (note $k_B T_c/J \approx 2.269$) for three values of the field H. Time averages $\bar{m}$ of $m(t)$ are then estimates of $\langle M \rangle_{T,H,L}$ and are indicated by arrows. From Binder and Landau [20].

is a slowly decaying function of the distance $\mathbf{r}_{ij}$ between the lattice sites, M_{rms} is even much larger. With periodic boundary conditions, $G(\mathbf{r}_{ij})$ is translationally invariant, and hence $M_{\text{rms}}(T) = (\sum_{i=1}^{N} \langle S_i S_j \rangle_T / N)^{1/2}$.

We now discuss this expression at T_c itself [21,23]. In an infinite system, the decay of correlations at T_c is described by

$$G(\mathbf{r}_{ij}) \approx \hat{G} |\mathbf{r}_{ij}|^{-(d-1+\eta)}, |\mathbf{r}_{ij}| \to \infty, \tag{43}$$

where $\hat{G}$ is a "critical amplitude" and η a critical exponent related to γ and ν via $\gamma/\nu = 2 - \eta$ [7]. We now approximate $G(\mathbf{r}_{ij})$ in the finite system of size L^d by taking $G(\mathbf{r}_{ij})$ also from (43), if $|\mathbf{r}_{ij}| \leq L/2$, and putting it to zero else. Then we obtain

$$\sum_{i=1}^{N} \langle S_i S_j \rangle_{T_c} \approx \int_0^{L/2} r_{ij}^{d-1} dr_{ij} \langle S_i S_j \rangle_{T_c} \propto \int_0^{L/2} r_{ij}^{1-\eta} dr_{ij} \propto L^{2-\eta} \tag{44}$$

and hence M_{rms} becomes ($N = L^d$)

$$M_{\text{rms}}^{(L)}(T = T_c) \propto (L^{2-d-\eta})^{1/2} \propto L^{-\beta/\nu}, \tag{45}$$

where we have used the scaling laws $2 - \eta = \gamma/\nu$ and $d = 2\beta/\nu + \gamma/\nu$. Equation (45) is exactly analogous to the corresponding result for the percolation problem (23).

Similarly, the fluctuation relation for the susceptibility

$$k_BT\chi(T) = \left\{ \sum_{i=1}^{N} (\langle S_i S_j \rangle_T - \langle S_i \rangle_T \langle S_j \rangle_T) \right\} \stackrel{H=0}{=} \sum_{i=1}^{N} \langle S_i S_j \rangle_T \tag{46}$$

yields, together with (44)

$$k_BT_c\chi(T_c) \propto L^{\gamma/\nu} \tag{47}$$

in analogy with (26). The exponents β_c and γ_c (5),(8) thus are $\beta_c = \beta/\nu$, $\gamma_c = \gamma/\nu$.

The analogy between the finite size results for the Ising model and for the problem of random percolation is no surprise at all: the Fortuin-Kasteleyn [24] mapping between bond percolation and the ferromagnetic 1-state Potts model [25,26] provides a description of the thermal order-disorder transition of the Ising model as a percolation problem of "physical clusters" [27-30]. Presently the main interest in this connection between thermal critical phenomena and geometrical critical phenomena lies in its usefulness as theoretical foundation for the Monte Carlo cluster algorithms [31-40] which are very effective because they greatly reduce "critical slowing down" [41-44]. Here we shall not discuss the details of this connection (this is the subject of other lectures [33] in this volume), but only quote the main points. In the ferromagnetic q-state Potts model [25], each lattice site may be in one out of q discrete states, and an energy J_P is won if two neighboring sites are in the same state. Thus the Hamiltonian is

$$\mathcal{H}_{\text{Potts}} = -J_P \sum_{\langle i,j \rangle} \delta_{\sigma_i,\sigma_j}, \qquad \sigma_i = 1,\ldots,q. \tag{48}$$

For $q = 2$ this model is equivalent to the Ising model (where we transform our variables from $\sigma_i = 1,2$ to $S_i = \pm 1$), (36), with $J_P = 2J$. Now it is clear that any state of the lattice for the Ising model can be described in terms of "geometrical clusters" of, say, "down spins", in a surrounding background of "up spins". However, these "geometrical clusters" [27,28] are not simply the "physical clusters": we postulate that the latter have a critical percolation transition at the point $T = T_c$, $H = 0$ only, while geometrical clusters percolate throughout the paramagnetic phase (if we vary the magnetic field H from strongly positive to negative values we somewhere encounter a percolation transition). Already Coniglio and Klein [29] have pointed out that one has to distinguish between "active" and "inactive" bonds in a geometrical cluster. The probability p for a bond to be active is

$$p = 1 - \exp(-2J/k_BT), \tag{49}$$

and only spins connected by "active bonds" form a "physical cluster". This rule can be proven [24] by deriving a percolation representation of the Potts model partition function Z_{Potts} as

$$Z_{\text{Potts}} = Tr_{\sigma_i} \exp(-\mathcal{H}_{\text{Potts}}/k_BT) = \sum p^b(1-p)^m q^{N_c}, \tag{50}$$

where b is the number of bonds, m is the number of missing bonds, N_c the number of clusters in a given bond configuration, and the sum is over all bond configurations.

In order to describe the variables of interest in the Ising model, it now is not sufficient to work with a single "cluster coordinate" (28),(38) ℓ as in the geometrical percolation problem, but rather we need two coordinates: the cluster magnetization $m_{c\ell} = \pm\ell$ (the sign gives the orientation of the spins inside the cluster denoted as $c\ell$), and the number of (active) bonds in a cluster which we denote as $u_{c\ell}p$. Denoting the number of clusters per lattice site with these properties as $p(m_{c\ell}, u_{c\ell})$, magnetization and energy per spin (at zero field H) are written, for a square lattice, as [38]

$$M = \frac{1}{N}\sum_i S_i = \sum_{m_{c\ell}}\sum_{u_{c\ell}} m_{c\ell}P(m_{c\ell}, u_{c\ell}) = \sum_{m_{c\ell}} m_{c\ell}p(m_{c\ell}), \tag{51}$$

$$E = \mathcal{H}_{\text{Ising}}/N = -2J\left(\sum_{m_{c\ell}}\sum_{u_{c\ell}} u_{c\ell}p(m_{c\ell}, u_{c\ell}) - 1\right) = -2J(p\frac{N_b}{N} - 1), \tag{52}$$

where $p(m_{c\ell})$ is the number of clusters per lattice site with magnetization $m_{c\ell}$, and N_b the total number of bonds. Since every site belongs to some cluster one has the normalization condition

$$\sum_{m_{c\ell}}\sum_{u_{c\ell}} |m_{c\ell}|p(m_{c\ell}, u_{c\ell}) = \sum_{m_{c\ell}} |m_{c\ell}|p(m_{c\ell}) = 1. \tag{53}$$

With some algebra one can also express the thermal averages of fluctuations in terms of the corresponding quantities of the percolation problem of physical clusters; for example, the specific heat C per lattice site becomes

$$\begin{aligned} C = \partial\langle E\rangle/\partial T &= [1/(Nk_BT^2)](\langle\mathcal{H}^2_{\text{Ising}}\rangle - \langle\mathcal{H}_{\text{Ising}}\rangle^2) = \\ &= 4J^2/(Nk_BT^2p^2)(\langle N_b^2\rangle - \langle N_b\rangle^2 - (1-p)\langle N_b\rangle) \end{aligned} \tag{54}$$

Splitting off from $p(m_{c\ell})$ the contribution of the largest cluster in the system, which we denote as $m_{c\ell}^{\infty}$,

$$p(m_{c\ell}) \equiv p'(m_{c\ell}) + \frac{1}{N}\delta_{m_{c\ell}, m_{c\ell}^{\infty}} \tag{55}$$

we write the absolute value of the magnetization as [38]

$$\langle|M|\rangle = \left\langle \left|\frac{m_{c\ell}^{\infty}}{N} + \sum_{m_{c\ell}} m_{c\ell}p'(m_{c\ell})\right| \right\rangle. \tag{56}$$

The susceptibility for $T > T_c$ is defined (for $H = 0$) as (using $\langle S_iS_j\rangle = 1$ if spins belong to the same cluster, $\langle S_iS_j\rangle = 0$, if they belong to different clusters, and $m_{c\ell}^2 = \ell^2$)

$$\begin{aligned} k_BT\chi = k_BT(\partial\langle M\rangle/\partial H)_{T,H=0} &= N\langle M^2\rangle = N^{-1}\sum_{ij}\langle S_iS_j\rangle = \\ &= \sum_{m_{c\ell}} m_{c\ell}^2\langle p(m_{c\ell})\rangle = \sum_{\ell} \ell^2 n_\ell. \end{aligned} \tag{57}$$

This quantity hence is related to the percolation susceptibility in the thermodynamic limit, where the omission of the largest cluster in (11) (for $p < p_c$) does not matter. Similarly, (56) is related to the percolation probability P_∞,

$$P_\infty = |m_{c\ell}^{\infty}|/N, \langle|M|\rangle \geq \langle P_\infty\rangle. \tag{58}$$

The equality sign holds for $N \to \infty$ while for finite N $\langle|M|\rangle$ exceeds $\langle P_\infty\rangle$ due to the contribution of the finite clusters included in (56): While $\langle\sum m_{c\ell} p'(m_{c\ell})\rangle = 0$, that is positive and negative values of the sum are equally probable, $|M|$ on the average is larger than P_∞ due to the absolute value sign in (56). In fact, for $T \gg T_c$ the dominating contribution to $|M|$ is no longer due to the (single) largest cluster (which itself is small in a finite system for $T \gg T_c$), but from the summation over many clusters: while in this limit $\langle P_\infty\rangle \propto 1/N$ we have $\langle|M|\rangle \propto M_{mrs}(T) \propto 1/\sqrt{N}$, as will be discussed below. This distinction between $\langle P_\infty\rangle$ and $\langle|M|\rangle$ is numerically unimportant for $T < T_c$, as numerical results show (Fig.7).

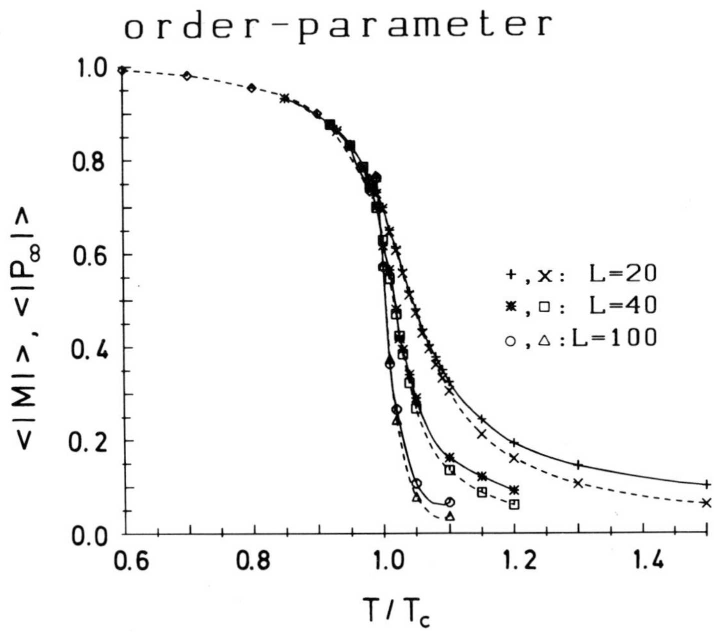

Fig. 7. Magnetization (full curves) and percolation probability (broken curves) for the two-dimensional nearest-neighbor ferromagnetic Ising model plotted vs. T/T_c for three system sizes: $\langle|M|\rangle$ for (+) $L = 20$, (∗) $L = 40$ and (circles) $L = 100$, and P_∞ is for (x) $L = 20$, (squares) $L = 40$, and (triangles) $L = 100$. Periodic boundary conditions were used throughout. All data were generated using the Swendsen-Wang cluster algorithm [31]. From D'Onorio De Meo et al. [38].

The fluctuation relation for the susceptibility below T_c is not given by (57), but rather by [1,18]

$$\chi' = (N/k_BT)(\langle M^2\rangle - \langle|M|\rangle^2). \tag{59}$$

The fact that different expressions for the susceptibility must be used above and below T_c in order to converge smoothly towards the thermodynamic limit is intimately related to the spontaneously broken symmetry, as will be discussed in more detail below. Using now again (55) to separate out the contribution of the largest cluster yields [38]

$$k_B T\chi' = \sum_{\ell}{}' \ell^2 n_\ell + N(\langle P_\infty^2\rangle - \langle |M|\rangle^2) \approx \sum_{\ell}{}' \ell^2 n_\ell + N(\langle P_\infty^2\rangle - \langle P_\infty\rangle^2). \tag{60}$$

This result has a very simple physical interpretation: the response function below T_c picks up a contribution not only from all the finite clusters (the term $\sum' \ell^2 n_\ell$, only considered in the percolation problem), but there is also a contribution present from the fluctuations in size of the largest (percolating) cluster. It turns out that for $T > T_c$ the second term asymptotically gets negligible against the first, since one can show [45]

$$N(\langle P_\infty^2\rangle - \langle P_\infty\rangle^2) \propto L^{-d}\log^2 L, L \to \infty, T > T_c,$$

but for $T < T_c$ the second term actually may be the dominating one, as numerical data show (Fig.8).

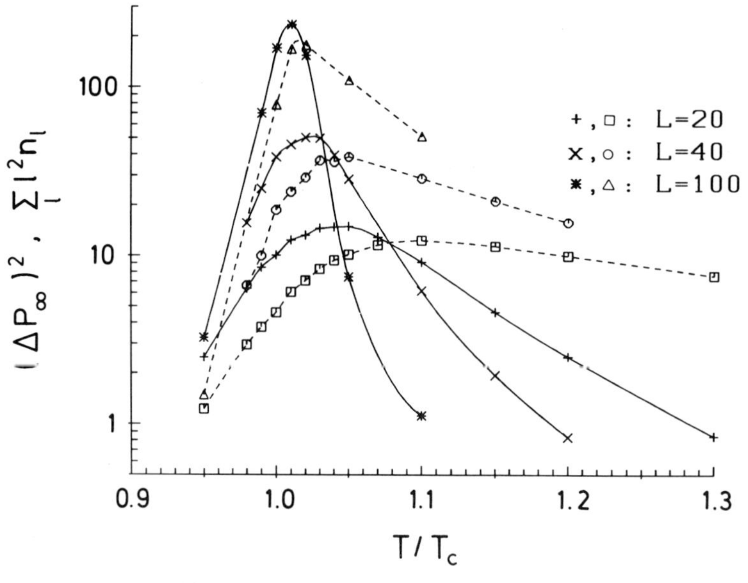

Fig. 8. Normalized fluctuation of the largest cluster, $N(\langle P_\infty^2\rangle - \langle P_\infty\rangle^2)$ [full curves] and second moment of the cluster distribution $\sum' \ell^2 n_\ell$ (broken curves) plotted vs. T/T_c, for the two-dimensional nearest-neighbor Ising ferromagnet on the square lattice. Three lattice sizes are shown. From D'Onorio De Meo et al.[38].

Thus the critical behavior in Ising ferromagnets (and also in Potts models) is associated with the percolation of physical clusters, but some care is necessary in relating the quantities of interest for the magnetic problem to the corresponding percolation probability, percolation susceptibility and so on. But the arguments of

Sect.2 which yielded the finite size scaling description, (29)-(34), carry over to the present problem as well. Corresponding quantities ($\langle |M| \rangle$ and $\langle P_\infty \rangle$, or χ, χ' and $\chi'_p = \sum' \ell^2 n_\ell$, respectively) have a scaling structure of the same type, only the scaling functions differ from each other. In terms of $\epsilon \equiv T/T_c - 1$ we have

$$\langle |M| \rangle = L^{-\beta/\nu} \tilde{M}(L/\xi) = L^{-\beta/\nu} \,\tilde{\tilde{M}}\,(\epsilon L^{1/\nu}), \tag{61}$$

$$\langle P_\infty \rangle = L^{-\beta/\nu} \tilde{P}(L/\xi) = L^{-\beta/\nu} \,\tilde{\tilde{P}}\,(\epsilon L^{1/\nu}). \tag{62}$$

The scaling functions $\tilde{M}$ and $\tilde{\tilde{M}}$ (or $\tilde{P}$ and $\tilde{\tilde{P}}$) simply differ by a transformation of their argument $L/\xi \propto L/|\epsilon|^{-\nu}$ to $L^{1/\nu}|\epsilon|$. Note the similarity between (61),(62) and (33): in the present case $\langle P_\infty \rangle$ is described by the Ising model exponents and not those of random percolation (β_p, ν_p, ...), of course.

The corresponding finite size scaling relations for the susceptibilities read

$$k_B T \chi = L^{\gamma/\nu} \tilde{\chi}(L/\xi) = L^{\gamma/\nu} \,\tilde{\tilde{\chi}}\,(\epsilon L^{1/\nu}), \tag{63}$$

$$k_B T \chi' = L^{\gamma/\nu} \tilde{\chi}'(L/\xi) = L^{\gamma/\nu} \,\tilde{\tilde{\chi}}'(\epsilon L^{1/\nu}), \tag{64}$$

$$\chi'_p \equiv \sum_\ell{}' \ell^2 n_\ell = L^{\gamma/\nu} \tilde{\chi}'_p(L/\xi) = L^{\gamma/\nu} \,\tilde{\tilde{\chi}}'_p\,(\epsilon L^{1/\nu}). \tag{65}$$

Fig.9 shows an example where such scaling relations are tested, using the exactly known T_c [46] and critical exponents [7,8]. It is seen that (64),(65) are reasonably well satisfied.

It is also possible to test the ideas expressed in (24)-(34) more precisely, by examining the cluster distribution n_ℓ in the finite system itself. In (24),(25) it was assumed that the main part of the size effect at the critical point is to cut off the power law $n_\ell \propto \ell^{-\tau}$ at a characteristic cluster size $\ell_L \propto L^{d_f} = L^{1/(\nu\sigma)}$. Consequently, $n_\ell \ell^\tau$ should be a function of the scaled variable $\ell/\ell_L = \ell/L^{1/(\nu\sigma)}$ or $\ell^\sigma/L^{1/\nu}$, respectively. Fig. 10 tests this type of finite size scaling behavior of the cluster size distribution at T_c. It is seen that indeed $n_\ell \ell^\tau$ is basically constant for $0.2 \lesssim \ell^\sigma/L^{1/\nu} \lesssim 0.6$, while for $\ell^\sigma/L^{1/\nu} \approx 0.7$ $n_\ell \ell^\tau$ rapidly drops to zero.

One distinction between percolation critical phenomena and thermal critical phenomena is the fact that in the cluster description of the Ising model each cluster has two signs of the cluster magnetization, $m_{c\ell} = \pm\ell$. While this fact does not make any difference as long as we consider the susceptibility χ above T_c (57), it has led to characteristic distinctions between $\langle |M| \rangle$ and $\langle P_\infty \rangle$ (56),(58) and between $k_B T \chi'$ and the percolation susceptibility $\sum_\ell' \ell^2 n_\ell$. The appearance of a broken symmetry means that there is a single largest cluster in the system, well separated in size from the next largest clusters. The sign of the magnetization of this largest cluster yields the sign of the spontaneous magnetization. In the Swendsen-Wang cluster algorithm [31] the cluster-updating steps do not conserve this sign, and so within one step of this algorithm the largest cluster changes at $H = 0$ its sign with probability 1/2. Thus the symmetry property (37) is readily established. In a single-spin flip algorithm [1], however, the sign of the magnetization is metastable, because a magnetization reversal needs a process where the largest cluster shrinks and the second-largest cluster (with a different sign of its magnetization) grows until it becomes the new largest

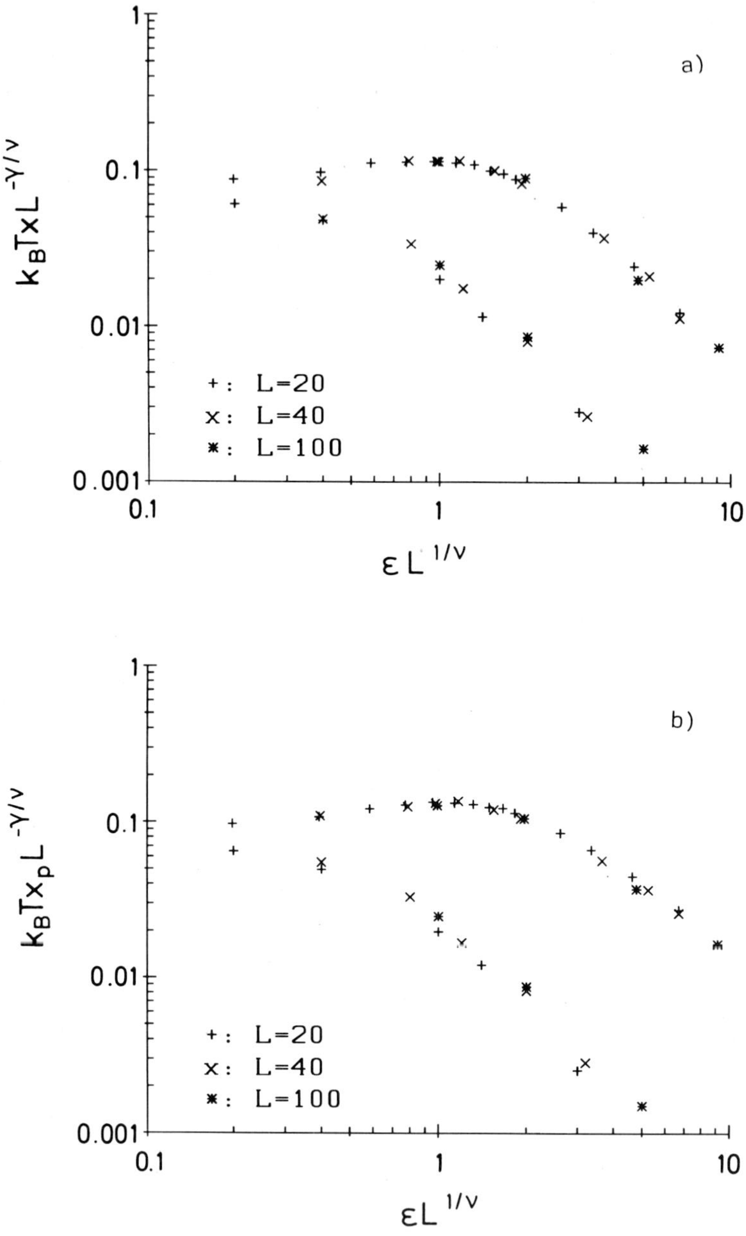

Fig. 9. Finite size scaling plot of $k_BT\chi'$ (a) and of χ'_p (b) versus $|\epsilon|L^{1/\nu}$, for the two-dimensional Ising model with nearest neighbor ferromagnetic interaction at the square lattice. The exactly known value of T_c [46] and of the critical exponents ($\nu = 1$, $\gamma = 7/4$) [7,8] were used. Three different lattice sizes are used as shown in the figure. From D'Onorio De Meo et al.[38].

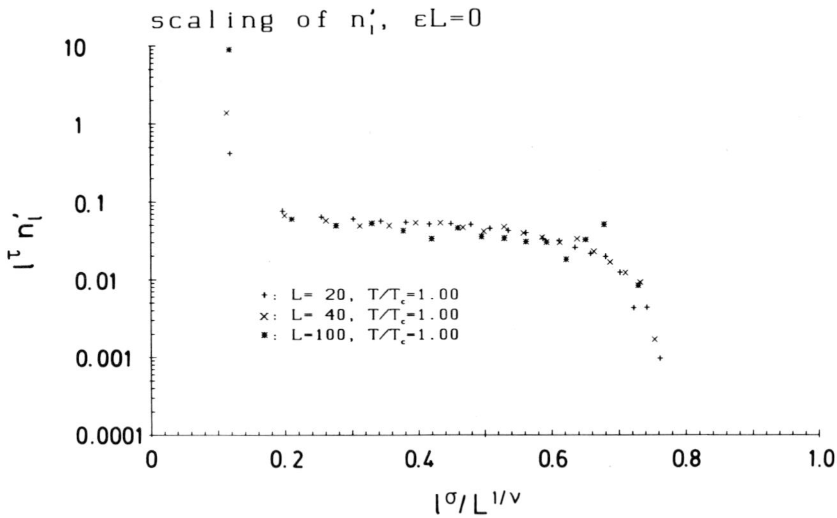

Fig. 10. Finite size scaling of the cluster size distribution $\ell^\tau n'_\ell$ (the largest cluster being omitted) in the two-dimensional nearest-neighbor ferromagnetic Ising model, using the exact T_c [46] and exactly known exponents ($\nu = 1$, $\sigma = 1/(\beta+\gamma) = 8/15$, $\tau = 2+\beta/(\gamma+\beta) = 31/15$). From D'Onorio De Meo et al. [38].

cluster. Again these considerations can be easily checked by inspection of the cluster size distribution generated in the Monte Carlo simulations (Fig.11). It is seen that for $T > T_c$ the distribution of the largest cluster and of the other clusters overlap over a significant range of sizes, while for $T < T_c$ they start to get well separated from each other.

4 The Order Parameter Probability Distribution in Finite Systems

In this section we develop an alternative description of finite size scaling and related "phenomenological renormalization" concepts [18]. For the Ising model, the results based on this approach basically are equivalent to those of the previous section. However, the merit of the present approach is its generality - it easily is generalized to more complicated cases, such as those where the "hyperscaling relation" between critical exponents $d\nu = \gamma + 2\beta$ [8] does not hold [47] and to anisotropic critical phenomena [48], as well as to first-order phase transitions [20,49,50]. By considering again the probability distribution $P_L(M)$ of the order parameter M for an Ising-like system in a finite volume $V = L^d$, which was already briefly alluded to in Fig.5, the question how the spontaneously broken symmetry appears in the thermodynamic limit although there is no such symmetry breaking in finite systems can be dealt with in a natural way.

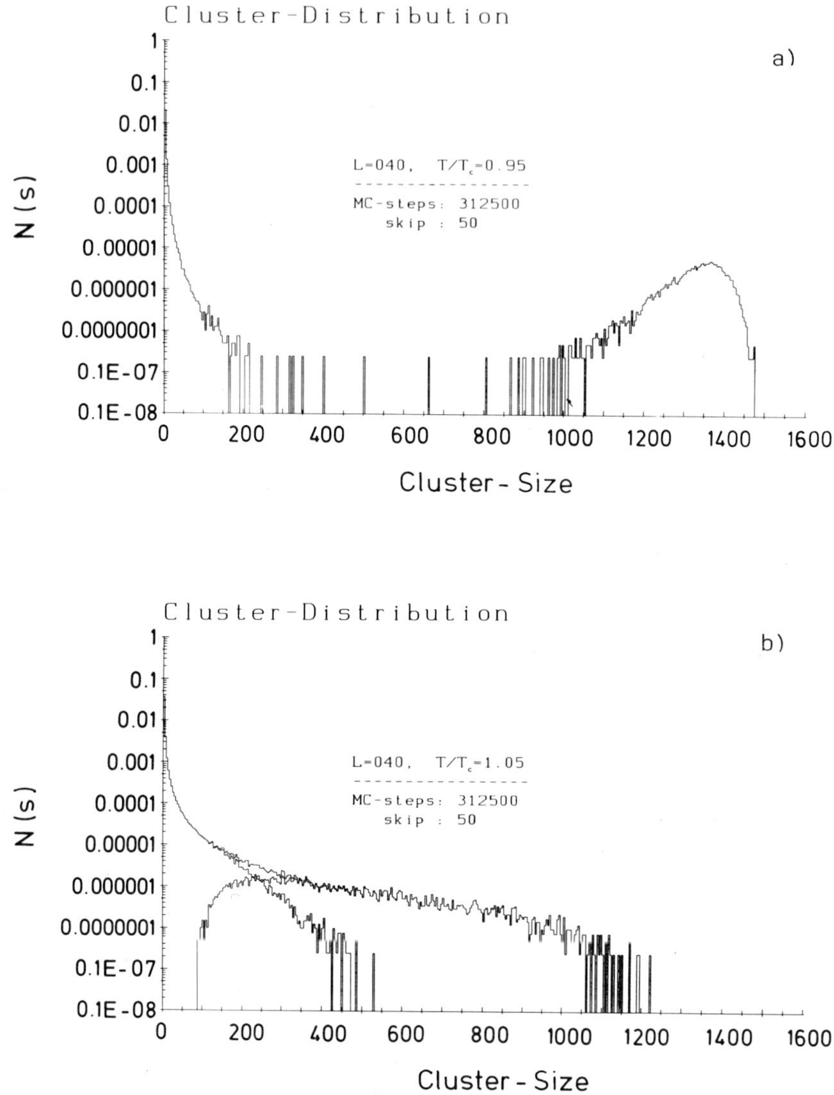

Fig. 11. Cluster size distribution n_ℓ plotted vs. cluster size ℓ for a ferromagnetic Ising square $L \times L$ lattice with $L = 40$ at $T/T_c = 0.95$ (a) and $T/T_c = 1.05$ (b). Four consecutive values of ℓ are averaged together. After omission of the first 1000 MCS, averages were based on 6250 configurations taken at "time" intervals of $\delta t = 50$ MCS, using the Swendsen-Wang [31] algorithm. In case (b), both the total distribution n_ℓ, the distribution of the largest cluster and the distribution n'_ℓ where the largest cluster is omitted are shown. From D'Onorio De Meo et al. [38].

The main ideas about this probability distribution $P_L(M)$ are summarized in Fig.12. For $T > T_c$ and linear dimensions L exceeding the correlation length ξ of order parameter fluctuations, this distribution should be a Gaussian [51]

$$P_L(M) = L^{d/2}[2\pi k_B T\chi^{(L)}]^{1/2} \exp\{-M^2 L^d/[2k_B T\chi^{(L)}]\}, \quad T > T_c, H = 0. \quad (66)$$

The "susceptibility" $\chi^{(L)}$ defined in (66) from the halfwidth of the distribution should smoothly tend towards the susceptibility χ of the infinite system as $L \to \infty$ (remember $\chi \propto |T - T_c|^{-\gamma}$). For $T < T_c$ but again $L \gg \xi$, the distribution is peaked at values $\pm M_L$ near $\pm M_{sp}$ (M_{sp} being the absolute value of the spontaneous magnetization). Near these peaks, again a description in terms of Gaussians applies (while a different behavior occurs near $M = 0$ [18,52], where the system configurations are dominated by mixed phase states involving domains with positive and negative magnetization, see Fig.5)

$$P_L(M) = \frac{L^{d/2}}{[2\pi k_B T\chi'^{(L)}]^{1/2}} \left\{ \frac{1}{2} \exp\left[-\frac{(M - M_L)^2 L^d}{2k_B T\chi'^{(L)}} \right] + \frac{1}{2} \exp\left[-\frac{(M + M_L)^2 L^d}{2k_B T\chi'^{(L)}} \right] \right\}, \quad T < T_c, \quad H = 0. \quad (67)$$

The small value of $P_L(M \approx 0) \propto \exp(-2L^{d-1} f_{\text{int}}/k_B T)$, where f_{int} is the interfacial tension of the system [52], measures the probability that the system moves away from the region near $+M_L$ by the spontaneous creation of two interfaces of size L^{d-1} to form a domain of negative magnetization spanning the system (Fig.5). By this mechanism the system can pass from states near $+M_L$ to states near $-M_L$ and vice versa. In a single spin-flip algorithm [1,2], the observation time needed to observe these transitions increases with L as $P_L(M = M_L)/P_L(M = 0)$; for large L thus such transitions cannot be seen during reasonable observation times.

In order to avoid problems resulting with this "tunneling" of the system from one peak of $P_L(M)$ to the other one, one does not sample the full symmetric distribution $P_L(M)$ as given by (67), for which

$$\langle M \rangle_L = \int_{-1}^{+1} M P_L(M) dM = 0, H = 0, \text{ all } T, \quad (68)$$

cf. (37), but rather only one half of it:

$$\langle M \rangle'_L = \int_0^1 M P_L(M) dM \ / \int_0^1 P_L(M) dM = \langle |M| \rangle_L. \quad (69)$$

When (67) describes the actual distribution $P_L(M)$ accurately, the restricted average $\langle M \rangle'_L$ coincides with the peak position M_L. But due to deviations of the actual distribution observed in a simulation from the double Gaussian form, M_L and $\langle M \rangle'_L$ may differ from each other. However, when extrapolated to the thermodynamic limit, all these quantities should yield the spontaneous magnetization [1,18,53]

$$\lim_{L\to\infty} M_L = \lim_{L\to\infty} \langle |M| \rangle_L = \lim_{L\to\infty} M_{\text{rms}} = M_{\text{sp}}. \quad (70)$$

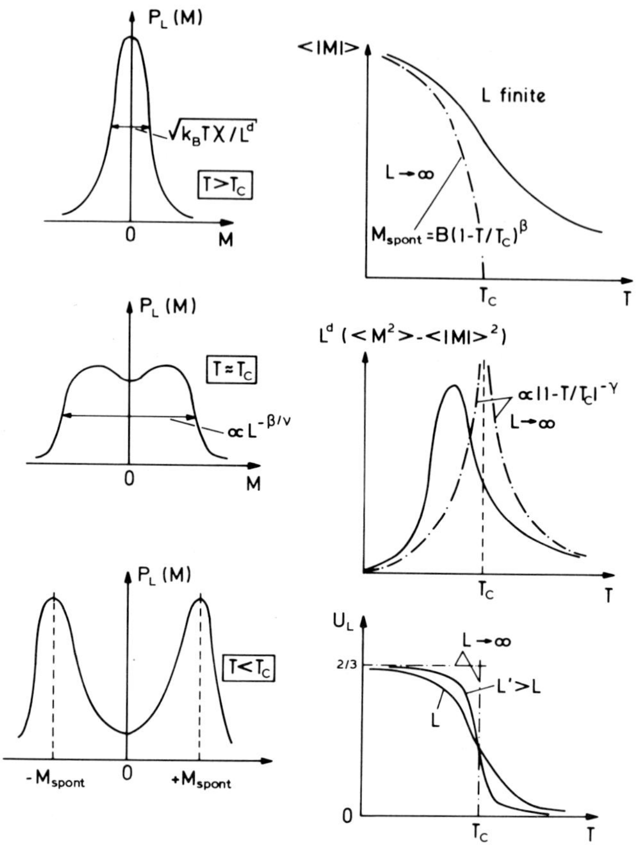

Fig. 12. Schematic evolution of the order parameter probability distribution $P_L(M)$ from $T > T_c$ to $T < T_c$ (from above to below, left part) and corresponding temperature dependence of the mean order parameter $\langle|M|\rangle$, the susceptibility $k_BT\chi'$, and the reduced fourth order cumulant U_L, $U_L = 1 - \langle M^4\rangle/[2\langle M^2\rangle^2]$. Dash-dotted curves indicate the singular variation which results in the thermodynamic limit, $L \to \infty$.

Of course, these relations are more convenient for the use in simulations rather than (38). Fig. 13 illustrates the use of these relations, where only a single ferromagnetic Ising system of size $N = 24^3$ was simulated but the order parameter distributions in *subsystems* of different linear dimensions $L \ll 24$ were recorded. The advantage of this finite size analysis of subsystems, of course, is that many values of L result from a single run.

Similarly, the susceptibilities can be estimated both from the fluctuation dissipation relations (relating it to magnetization fluctuations), and the half widths ΔM, or the heights $P_L(0)$, $P_L(M_L)$ of the peaks:

$$\lim_{L\to\infty} \frac{\langle M^2\rangle L^d}{k_BT} = \lim_{L\to\infty} \frac{[P_L(0)]^{-2}L^d}{2\pi k_BT} = \lim_{L\to\infty} \frac{(\Delta M)^2 L^d}{8k_BT\ln 2} = \chi,$$

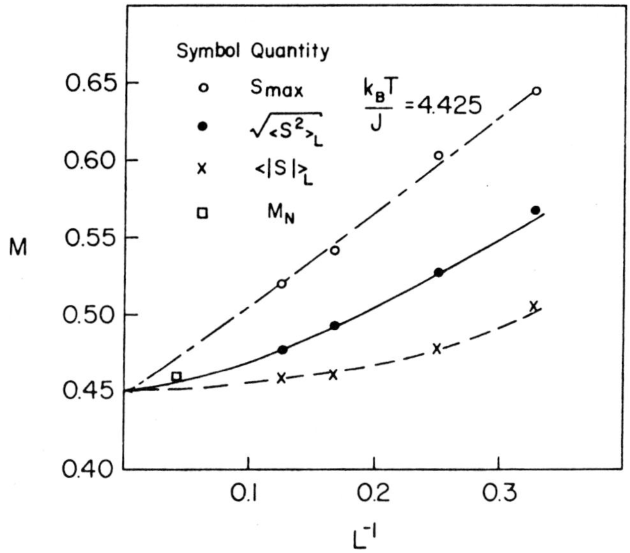

Fig. 13. Estimates of the spontaneous magnetization of the three-dimensional Ising model with nearest-neighbor interaction J on the simple cubic lattice at a temperature $k_BT/J = 4.425$ slightly below criticality ($k_BT_c/J \approx 4.51$). Circles (denoted by $s_{\max}$ in the figure) refer to the position M_L of the maximum of the distribution, $M_{\rm rms}$ is denoted by full dots, $\langle|M|\rangle_L$ by crosses and the direct estimation for the magnetization of the total system of $N = 24^3$ spins is denoted by M_N (square). The curves indicating the extrapolation to $L^{-1} \to 0$ are guides to the eye only. From Kaski et al. [53].

$$T > T_c, \tag{71}$$

$$\lim_{L\to\infty} \frac{\langle M^2\rangle - \langle |M|\rangle^2}{k_BT} L^d = \lim_{L\to\infty} \frac{[P_L(M_L)]^{-2} L^d}{8\pi k_B T} = \lim_{L\to\infty} \frac{(\Delta M)^2 L^d}{8k_BT\ln 2} = \chi',$$

$$T < T_c. \tag{72}$$

Equations (71),(72) hold for $L \gg \xi$. However, in practical applications the correlation length ξ is not known a priori. The Gaussian character of the distribution then is conveniently checked by calculating the fourth-order cumulant U_L [18] or the equivalent quantity $g_L = -3U_L$ called the renormalized coupling constant [47],

$$U_L = 1 - \langle M^4\rangle_L/(3\langle M^2\rangle_2^2). \tag{73}$$

For a single Gaussian centered at $M = 0$ (66), $U_L \equiv 0$. One can show [18] that for $T > T_c$ and $L \gg \xi$ the cumulant decreases towards zero as $U_L \propto L^{-d}$. For $T < T_c$ and $L \gg \xi$, one can show from (67) that U_L tends to $U_\infty = 2/3$ as $L \to \infty$. For $L \ll \xi$, on the other hand, U_L varies only weakly with temperature and linear dimension, it stays close to the (universal but nontrivial) "fixed-point" value $U^\star$, as will be discussed further below.

This behavior of the cumulant makes U_L very useful for obtaining estimates of T_c itself which are not biased by any assumptions about critical exponents. One may plot U_L versus T for various linear dimensions L and estimate T_c from the common intersection point of these curves. Due to corrections to finite size scaling, there may be some scatter in the intersection points for different pairs (L, L') if one works with very small linear dimensions (Fig.14). Of course, the method is by no means restricted to models with discrete degrees of freedom, and thus Fig.14 presents as an example data for the ϕ^4- model on the square lattice [54]. The Hamiltonian of this model is given by

$$\mathcal{H}_{\phi^4} = \sum_i (\frac{1}{2}A\phi_i^2 + \frac{1}{4}B\phi_i^4) + \sum_{\langle i,j\rangle} \frac{1}{2}C(\phi_i - \phi_j)^2, -\infty < \phi_i < +\infty. \quad (74)$$

Data such as those shown in Fig.14 can be generated with modest effort (less than 100 hours CPU time on a workstation), clearly for higher accuracy more substantial CPU effort, more sophisticated algorithms and data analysis techniques, and more powerful computers are useful. The "cumulant intersection method" [18] and variants and extensions thereof have been applied to many different models [18,47,48,53-68], including sophisticated examples such as spin glasses [59], Potts glasses [60], alloy models with many competing order parameters [61,62], polymer mixtures [63,64], XY models with cubic anisotropy [55], three-dimensional Heisenberg ferromagnets [65] and antiferromagnets [68] and even the gas-liquid transition of the off-lattice Lennard-Jones fluid [67]. As an example of the high accuracy that today is possible when one uses sophisticated programs (such as vectorizing multispin coding programs for the Ising model [69] on the Cyber 205 supercomputer, together with the "histogram techniques" [70], Fig.15 shows high accuracy data on the three-dimensional Ising model [66]. Here "effective critical temperatures" for finite L are defined by searching the temperatures where various quantities in the finite system have their maximum slope. Taking all these estimates together, one obtains the final estimate for the critical coupling as $K_c = 0.2216595 \pm 0.0000026$. The accuracy of this estimate is believed to be at least competitive (and perhaps better) than sophisticated Monte Carlo renormalization group [71,72] or high temperature series expansion techniques [73,74]. Very accurate critical exponent estimates have been extracted as well [65,66].

In order to discuss how one extracts critical exponents and T_c in the best way from finite size systems, it is useful to discuss the description of the order parameter distribution function $P_L(M)$ in the region where L is not necessarily larger than the correlation length ξ. The key idea is that $P_L(M)$, which is a function of L, M and ξ (which expresses the temperature dependence via $\xi \propto |1 - T/T_c|^{-\nu}$, as noted above) does not depend separately on these three variables, but only on two scaled combinations, L/ξ and $M\xi^{\beta/\nu}$ [18]:

$$P_L(M) = \xi^{\beta/\nu} P(L/\xi, M\xi^{\beta/\nu}) = L^{\beta/\nu}\tilde{P}(L/\xi, ML^{\beta/\nu}), L \to \infty, \xi \to \infty. \quad (75)$$

Here the power-law prefactors are needed to ensure the proper normalization,

$$\int dM P_L(M) = 1, \text{for all } L, \quad \xi, \quad (76)$$

and in the last part of (75) we have used $ML^{\beta/\nu} = (M\xi^{\beta/\nu})(L/\xi)^{\beta/\nu}$ instead of $M\xi^{\beta/\nu}$ as our second scaling variable. From (75) it is straightforward to derive the

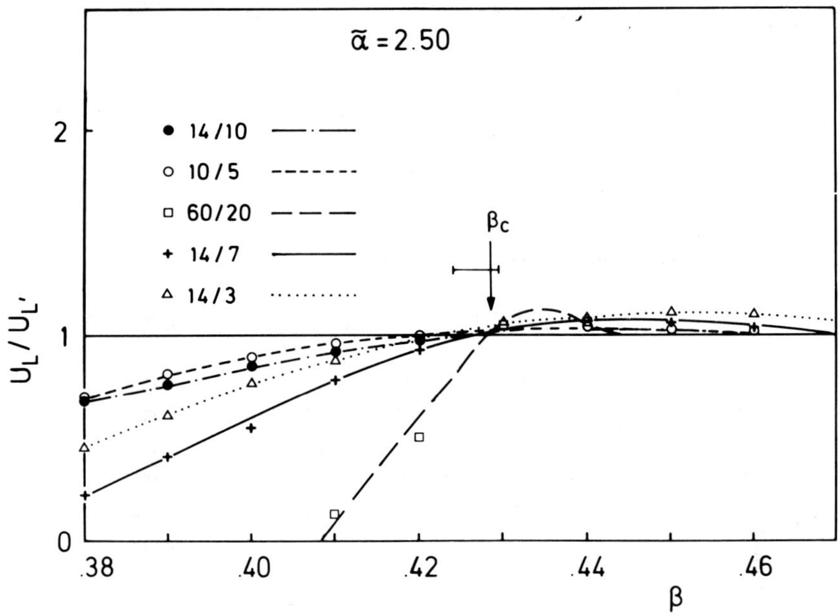

Fig. 14. Determination of the critical coupling $\tilde{\beta}_c$ of the ϕ^4 model on the square lattice (74), where parameters are renormalized as $\tilde{\alpha} = (A + 2dC)^2/B$, $\tilde{\beta} = -C(A + 2dC)/B$, for the particular choice $\tilde{\alpha} = 2.50$, and various ratios $U_L/U_{L'}$, for various pairs L/L' as indicated in the figure. From Milchev et al. [54].

standard finite size scaling relations, (61),(63) and (64), by taking suitable moments of the distribution,

$$\langle |M| \rangle = \int |M| P_L(M) dM, \ \langle M^2 \rangle = \int M^2 P_L(M) dM. \tag{77}$$

For the fourth order cumulant U_L considered in (73), we similarly obtain

$$U_L = 1 - \tilde{\chi}_4(L/\xi)/\{3[\tilde{\chi}(L/\xi)]^2\}, \tag{78}$$

where $\tilde{\chi}_4(L/\xi)$ is related to the 4th moment,

$$\langle M^4 \rangle_L = L^{-4\beta/\nu} \tilde{\chi}_4(L/\xi). \tag{79}$$

In (78), the power law prefactors of L have cancelled out, since $\langle M^2 \rangle_L = L^{-2\beta/\nu}\tilde{\chi}(L/\xi)$, cf. (63), (57). Note that $k_B T \chi = L^d \langle M^2 \rangle = L^{d-2\beta/\nu}\tilde{\chi}(L/\xi) = L^{\gamma/\nu}\tilde{\chi}(L/\xi)$ requires the hyperscaling relation, $d\nu = \gamma + 2\beta$, to be valid. Critical phenomena for which hyperscaling does not hold will be considered later (Sect.6). From (78) we see that for $T \to T_c$, that is $\xi \to \infty$, we simply have a universal [6,18] constant independent of L

$$U_L(T = T_c) = 1 - \tilde{\chi}_4(0)/\{3[\tilde{\chi}(0)]^2\} = U^\star. \tag{80}$$

This observation justifies the identification of T_c from the intersection point of different size cumulants (Fig.14).

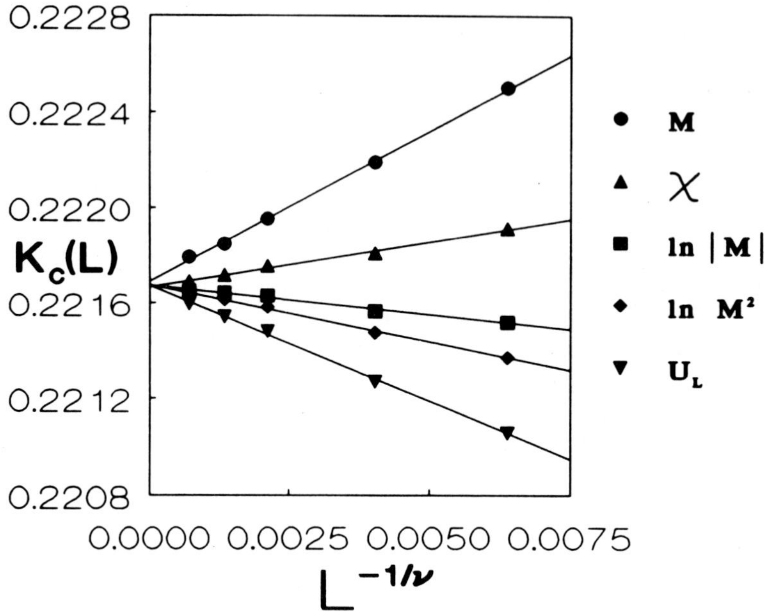

Fig. 15. Size dependence of the finite-lattice effective critical couplings $K_c(L) \equiv J/k_BT_c(L)$ for the three-dimensional nearest-neighbor ferromagnetic Ising model at the simple cubic lattice, estimated from several quantities for $L = 24, 32, 48, 64$ and 96. Here $T_c(L)$ is defined for each quantity as the temperature where the slope of the quantity has its maximum. The lines are fits to the equation $K_c(L) = K_c + \lambda L^{-1/\nu}$, with $\nu = 0.6289$ and K_c, λ being adjusted. From Ferrenberg and Landau [66].

The order parameter distribution $P_L(M)$ also is useful to elucidate the distinction between $k_BT\chi = L^d\langle M^2\rangle$ and $k_BT\chi' = L^d\left(\langle M^2\rangle - \langle|M|\rangle^2\right)$, which has already been emphasized above (57),(59). Note that for $T > T_c$ we have $\chi \equiv \lim_{H\to 0}\lim_{L\to\infty} (\partial\langle M\rangle_L/\partial H)_T$ while for $T < T_c$ we have $\chi' \equiv \lim_{H\to 0}\lim_{L\to\infty} (\partial\langle M\rangle_L/\partial H)_T$, so the response function above and below T_c must be defined differently. From (66), it is straightforward to show that for $T > T_c$ χ' never approaches χ, but rather is smaller by a constant factor $1 - 2/\pi$:

$$\lim_{L\to\infty} \chi'(L,T) = \lim_{L\to\infty} L^d(\langle M^2\rangle_L - \langle|M|\rangle_L^2)/k_BT = \chi(1-2/\pi), T > T_c. \qquad (81)$$

Thus χ' diverges with the same exponent γ as χ does, but the prefactor (critical amplitude) is reduced by a factor $1 - 2/\pi$. This point is sometimes confused in the literature, where sometimes $\lim_{L\to\infty}\chi'(L,T)$ is taken as an estimate for χ above T_c as well, and sometimes the formula $\chi = \lim_{L\to\infty} L^d(\langle M^2\rangle_L - \langle M\rangle^2)/k_BT$ is used at all temperatures. This latter relation, however, in the absence of a symmetry-breaking field makes sense only for $T > T_c$, where $\langle M\rangle_L \equiv 0$ (37) applies, and then it reduces to (57). For $T < T_c$, on the other hand, in a Monte Carlo simulation (using single spin flip algorithms) for a finite system near T_c, $\langle M\rangle_L^2$ is not a useful quantity. For observation times which are much smaller than the ergodic time τ_e, $\langle M\rangle_L^2 \approx \langle|M|\rangle_L^2$, which is close to the squared spontaneous magnetization $M_{\rm sp}^2$, and then $\chi \approx \chi'$ even below T_c. However, for observation times which are much larger than τ_e, $\langle M\rangle_L^2 \equiv 0$, and $\chi = L^d\langle M^2\rangle/k_BT \approx L^d\langle|M|\rangle^2/k_BT \approx L^dM_{\rm sp}^2/k_BT \gg \chi'$. In fact, since $M_{\rm sp}(T)$ steadily increases for $T < T_c$, $k_BT\chi$ is a monotonously increasing function as T is lowered and $k_BT\chi(T \to 0) = L^d$, the volume of the system, since at $T = 0$ the spontaneous magnetization reaches its saturation value $M_{\rm sp}(T = 0) = 1$. For observation times which are of the same order as τ_e itself, one may get a rather erratic behavior, where $\langle M\rangle_L^2$ can take any value between zero and $\langle|M|\rangle_L^2$, when the magnetization has jumped back and forth between $+\langle|M|\rangle_L$ and $-\langle|M|\rangle_L$ just a few times. Now τ_e at T_c is of the order of $\tau_{\max}$, $\tau_{\max}$ being the maximum "intrinsic" relaxation time, while for $T < T_c$ $\tau_{\max}$ steadily decreases with decreasing temperature while τ_e increases. In order to have reasonably small statistical errors, one has to work with observation time $t_{\rm obs} \gg \tau_{\max}$. Below T_c one will thus necessarily encounter a temperature region where $\tau_{\rm obs} \approx \tau_e$, and there $\langle M\rangle_L^2$ is meaningless for $H = 0$! Thus it is an inevitable consequence of symmetry breaking at a thermal phase transition, that above T_c and below T_c different fluctuation formulae must be used to estimate the susceptibility of the infinite system properly. This behavior of $k_BT\chi$ and $k_BT\chi'$ is schematically illustrated in Fig.16.

Thus we have justified the finite size scaling results, (61)(63) and (64) from the consideration of the order parameter probability distribution as well. These finite size scaling properties form the basic ground for many numerical investigations of critical phenomena [6]. Originally one has tried to estimate critical exponents β/ν, γ/ν, $1/\nu$ from "data collapsing": Plotting $\langle|M|\rangle L^{\beta/\nu}$ and $\chi' L^{-\gamma/\nu}$ versus $\epsilon L^{1/\nu}$ as done in Figs.4,9 one adjusts the exponents until the "family of curves" $\langle|M|\rangle_{T,L}$ or $\chi(T,L)$ collapses best on a "master curve", which then is an estimate of the respective scaling function [1,2,6,54,75-78]. This master curve in Figs.4,9 has two branches, one for $T > T_c$ (or $p < p_c$) and one for $T < T_c(p > p_c)$, respectively. The figures already show the disadvantages of this "data collapsing" method: one simultaneously has to fit several parameters (γ/ν, $1/\nu$, sometimes T_c is estimated from the same fit) and since there always is some scatter due to statistical errors and due to systematic corrections to scaling, the superposition on a master curve is never perfect and the judgement for which parameters optimal "data collapsing"occurs is nontrivial and somewhat subjective. In fact, for the sake of saving computer time one likes to work with Monte Carlo "data" for which neither L nor ξ are very large, and thus systematic deviations from finite size scaling must be expected: finite size scaling only is true in the asymptotic limit $L \to \infty$, $\xi \to \infty$, L/ξ finite. These systematic deviations are not accounted for by the analysis that has been presented so far.

Thus there is a need for alternative methods for the estimation of $\frac{1}{\nu}$, $\frac{\gamma}{\nu}$, $\frac{\beta}{\nu}$ and so on from finite size scaling, where these exponents can be estimated independently from each other, and where effects due to corrections to finite size scaling can be systematically considered. Motivated by the concept of "phenomenological renormalization", it has been suggested [18] to estimate exponents from pairs of sizes L, $L' = bL$ with a scale factor $b > 1$ as follows

$$\frac{1}{\nu} = \frac{\ln(\partial U_{bL}/\partial U_L)}{\ln b}|_{U^*}, \tag{82}$$

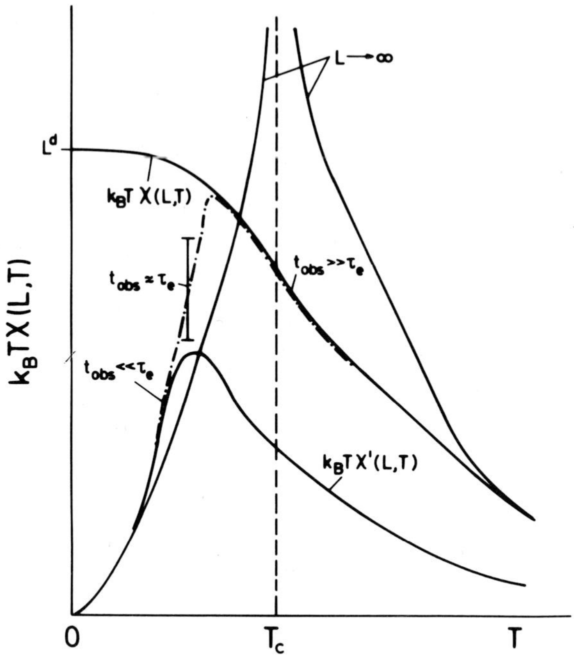

Fig. 16. Schematic temperature variation of the normalized susceptibilities $k_BT\chi(L,T)$ and $k_BT\chi'(L,T)$ as defined in (57), (59). The dash-dotted curve illustrates the standard susceptibility formula $k_BT\chi = L^d(\langle M^2\rangle_L - \langle M\rangle_L^2)$, which is not well defined for $H = 0$ for the temperature region where $t_{\rm obs} \approx \tau_e$. For $L \to \infty$, of course, this temperature region shrinks and ultimately coincides with T_c: symmetry breaking simply appears via "ergodicity breaking". For further explanations cf. text. From Binder and Heermann [1].

$$\frac{\gamma}{\nu} = \frac{\ln[\chi'_{\max}(bL)/\chi'_{\max}(L)]}{\ln b} = \frac{\ln[\chi(bL,T_c)/\chi(L,T_c)]}{\ln b} \tag{83}$$

$$\frac{\beta}{\nu} = -\frac{\ln[\langle|M|\rangle_{bL,T_c}/\langle|M|\rangle_{L,T_c}]}{\ln b}. \tag{84}$$

Note that the first of these relations immediately follows when we expand $U_L = \tilde{U}(\epsilon L^{1/\nu})$ (78) at the intersection point $U^\star$,

$$\tilde{U}(\epsilon L^{1/\nu}) \approx U^\star + U_1^\star \epsilon L^{1/\nu} + o\{(\epsilon L^{1/\nu})^2\}, \tag{85}$$

which holds because thermal properties of finite systems are analytic in T, and hence $U_1^\star$ is a constant, and hence the slope $(\partial\tilde{U}_L/\partial\epsilon)_{U^\star} = U_1^\star L^{1/\nu}$ and therefore $(\partial\tilde{U}_{bL}/\partial U_L)_{U^\star} = (\partial\tilde{U}_{bL}/\partial\epsilon)_{U^\star}/(\partial\tilde{U}_L/\partial\epsilon)_{U^\star} = b^{1/\nu}$. Equations (83),(84) are a straightforward consequence of (61),(63) and (64), of course.

Now systematic corrections to finite size scaling are easily considered [18]. At T_c (or at the temperature where $\chi'(bL,T)$ has its maximum) such corrections are, to leading order, additive terms with less strongly divergent exponents:

$$\chi(L,T_c) = L^{\gamma/\nu}\tilde{\chi}(0)(1 + \chi^{\rm corr}L^{-x_{\rm corr}} + \ldots), \tag{86}$$

χ^{corr} being another amplitude factor and x_{corr} the leading correction exponent. Now (83) gets replaced by

$$\frac{\ln[\chi(bL,T)/\chi(L,T_c)]}{\ln b} = \frac{\gamma}{\nu} - \frac{\chi^{\text{corr}} L^{-x_{\text{corr}}}}{\ln b}(1 - b^{-x_{\text{corr}}}) + \dots \qquad (87)$$

Thus the recipe is to obtain $\chi(bL,T_c)/\chi(L,T_c)$ for a variety of choices of L and b. Plotting then the left hand side of (87) at fixed L versus $(\ln b)^{-1}$, one should obtain for each choice of L a different curve, which should extrapolate linearly to the same value of γ/ν as $(\ln b)^{-1} \to 0$. The same procedure applies for the estimation of β/ν and $1/\nu$. The disadvantage of this method, of course, is that one needs several choices of L and many choices of b, as well as a very good statistical accuracy of data. As an example of a nontrivial application to a two-dimensional model, we discuss the Ising antiferromagnet with nearest-neighbor interactions J_{nn}, J_{nnn} [57] (Fig.17). This model exhibits a transition to a layered antiferromagnetic structure for $J_{nnn}/J_{nn} > 1/2$, belonging to the "universality class" [8] of the XY model with cubic anisotropy [79], with *nonuniversal* critical exponents. The exponent estimates which can be obtained from the analysis outlined above are comparable in accuracy [57] to results obtained from Monte Carlo renormalization group (MCRG) [80], Monte Carlo data collapsing [77], high-temperature series extrapolations [81] and transfer matrix calculations for finite strips [82], see Fig. 18. It is seen that standard real space renormalization group techniques [83] and also high temperature series analysis [81] is less reliable, while MCRG [80] and transfer matrix renormalization [82] yield results of comparable accuracy. However, the transfer matrix method clearly is limited to two-dimensional problems, while the Monte Carlo method works in three and four dimensions as well. At this point, we draw again attention to recent high accuracy studies of the three dimensional Ising [66] and Heisenberg [65] models, where an accuracy of exponent estimates competitive with the best MCRG [71,72] and field-theoretic renormalization methods [84] was reached.

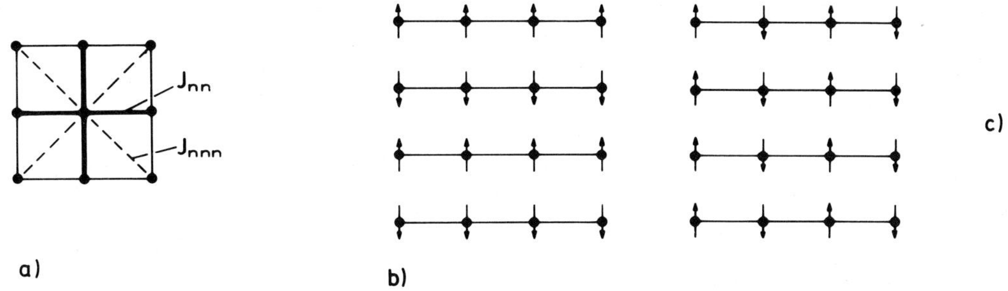

Fig. 17. (a) Square lattice with exchange J_{nn} between nearest neighbors (full lines) and exchange J_{nnn} between next-nearest neighbors (broken lines) (b,c) Spin arrangements in the layered antiferromagnetic structure, where ferromagnetically aligned rows of up spins alternate with ferromagnetically aligned rows of down spins. These rows can be oriented in the x-direction (b) or in the y-direction (c). These two orderings are always degenerate with each other and hence should be considered as the two components of a two-component order parameter (ψ_x, ψ_y). From Binder and Heermann [1].

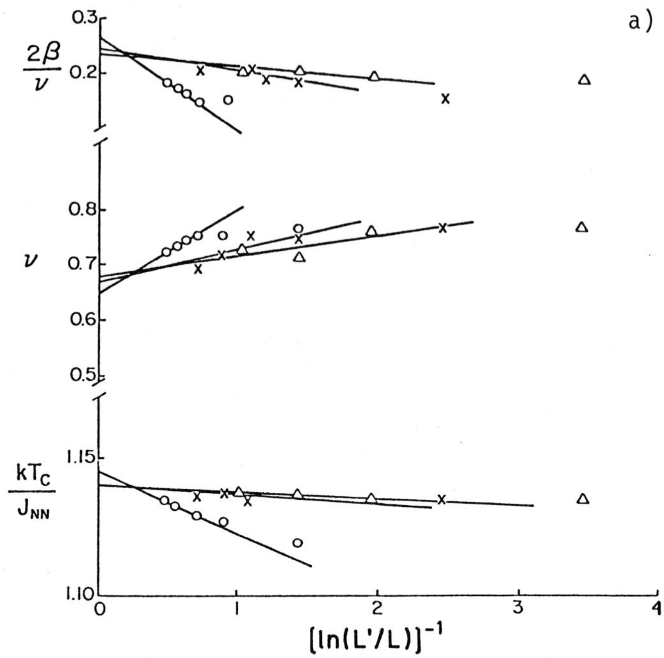

Fig. 18. (a) Variation of critical parameters with $(\ln b)^{-1}$, for the Ising antiferromagnet on the square lattice with $R = J_{nnn}/J_{nn} = 0.65$. Data are for $L = 4$ (circles), $L = 8$ (crosses) and $L = 12$ (triangles), $L' = bL$. Finite blocks with periodic boundary conditions were used throughout.
(b) Variation of the correlation length exponent (upper part) and critical temperature T_c (lower part) with R. Circles show results of phenomenological renormalization, as explained in (a); triangle in the upper part denotes the Monte Carlo "data collapsing" result [77], triangles in the lower part are real-space renormalization results [83]. Crosses denote MCRG results [80], full dots series expansion results [81], and squares show transfer-matrix renormalization data [82]. From Landau and Binder [57].

5 Finite Size Rounding of Critical Slowing Down

So far, we have considered only static quantities and analyzed the respective finite-size effects. Now it is well-known that the Metropolis importance sampling Monte Carlo method [85] amounts to a numerical realization of a Markovian master equation which can be interpreted as a description of a dynamical stochastic process [1-5,86,87]. E.g., the single spin-flip Monte Carlo method applied to the Ising model can be interpreted [86] as a numerical study of the Glauber [88] Kinetic Ising model.

Studies of dynamic relaxation processes with Monte Carlo methods is a subject of its own interest [1-4] but outside of consideration here. Thus we summarize only those points which need to be considered when one wishes to apply computer simulation techniques in conjunction with finite size scaling to analyze phase transitions.

We start by noting that the "intrinsic" relaxation time τ of the single spin flip Kinetic Ising model or related stochastic models without conservation laws [89] in the

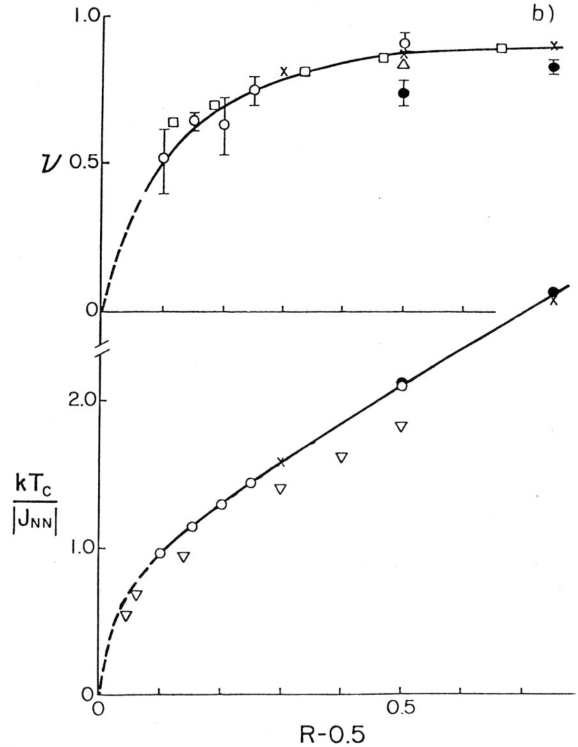

Fig. 18. Cont.

thermodynamic limit ($L \to \infty$) exhibits a critical singularity, described by a dynamic exponent z (with $z \geq 2 - \eta$)

$$\tau \propto \xi^z \propto |1 - T/T_c|^{-\nu z}. \tag{88}$$

In a finite system, this critical divergence exhibits a finite size rounding, again controlled by the rule that the rounding sets in when L is comparable to ξ. Hence the maximum "intrinsic" relaxation time is

$$\tau_{\max} \propto L^z \qquad (T = T_c). \tag{89}$$

From (89) and the result for the statistical error δA of a quantity A [1,2,86] based on n observations a "time interval" δt apart

$$\begin{aligned} \langle(\delta A)^2\rangle &= \frac{1}{n}[\langle A^2\rangle - \langle A\rangle^2]\left(\frac{\tau_A}{\delta t}\right) \approx \frac{2\tau_A}{n\delta t}\left[\langle A^2\rangle - \langle A\rangle^2\right] = \\ &= 2\frac{\tau_A}{t_{\text{obs}}}\left[\langle A^2\rangle - \langle A\rangle^2\right], \qquad t_{\text{obs}} = n\delta t \gg \tau_A, \end{aligned} \tag{90}$$

τ_A being the relaxation time of the quantity A, we realize that it will be difficult to reach a high accuracy near T_c, since τ_A there gets large. Let us analyze this problem for the order parameter $\langle |M| \rangle$: using (90) we get

$$\langle (\delta M)^2 \rangle_{T_c} = \frac{2\tau_{\max}}{t_{\text{obs}}} (\langle M^2 \rangle_{T_c} - \langle |M| \rangle^2_{T_c}) \tag{91}$$

and combining this with (59),(64),(89) yields

$$\langle (\delta M)^2 \rangle_{T_c} = \frac{2\tau_{\max} \chi'_{\max} k_B T_c}{t_{\text{obs}} L^d} \propto \frac{L^{z+\gamma/\nu - d}}{t_{\text{obs}}}. \tag{92}$$

Since *on a serial computer* the CPU time for one Monte Carlo step per site increases as L^d, the CPU time needed to obtain data for $\langle |M| \rangle$ at T_c with a given constant accuracy increases with L as $L^{z+\gamma/\nu} \approx L^4$, using the fact that $z \approx 2$ [90,91], independent of d for $d \leq 4$. Thus an increase by a factor of 10 in the linear dimension would require 10^4 more computing time. Thus we see why vector processors and parallel computers, for which there is less increase in CPU time needed for one Monte Carlo step per site with increasing linear dimension, are indispensable for the Monte Carlo simulation of very large lattices at T_c. Obviously, for this problem it is very important to develop algorithms that reduce critical slowing down, such as the Swendsen-Wang [31,32] and Wolff [33,34,36] cluster algorithms. This is outside of consideration here.

The fact that successive configurations generated by Monte Carlo methods (other than simple sampling) are highly correlated, and the correlation time τ increases dramatically with increasing linear dimension at T_c (89), is particularly cumbersome when one samples the specific heat C from internal energy fluctuations, or the susceptibility χ from magnetization fluctuations, and so on [92]. This is because these quantities are systematically underestimated if the observation time on which the analysis is based is too short. Suppose we divide a very long run into N_b bins of N observations of magnetization, energy and so on each. For each bin the specific heat C and susceptibility χ are now estimated from the corresponding fluctuation relation, and the results are averaged over all the bins. The results of such a procedure, for the three-dimensional Ising model at $J/k_B T_c = 0.221654$ using $\delta t = 10 MCS/\text{site}$ and a total effort of at least 3×10^6 MCS/site for each lattice size are shown in Fig.19 as a function of "bin length" N. One expects that both χ and C are reduced for finite N in comparison to their expectation value $\chi_\infty(L)$, $C_\infty(L)$ simply by a factor $(1 - 1/n)$, n being the number of statistically independent observations [93]. Using the fact that the number of independent "measurements" obtained from N correlated measurements is [86,87] $n = N/[2(\tau/\delta t) + 1]$, we have [92]

$$\chi_N(L) = \chi_\infty(L) \left[1 - \frac{2(\tau/\delta t) + 1}{N}\right], C_N(L) = C_\infty(L) \left[1 - \frac{2(\tau/\delta t) + 1}{N}\right], \tag{93}$$

for $n \gg 1$. Fig. 19 shows that for large L due to the dramatic increase of τ with L for $T = T_c$ (89) N must be huge in order to avoid a dramatic error in either χ or C!

For $T \geq T_c$, the "intrinsic" order parameter relaxation time considered in (88),(89) is in fact the absolutely largest relaxation time in the system. For $T < T_c$, this is not true due to the existence of the "ergodic" time τ_e. As was already mentioned, the ergodic time is proportional to (cf. Fig.5) [52]

$$\tau_e \propto P_L(M = M_L)/P_L(M = 0) \propto \exp(2L^{d-1} f_{\text{int}}/k_B T). \tag{94}$$

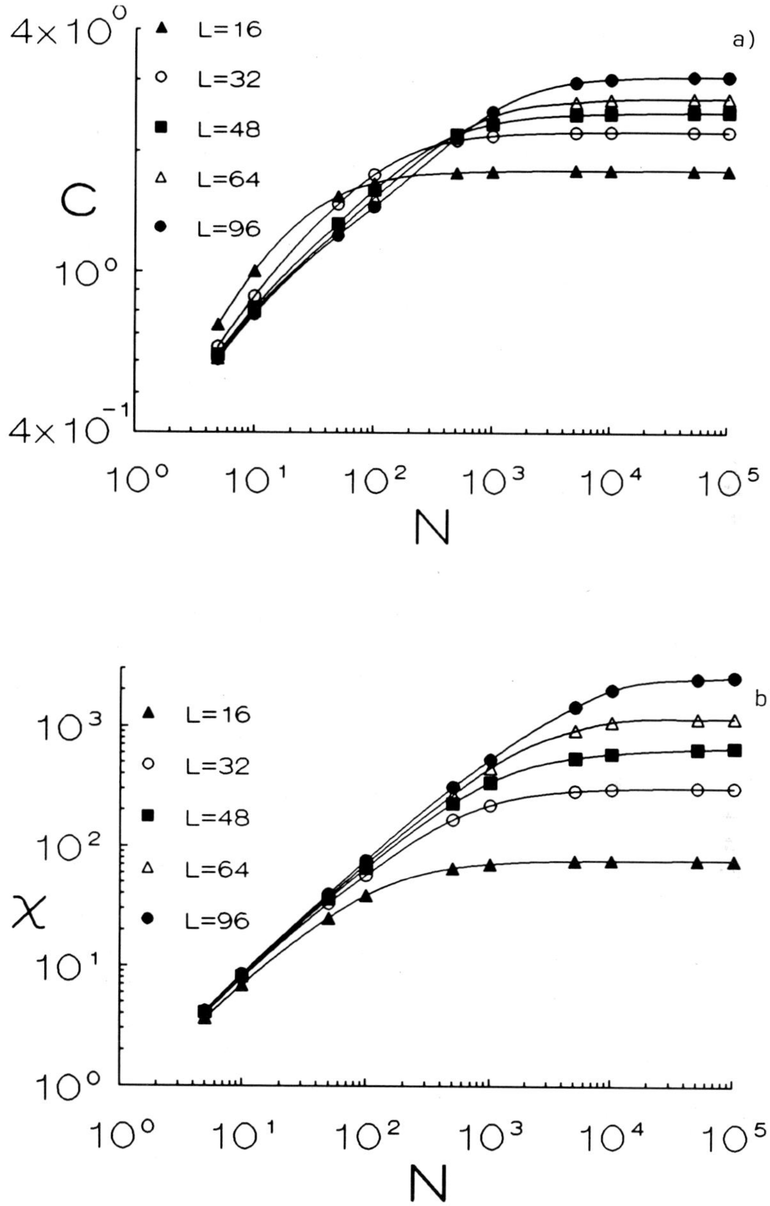

Fig. 19. Specific heat (a) and susceptibility (b) plotted vs. the number of measurements N in a bin, for $\delta t = 10$ MCS and the simple cubic nearest neighbor Ising model right at T_c. Different symbols denote different linear dimensions, as indicated in the figure. From Ferrenberg, Landau and Binder [92].

Since near T_c we have $f_{\text{int}}/k_BT \propto \xi^{-(d-1)}$, one can write a finite size scaling relation for the ergodic time as follows

$$\tau_e \propto L^z \exp[\text{const}(L/\xi)^{d-1}], \qquad T < T_c. \tag{95}$$

Equation (95) was studied by simulations [94], and in fact also some of the most precise estimates of z are due to simulations [90,91].

In addition to the intrinsic relaxation time τ, (88), which describes how a small deviation δM from the spontaneous magnetization M_{sp} relaxes, and to the ergodic time τ_e, describing the time needed to flip over the sign of the total magnetization (Fig.5), various nonlinear relaxation times also are of interest, particularly below T_c. E.g., if we start the system in a fully aligned ferromagnetic state, the relaxation times $\tau_M^{(n\ell)}$, $\tau_E^{(n\ell)}$ of magnetization, energy and so on needed to relax to their equilibrium also are expected to diverge near T_c. The resulting power laws, however, differ from (88); rather one has [95,96]

$$\tau_M^{(n\ell)} \propto (1 - T/T_c)^{-(\nu z - \beta)}, \quad \tau_E^{(n\ell)} \propto (1 - T/T_c)^{-(\nu z - 1 + \alpha)}, \tag{96}$$

α being the critical exponent of the specific heat [7]. Writing (96) in terms of ξ it is again easy to predict the finite size rounding of these quantities as

$$\tau_M^{(n\ell)}|_{T_c} \propto L^{z-\beta/\nu}, \quad \tau_E^{(n\ell)}|_{T_c} \propto L^{z-(1-\alpha)/\nu}. \tag{97}$$

Equation (97) clearly demonstrates that one must not rely on the relaxation of the energy when a judgement is desired whether after the start of a simulation the system "has reached equilibrium" and averaging can be started, because the order parameter relaxes slower than the energy does.

If one starts a simulation in a disordered system and no symmetry breaking field is present, ordered domains form whose linear dimensions $\ell(t)$ grow according to a power law [97,98] (Figs. 20,21)

$$\ell(t) = (\Omega t)^x, \qquad \Omega \propto (1 - T/T_c)^{\nu z - \nu/x}. \tag{98}$$

Now with such a "disordered start" in the regime of the ordered phase the times to reach equilibrium are at least of the order of t_L defined by

$$\ell(t_L) \equiv L; \text{ that is } t_L = \Omega^{-1} L^{1/x} \propto \tau (L/\xi)^{1/x} \propto \xi^z (L/\xi)^{1/x}. \tag{99}$$

While the times $\tau_M^{(n\ell)}$, $\tau_E^{(n\ell)}$ are small for T distinctly below T_c, the time t_L is always rather large, since $x = 2$ in the nonconserved kinetic Ising model [97], and hence

$$t_L \propto L^2. \tag{100}$$

It should be noted that this time t_L is only a lower bound to establish true equilibrium: it only says that domains with sizes comparable to the system linear dimension are present. There is no guarantee that the system exists in a monodomain state. Often one finds that [99,100] the system after the time t_L is in a state with a few frozen-in domain walls oriented parallel to a surface of the system, such that each wall is connected to itself via the periodic boundary condition (Fig.5). It then may take a very long time for such frozen-in domain walls to disappear from the system.

a)

Fig. 20. A series of snapshot pictures of a time evolution of an Ising system of size 120 x 120 with $J_{nnn} = J_{nn} < 0$ at $k_BT/|J_{nn}| = 1.33$ (note $k_BT_c/|J_{nn}| \approx 2.07$). Sites with spin down are denoted by a circle, triangle, x or +, if they belong to one of the four domains possible for the structures shown in Fig. 17b,c. Sites belonging to domain walls are not shown. Times shown are $t = 10$ (a), 20 (b), 40 (c), 60 (d), 100 (e) and 140 (f), measured in MCS/site after a start in a completely random configuration. From Sadiq and Binder [98].

6 Finite Size Scaling without Hyperscaling

The finite size scaling theory as it has been described up to now, where one scales the linear dimension L with the correlation length ξ, rests on the validity of the hyperscaling relation between critical exponents, $d\nu = \gamma + 2\beta$ [18,47]. As is well-known, the physical interpretation of hyperscaling is that there is only one characteristic length that diverges as $T \to T_c$, and hence it is clear that this length needs to be scaled with L. In contrast, when hyperscaling does not hold, several divergent lengths can be defined, and then the situation is not so simple.

As is well-known, hyperscaling is violated in systems above their marginal dimensionality $d^\star$ where mean-field predictions [7] for critical exponents become valid (for example, $d^\star = 4$ for Ising models [8], $d^\star = 6$ for percolation [14]). Then with fully periodic boundary conditions in an L^d geometry a simple modified form of finite size scaling holds [47], the correlation length ξ being replaced by a *thermal length* ℓ_T defined by [47,101]

$$\ell_T^d = k_BT\chi' M^{-2} \propto |1 - T/T_c|^{-(\gamma+2\beta)} = |1 - T/T_c|^{-2}, \tag{101}$$

where in the last equality the mean field exponents $\gamma = 2\beta = 1$ were inserted.

b)

c)

Fig. 20. Cont.

d)

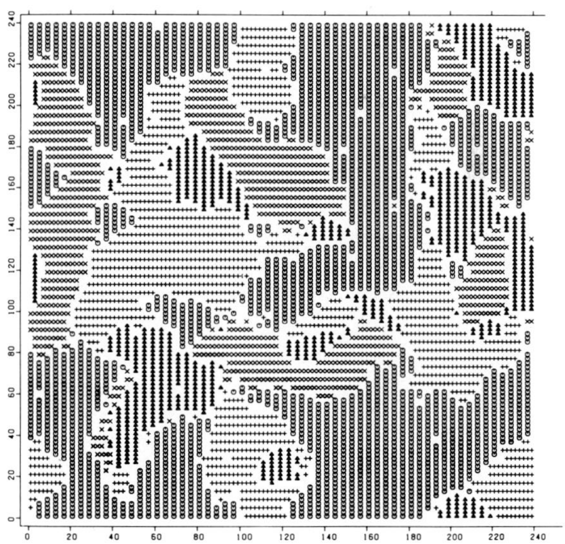

e)

Fig. 20. Cont.

f)

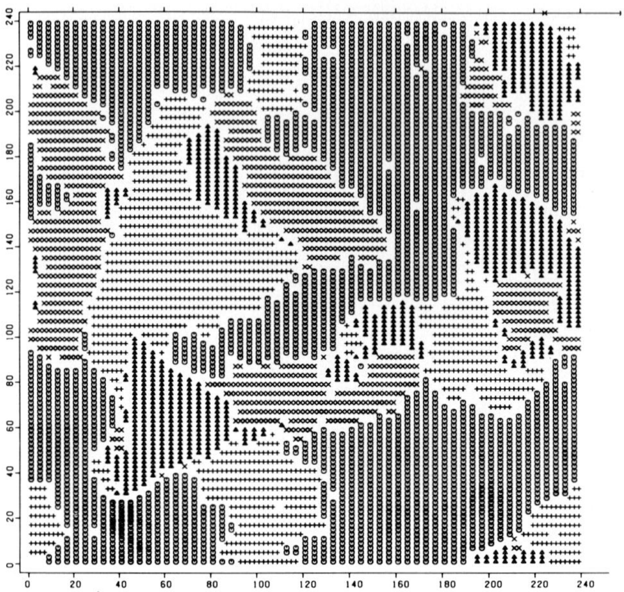

Fig. 20. Cont.

Equation (101) is understood by noting that (67) is true also for $d > d^\star$, and the argument of the exponential function can be written as

$$\frac{(M \pm M_{\rm sp})^2 L^d}{2k_BT\chi'} = \frac{(M/M_{\rm sp} \pm 1)^2}{2} \frac{L^d}{k_BT\chi' M^{-2}} = \frac{(M/M_{\rm sp} \pm 1)^2}{2} \left(\frac{L}{\ell_T}\right)^d , \quad (102)$$

which shows that L scales with ℓ_T. For $d < d^\star$, $\ell_T \propto \xi$ but for $d > d^\star$ ξ has a temperature dependence that differs from that of ℓ_T. The general scaling behavior replacing (75) is

$$\begin{aligned} P_L(M) &= (1 - T/T_c)^{-\beta} P'(L/\ell_T, M(1 - T/T_c)^{-\beta}) = \\ &= L^{d\beta/(\gamma+2\beta)} \tilde{P}'(L/\ell_T, ML^{d\beta/(\gamma+2\beta)}). \end{aligned} \quad (103)$$

This means that the exponent $1/\nu$ in (61)-(65) is replaced by the exponent $d/(\gamma+2\beta)$, the inverse of the exponent of the thermal length ℓ_T. Equation (103) has been verified for the five-dimensional Ising model [47,101], where $1/\nu = 2$ while $d/(\gamma + 2\beta) = 5/2$, see Fig. 22.

7 Finite Size Scaling for First Order Phase Transitions

Next we turn to finite size effects at first order phase transitions [20,49,50]. In an infinite system, a first order transition is characterized by a delta-function singularity,

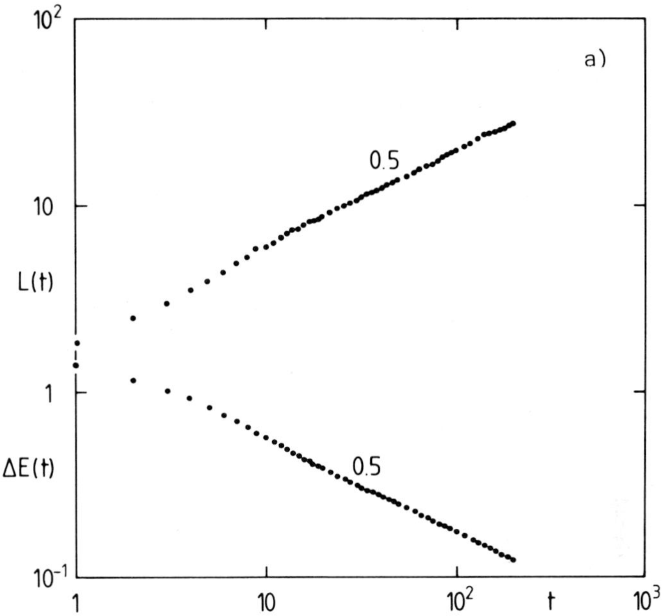

Fig. 21. Log-log plot of the average domain linear dimension $\ell(t)$ vs. times (upper curve) and of the excess energy $\Delta E(t)$ contained in domain walls (lower curve), for $k_B T/|J_{nn}| = 1.33$ (a) and 0.75 (b). Time is measured in units of Monte Carlo steps per site (using Glauber-type dynamics as in case of Fig. 20). Both figures represent averages over 45 samples of 80×80 lattices with periodic boundary conditions. From Sadiq and Binder [98].

see Fig.23. For example, if the transition is driven by temperature this singularity is the latent heat; in an Ising magnet, a first order transition occurs for $T < T_c$ at $H = 0$ on variation of the magnetic field, and hence we get a delta function in the susceptibility. In finite systems, of course, these delta functions are again rounded off [20,49,50,102-119].

One can understand this behavior most simply by generalizing (67) to include the dependence on the magnetic field [20]: the weights of the two Gaussians are no longer equal ($= \frac{1}{2}$), but rather weighted according to the Zeeman energy Boltzmann factors $\exp(\pm H M_L L^d / k_B T)$. In addition, one must take into account that for $H \neq 0$ the Gaussian peaks no longer occur at $M = \pm M_L$ but rather at $M = \pm M_L + \chi^{(L)} H$. This yields

$$P_L(M) = L^{d/2} [2\pi k_B T \chi^{(L)}]^{1/2}$$
$$\left\{ \exp\left(\frac{H M_L L^d}{k_B T}\right) \exp\left(-\frac{(M - M_L - \chi^{(L)} H)^2 L^d}{2 k_B T \chi^{(L)}}\right) \right.$$
$$\left. + \exp\left(-\frac{H M_L L^d}{k_B T}\right) \exp\left(\frac{(M + M_L - \chi^{(L)} H)^2 L^d}{2 k_B T \chi^{(L)}}\right) \right\} /$$

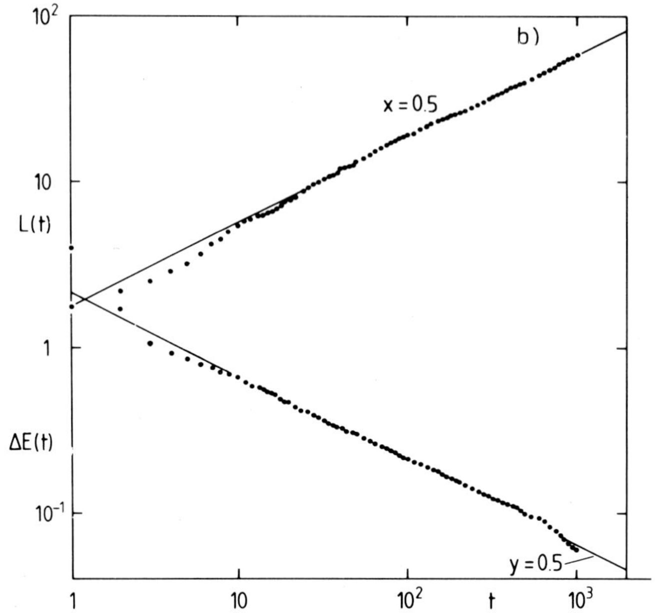

Fig. 21. Cont.

$$\left\{\exp\left(\frac{HM_LL^d}{k_BT}\right)+\exp\left(-\frac{HM_LL^d}{k_BT}\right)\right\} \tag{104}$$

This approach yields for the magnetization

$$\langle M\rangle_L = \chi^{(L)}H + M_L\tanh\left(\frac{HM_LL^d}{k_BT}\right) \tag{105}$$

and the susceptibility

$$\chi(H,T,L)=\left(\frac{\partial\langle M\rangle_L}{\partial H}\right)_T=\chi^{(L)}+M_L^2\frac{L^d}{k_BT}\cosh^{-2}(HL^dM_L/k_BT). \tag{106}$$

Equation (106) shows that the delta function singularity which occurs for $H=0$ in the limit $L\to\infty$ for finite L is smeared out into a peak of height proportional to L^d and of width ΔH proportional to L^{-d}.

Of course, (104)-(106) apply only if one records the true equilibrium behavior of the system (that is, one must apply observation time $t_{\rm obs}$ exceeding the ergodic time τ_e, see Fig. 23). Working in the canonical ensemble with a single spin flip-type algorithm, one would observe for too short observation time a single-peak structure for $P_L(M)$ near $H=0$ rather than the correct double-peak structure: the ordered state of the Ising ferromagnet then is metastable even in a weak field of direction opposite to the magnetization. Such metastable states and the mechanism of their decay is a problem of widespread interest [49,120-132] but outside of consideration here. In

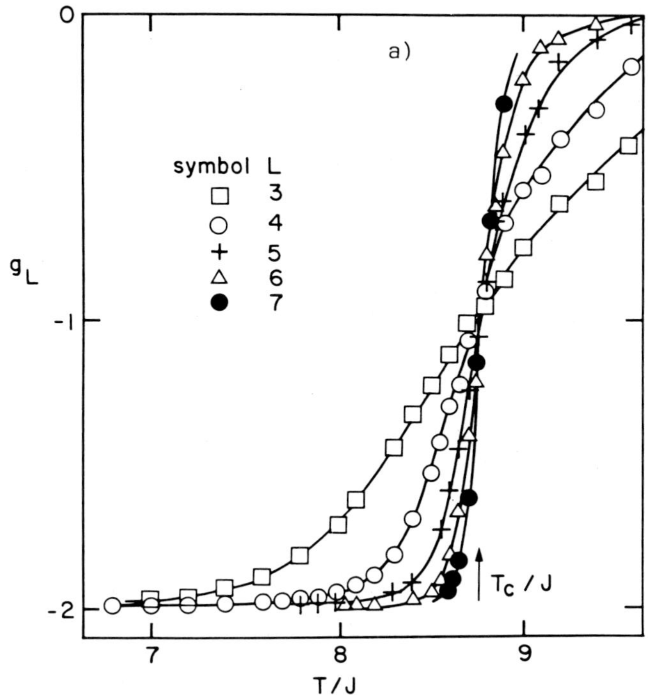

Fig. 22. (a) Plot of the "renormalized coupling constant" $g_L = -3U_L$ of the five dimensional nearest-neighbor hypercubic Ising model versus temperature. Curves for different lattice size intersect in a common intersection, which agrees with the critical temperature T_c predicted from high-temperature series expansions. (b) Same data as in (a) but replotted as $(T/T_c - 1)L^{5/2} \propto (L/\ell_T)^{5/2}$, as suggested from (101)-(103). From Binder et al. [47].

the canonical ensemble, (104)-(106) (and their generalization to temperature-driven first-order transitions) thus are useful only for very small systems, see for example, Fig. 24. However, one can avoid to a large extent the fight against the slowing down at first-order transitions (described by the time τ_e) by working in different statistical ensembles, for example, the gaussian ensemble [115,133], "microcanonical" ensembles where a suitably invented kinetic energy term is added [134], "multicanonical ensembles" [118,135,136], and so on. This is a field of intense current research, but out of consideration here. But Fig. 24 shows that the present description is in quantitative agreement with simulation results for the two-dimensional nearest-neighbor Ising square lattice, when the spontaneous magnetization M_{sp} (M_L is taken equal to the spontaneous magnetization M, independent of L, for $L \gg \xi$) is known exactly [137], and no adjustable parameters exist whatsoever.

The field-driven transition in the Ising model is particularly simple, since the symmetry $\chi(H,T,L) = \chi(-H,T,L)$ holds exactly for all L: thus the transition is rounded *but not shifted* by the finite size of the system. The above symmetry can be broken either for example, by multispin-interactions [20,138], by boundary fields in

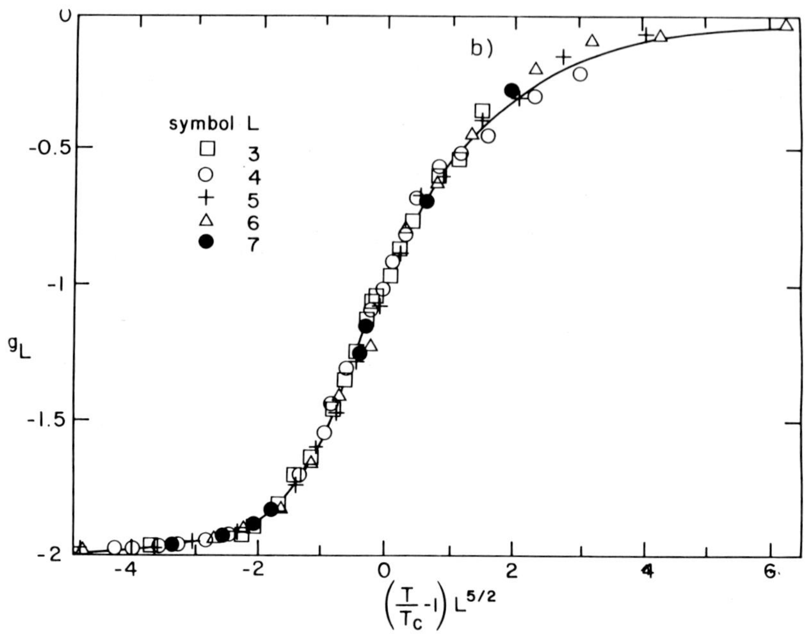

Fig. 22. Cont.

systems with free surface boundary conditions [139-141] and so on. While the first case (asymmetric bulk transition) leads to a shift exponent [20] $\lambda_H = d$ (10), the second case (symmetry-breaking boundary fields) yields [139-141] $\lambda_H = 1$. In both cases, the rounding exponent is $\theta_H = d$ at field driven first-order transitions and the exponent γ_c describing the maximum of the response function (5) also is $\gamma_c = d$. In the purely symmetrical case, $\chi(H,T,L) = \chi(-H,T,L)$, the energy is regular through the transition at $H = 0$ in the thermodynamic limit, and then the specific heat also in the finite system does not develop any singular behavior as $L \to \infty$. Only for the asymmetric case, where in the thermodynamic limit the transition does not occur at $H = 0$ but at a nontrivial critical field $H_c(T)$, do we expect a latent heat singularity in the specific heat as well (cf. Fig.23), and then (6) applies with $\alpha_c = d$ as well. While in the symmetrical case the order parameter at $H = 0$ is exactly zero for all finite L for symmetry reasons (cf. Fig.23), for all asymmetrical cases it is of order unity at the transition field $H_c(T)$ and hence β_c (8) does not occur ($\beta_c = 0$).

A more general case, also more interesting for applications, are first order transitions driven by temperature from an ordered phase at low temperature to a disordered phase at higher temperatures. Obviously, there is never a symmetry between the high-temperature and the low temperature phases, and also the latter is degenerate while the former is not. In this case, hence, there always is a shift of the effective transition temperature $T_c(L)$, where the rounded peak of the specific heat representing the smeared delta function of the latent heat has its maximum $C_L^{\max}$ relative to the true transition temperature $T_c(\infty)$. Finite size effects can again be understood by a

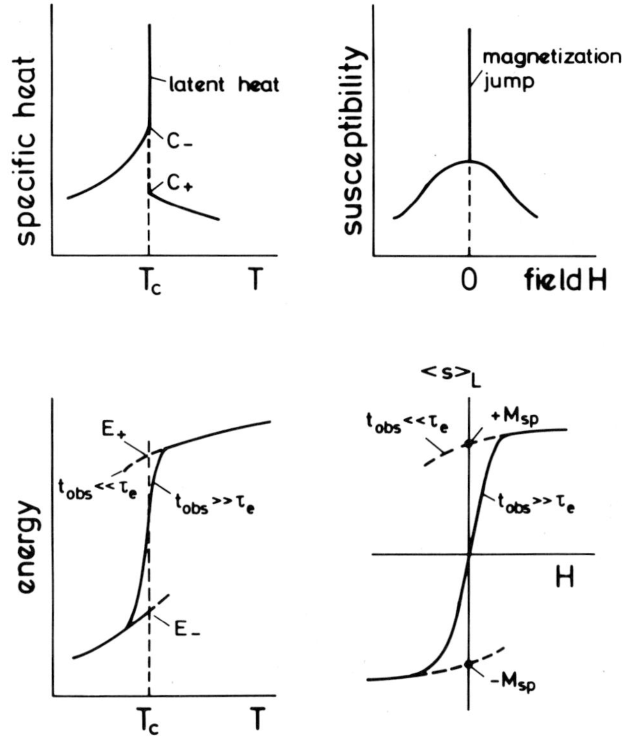

Fig. 23. Schematic variation of the specific heat and internal energy with temperature near a temperature-driven first order transition at T_c (left part). The delta function singularity of the specific heat C represents the latent heat $E_+ - E_-$. Full curve for the energy shows the equilibrium behavior (observation time $t_{\text{obs}} \gg \tau_e$, the ergodic time), while broken curve shows hysteresis (metastable states occur for $t_{\text{obs}} \ll \tau_e$). In the right part, the variation of the susceptibility and magnetization at a field-driven transition at $H = 0$ is shown. Now the delta function singularity of χ represents the magnetization jump M_{sp}. Again for $M(H)$ the full curve shows the equilibrium behavior ($t_{\text{obs}} \gg \tau_e$) while broken curves indicate metastable states (for $t_{\text{obs}} \ll \tau_e$). From Binder and Heermann [1].

discussion in terms of thermodynamic fluctuation theory [51], similar to (67),(104). We just note the fluctuation relation linking energy fluctuations in the canonical ensemble to the specific heat to write Gaussians for the energy distribution $P_L(E)$ for each pure phase and superpose them with the proper but unnormalized weights a_+, a_- for the high temperature (+) and low temperature (-) phase ($\pm$ refers to the sign of $\Delta T \equiv T - T_c(\infty)$) [50]

$$P_L(E) \propto \frac{a_+}{\sqrt{C_+}} \exp\left\{-\frac{(E - E_+ - C_+\Delta T)^2 L^d}{2k_B T^2 C_+}\right\} +$$

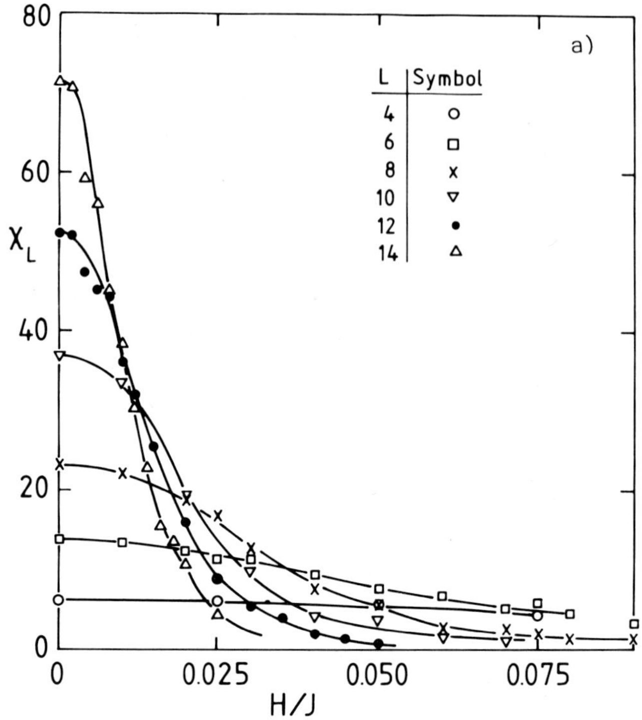

Fig. 24. (a) Susceptibility $\chi(H,T,L)$ of nearest-neighbor Ising square lattices at $k_BT/J = 2.1$ plotted versus magnetic field for various $L's$. Curves are guides to the eye only. (b) Same data replotted in scaled form, $\chi(H,T,L)/L^2$ plotted versus scaled field HL^2/J. Arrow indicates the asymptotic value $M_{\rm sp}^2 J/k_BT$ calculated from the exact solution [137]. Note that $k_BT_c/J \cong 2.269$ for the Ising model [46]. Broken curve is the scaling function of (106), the additive correction term $\chi^{(L)}$ being omitted. From Binder and Landau [20].

$$+\frac{a_-}{\sqrt{C_-}}\exp\left\{-\frac{(E-E_--C_-\Delta T)^2L^d}{2k_BT^2C_-}\right\}. \tag{107}$$

Here the specific heat in the infinite system near T_c behaves as

$$\lim_{T\to T_c^-} C(T) = C_-, \quad \lim_{T\to T_c^+} C(T) = C_+, \tag{108}$$

and one should remember that (107) is true at best in the limit $L \gg \xi$ and contributions due to interfaces between coexisting phases are disregarded again. If L is not very much larger than ξ, (107) can still be used but $E_\pm$, $C_\pm$ in (107) should be taken as L-dependent quantities, $E_\pm^{(L)}$, $C_\pm^{(L)}$. Since we expect [9,106]

$$E_\pm^{(L)} - E_\pm^{(\infty)} \propto \exp(-L/\xi_\pm), \quad C_\pm^{(L)} - C_\pm^{(\infty)} \propto \exp(-L/\xi_\pm), \quad L \gg \xi_\pm, \tag{109}$$

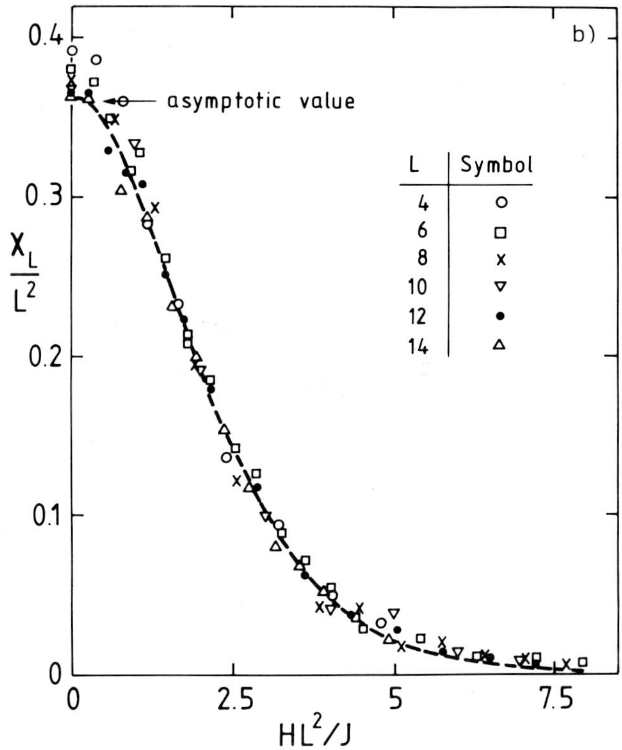

Fig. 24. Cont.

where $\xi_\pm$ are the correlation lengths of fluctuations in the pure phases at phase coexistence for $T = T_c(\infty)$ in the thermodynamic limit which stay finite, we again anticipate a rapid approach of $E_\pm^{(L)}$, $C_\pm^{(L)}$ to $E_\pm^{(\infty)}$, $C_\pm^{(\infty)}$ and ignore the L-dependence in (109) in the following throughout.

The weights a_+, a_- are expressed in terms of the degeneracies q_+, q_- of the two phases (normally $q_+ = 1$, $q_- = q$) and their free energy differences $\Delta F = F_+ - F_-$ as

$$a_+ = q_+ \exp\left\{-\frac{\Delta F L^d}{2k_B T}\right\}, \quad a_- = q_- \exp\left\{-\frac{\Delta F L^d}{2k_B T}\right\} \tag{110}$$

While originally [50] it was suggested to use $a_\pm = q_\pm \sqrt{C_+} \exp[\mp(\Delta F L^d)/(2k_B T)]$, based on a Landau theory suggesting that at T_c all peaks of the order parameter distribution $P_L(\mathbf{M})$ of the coexisting pure phases should have equal *height*, one now believes, based on rigorous arguments [108-110] that this is incorrect, and rather at $T = T_c$ all coexisting pure phases have equal *weight*. Using then $F_\pm = E_\pm - TS_\pm$, $\Delta F(T_c) = 0$ one finds $\Delta F \approx -(E_+ - E_-)\Delta T/T_c$ near $T_c = T_c(\infty)$ and thus

$$a_\pm = q_\pm \exp\left[\pm\frac{\Delta T(E_+ - E_-)L^d}{2k_B T T_c}\right] \tag{111}$$

From (107),(111) it is again straightforward to obtain the energy $\langle E\rangle_L$ and specific heat $C(T,L)$ explicitly,

$$\langle E\rangle_L = \frac{a_+E_+ + a_-E_-}{a_+ + a_-} + \Delta T\frac{a_+C_+ + a_-C_-}{a_+ + a_-}, \tag{112}$$

$$C(T,L) = \frac{\partial\langle E\rangle_L}{\partial T} = \frac{a_+C_+ + a_-C_-}{a_+ + a_-} + \frac{a_+a_-}{k_BT^2}L^d \frac{[(E_+ - E_-) + (C_+ - C_-)\Delta T]^2}{(a_+ + a_-)^2}. \tag{113}$$

From (113) one finds that the maximum of the specific heat occurs at [109]

$$\frac{T_c(L) - T_c}{T_c} = \frac{k_BT_c\ln(q_+/q_-)}{E_+ - E_-}\frac{1}{L^d}, \tag{114}$$

and has a height [50,109]

$$C_L^{\max} \approx \frac{(E_+ - E_-)^2L^d}{4k_BT_c^2} + \frac{C_+ + C_-}{2}. \tag{115}$$

Since the temperature region δT over which the rounding of the delta function peak occurs is given just by taking the argument of the exponential functions in (111) equal to order unity, $\delta T \approx 2k_BT_c^2/[(E_+ - E_-)L^d]$, we conclude that the exponents λ_T, α_c, θ_T defined in (1),(2) and (6) are all equal to the dimensionality

$$\lambda_T = \alpha_c = \theta_T = d. \tag{116}$$

Thus, since there is no diverging characteristic length to which the linear dimension L could be compared at a first-order transition, it is simply the volume L^d that controls the size effects [20,50,102-119].

Figs.25,26 present the first examples [50] (meanwhile considerably more accurate work is available [111-114,116-119]) for the q-state Potts model (48), for the case $q = 10$, $d = 2$. The square lattice is a good test case since T_c, E_+, E_- is known from exact solutions [142] and hence a nontrivial test of (107)-(116) is possible. It is seen already from the limited data of Ref.50 that the above phenomenological theory accounts for the behavior of the Monte Carlo results nicely (Fig. 25).

A quantity which is useful for distinguishing first order transitions from second order transitions is [50]

$$V_L = 1 - \langle E^4\rangle_L/[3\langle E^2\rangle_L^2] \tag{117}$$

One finds that V_L converges for $L \to \infty$ towards $V_\infty = 2/3$ in all cases except for first order transitions where it is related to the latent heat $|E_+ - E_-|$,

$$V_L^{\min} = \frac{2}{3} - \frac{1}{3}(E_+^2 - E_-^2)^2/(2E_+E_-)^2 + O(L^{-d}). \tag{118}$$

This minimum occurs at a temperature [109]

$$\frac{T_v(L) - T_c}{T_c} = \frac{k_BT_c}{E_+ - E_-}\ln\left(q\frac{E_-^2}{E_+^2}\right)\frac{1}{L^d}, \tag{119}$$

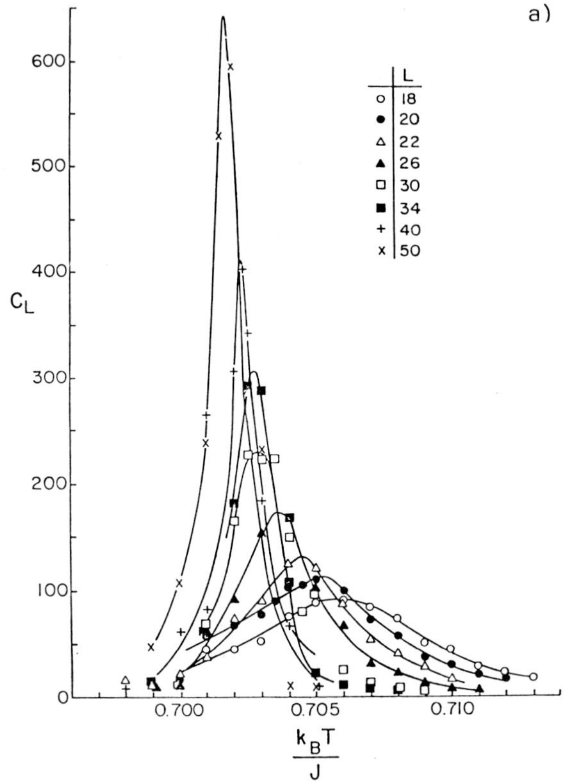

Fig. 25. (a) Temperature variation of the specific heat for various lattice linear dimensions L for the 10-state Potts model on the square lattice. Data for some lattice sizes were omitted to preserve the clarity of the figure. Curves are guides to the eye only. (b) Scaling of the specific heat data, plotting C_L/L^2 versus $[T - T_c(\infty)]L^2$. In this case $q_- = q = 10$, $q_+ = 1$. The solid curve results from (107)-(113), as described in Ref.[50]. Note that the additive correction $[a_+C_+ + a_-C_-]/(a_+ + a_-)$ in (113) has been omitted. From Challa et al. [50].

and hence $T_v(L)$ can also be used as a criterion to locate the first order transition. While both $T_v(L)$ and the temperature $T_c(L)$ of the specific heat maximum differ from $T_c = T_c(\infty)$ by terms of order $1/L^d$, Borgs et al. [109,110] show that a temperature $T_i(L)$ found from an intersection of the curve $E(T, L) \equiv \langle E \rangle_L$ and $E(T, 2L)$

$$E(T_i, L) = E(T_i, 2L) \tag{120}$$

would differ from T_c only by exponentially small corrections

As another example of these concepts, Fig. 26 compares the behavior of V_L for $q = 2$ (Ising model), $q = 3$ (where the transition in $d = 2$ is also second order, but with different critical exponents [26]) and $q = 10$. It is seen that for $q = 2, 3$ V_ℓ converges smoothly towards $2/3$, a shallow minimum disappears as L increases, while for $q = 10$ the minimum sharpens up as L increases and is in fact consistent with the theoretical value in this case ($V_\infty^{\min} = 0.559$).

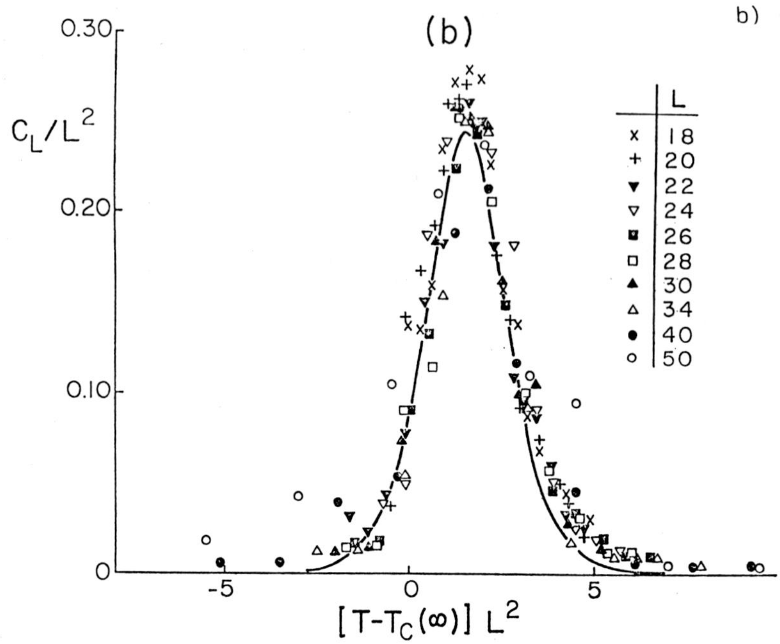

Fig. 25. Cont.

While finite size effects at first order transitions are now fairly well understood for models with a discrete number of states such as the Ising model transition as function of the field [20,106] or the Potts model transition as function of temperature [50,109-119], much less is known about first order transitions in systems where in the ordered phase a continuous symmetry is broken, such as the field-driven transition of *XY*- and Heisenberg ferromagnets [107,143-146].

8 Finite Size Effects in Anisotropic Geometries and for Anisotropic Critical Behavior

In this section, we briefly mention extensions of finite size scaling concepts [48] to the case where in one lattice direction one has a linear dimension ($L_\parallel$) that differs from the linear dimension $L_\perp$ in all other $(d-1)$ directions. In particular, for $L_\parallel \to \infty$ the system behaves quasi-one dimensional and this leads to some simplifications that can be utilized [10,147].

Here we summarize the main points only. In a finite system with isotropic critical behavior, the generalization of (75) simply is [48]

$$P_{L_\perp,L_\parallel}(M) = L_\perp^{\beta/\nu}\tilde{P}(L_\perp/\xi, ML_\perp^{\beta/\nu}, L_\parallel/\xi) = \\ = L_\perp^{\beta/\nu}\,\tilde{\tilde{P}}\,(L_\perp/\xi, ML_\perp^{\beta/\nu}, L_\parallel/L_\perp). \tag{121}$$

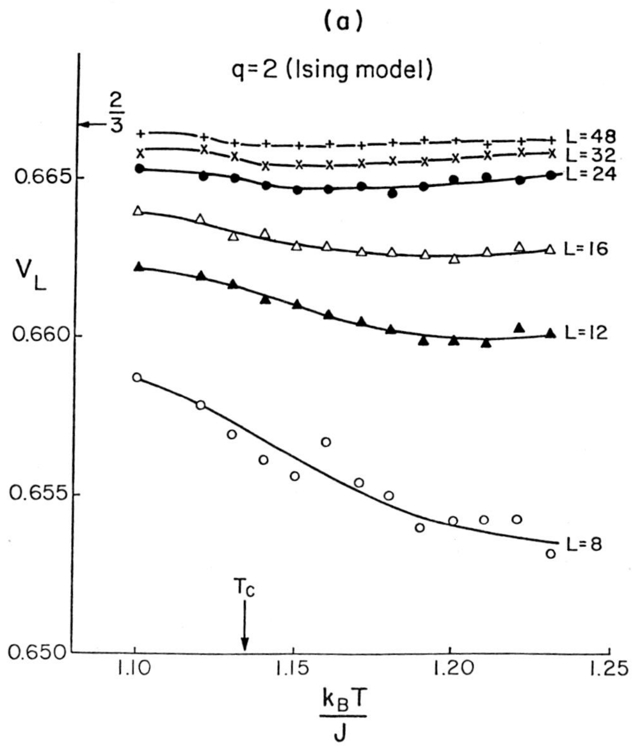

Fig. 26. Temperature variation of the energy cumulant V_L (117) for the q-state Potts model on the square lattice for various choices of L and the cases $q = 2$ (a), $q = 3$ (b) and $q = 10$ (c). Exactly known transition temperatures and the trivial limit of V_L are indicated by arrows. Curves are guide to the eye only. Note that J here means the coupling constant J_p of the Potts Hamiltonian (48) throughout. From Challa et al. [50].

Here we have asserted that *all* linear dimensions scale with ξ, and in the last step we have redefined the scaling function by choosing $(L_\parallel/\xi)/(L_\perp/\xi) = L_\parallel/L_\perp$ rather than $L_\parallel/\xi$ as third variable. This variable $L_\parallel/L_\perp$ then also appears in the resulting moments, for example,

$$\langle|M|\rangle = L_\perp^{-\beta/\nu}\tilde{M}(L_\perp/\xi, L_\parallel/L_\perp), \tag{122}$$

$$\chi = L_\perp^{\gamma/\nu}\tilde{\chi}(L_\perp/\xi, L_\parallel/L_\perp). \tag{123}$$

These scaling forms show that a simple scaling holds only if varying $L_\perp$ one keeps the "aspect ratio" $L_\parallel/L_\perp$ fixed. For the two-dimensional Ising model, this fact has been checked by explicit calculation [148,149], see Fig. 27. Equation (123) also has been suggested using conformal invariance [150].

Let us consider now the two-dimensional case with $L_\parallel \gg L_\perp$ and $T = T_c$. Then the growth of critical correlations is limited by $L_\perp$ and actually we expect a behavior

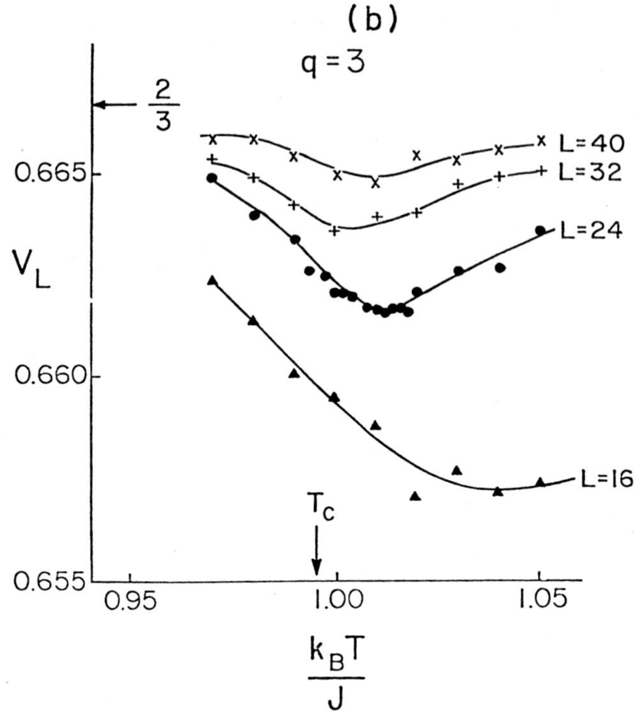

Fig. 26. Cont.

of the correlation function $\langle S(0,0)S(x,y)\rangle$ in the direction x along the strip to decay as [48]

$$\langle S(0,0)S(x,y)\rangle_{T_c} \propto L_\perp^{-\eta} \exp\{-\text{const}(x/L_\perp)\}, \quad L_\perp \ll x \ll L_\parallel. \tag{124}$$

Using this in the fluctuation relation

$$k_BT_c\chi(T_c) = \sum_j \langle S_iS_j\rangle \sim \int_0^{L_\parallel} dx \int_0^{L_\perp} dy \langle S(0,0)S(x,y)\rangle_{T_c} \tag{125}$$

readily yields a result independent of $L_\parallel$

$$k_BT_c\chi(T_c) \propto L_\perp^{1-\eta} \int_0^\infty dx \exp\{-\text{const}(x/L_\perp)\} \propto L_\perp^{2-\eta} = L_\perp^{\gamma/\nu}. \tag{126}$$

From this one estimates the order parameter at T_c as

$$\langle |M|\rangle_{T_c} \propto \sqrt{\langle M^2\rangle_{T_c}} = \sqrt{k_BT_c\chi(T_c)/(L_\parallel L_\perp)} \propto L_\perp^{-\beta/\nu}(L_\perp/L_\parallel)^{-1/2}. \tag{127}$$

Fig.28 provides numerical evidence for this relation.

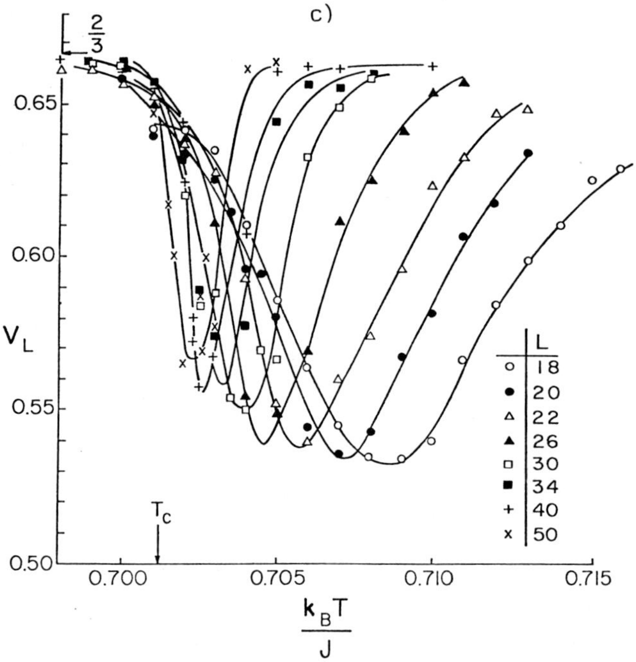

Fig. 26. Cont.

We now briefly discuss the generalization to truly anisotropic critical phenomena, where in the thermodynamic limit the correlation length diverges with different exponents in different directions of the lattice,

$$\xi_{\parallel} \propto t^{-\nu_{\parallel}}, \quad \xi_{\perp} \propto t^{-\nu_{\perp}}. \tag{128}$$

Let us assume that a generalized hyperscaling relation holds

$$(d-1)\nu_{\perp} + \nu_{\parallel} = \gamma + 2\beta. \tag{129}$$

Equations (128),(129) are true, for instance, for uniaxial Lifshitz points [151]. Then the obvious generalization of (122),(123) is [147]

$$\langle |M| \rangle = L_{\perp}^{-\beta/\nu_{\perp}} \tilde{M}(L_{\perp}/\xi_L, L_{\parallel}/\xi_{\parallel}) = L_{\perp}^{-\beta/\nu_{\perp}} \; \tilde{\tilde{M}} \; (L_{\perp}/\xi_{\perp}, L_{\parallel}^{\nu_{\perp}/\nu_{\parallel}}/L_{\perp}) \tag{130}$$

$$k_B T \chi = L_{\perp}^{\gamma/\nu_{\perp}} \tilde{\chi}(L_{\perp}/\xi_{\perp}, L_{\parallel}^{\nu_{\perp}/\nu_{\parallel}}/L_{\perp}). \tag{131}$$

In the last equality of (130), we again transformed the scaling function by choosing instead of the argument $L_{\parallel}/\xi_{\parallel}$ an argument $(L_{\parallel}/\xi_{\parallel})^{\nu_{\perp}/\nu_{\parallel}}(\xi_{\perp}/L_{\perp}) \propto (L_{\parallel}^{\nu_{\perp}/\nu_{\parallel}} t^{\nu_{\perp}})$ $(\xi_{\perp}/L_{\perp}) \propto L_{\parallel}^{\nu_{\perp}/\nu_{\parallel}}/L_{\perp}$. Thus a simple finite size scaling structure no longer results

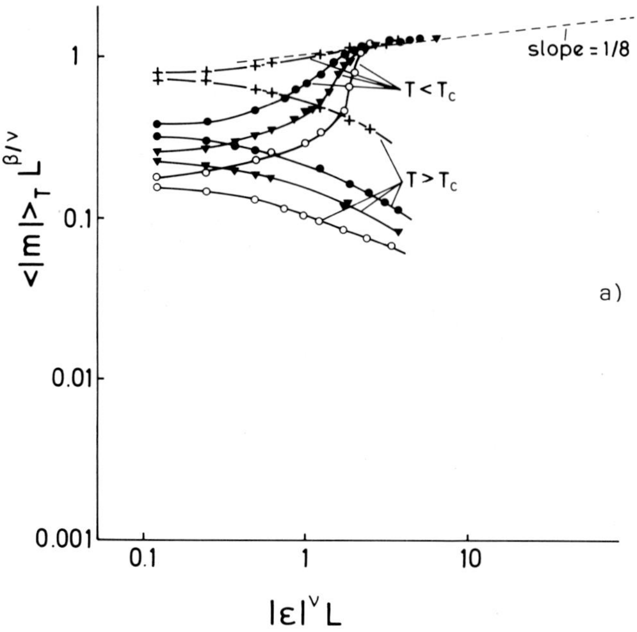

Fig. 27. (a) Finite size scaling plot of the scaled absolute value of the magnetization for an Ising square lattice in the geometry $L_\perp \times L_\parallel$ with $L_\perp = 12$ fixed, including data for $L_\parallel = 12$ (crosses), 72 (dots), 144 (triangles) and 288 (circles). Curves are only guides to the eye. Broken straight line indicates the behavior of the scaling function for large values of the abscissa. Note that $L_\perp$ is denoted as L in this figure. (b) Same as (a) but for fixed aspect ratio $L/L_\parallel = 1/6$. Dots refer to $L = 12$, crosses to $L = 24$ and triangles to $L = 36$, respectively. Note that $\beta = 1/8$, $\nu = 1$ and T_c are taken from the exact solution [137], $\epsilon = 1 - T/T_c$. From Albano et al. [149].

for fixed aspect ration $L_\parallel/L_\perp$, but rather now the ratio $L_\parallel^{\nu_\perp/\nu_\parallel}/L_\perp$ should be kept constant!

An even more complicated behavior results if hyperscaling relations do not hold. This can be demonstrated for simple Landau theory [48,147]: Writing the partition function as a path integral (t is the relative distance from the transition temperature in Landau theory, u_0 a coefficient)

$$Z = \int D\psi \exp[-\mathcal{H}_{\text{eff}}\{\psi\}],\ \mathcal{H}_{\text{eff}}\{\psi\} = \int d\mathbf{r} \left\{\frac{1}{2}(\nabla\psi)^2 + \frac{1}{2}t\psi^2 + \frac{u_0}{4!}\psi^4\right\}, \qquad (132)$$

we can treat the problem for $L_\parallel \gg L_\perp$ as quasi-one-dimensional:

$$\mathcal{H}_{\text{eff}}(\psi) = L_\perp^{d-1} \int_0^{L_\parallel} dx \left[\frac{1}{2}\left(\frac{d\psi}{dx}\right)^2 + \frac{1}{2}t\psi^2 + \frac{u_0}{4!}\psi^4\right]. \qquad (133)$$

In (132),(133) a factor $1/k_BT$ has been absorbed in $\mathcal{H}_{\text{eff}}\{\psi\}$ already. Now it is clear that the main contribution to the path integral comes from such paths for which

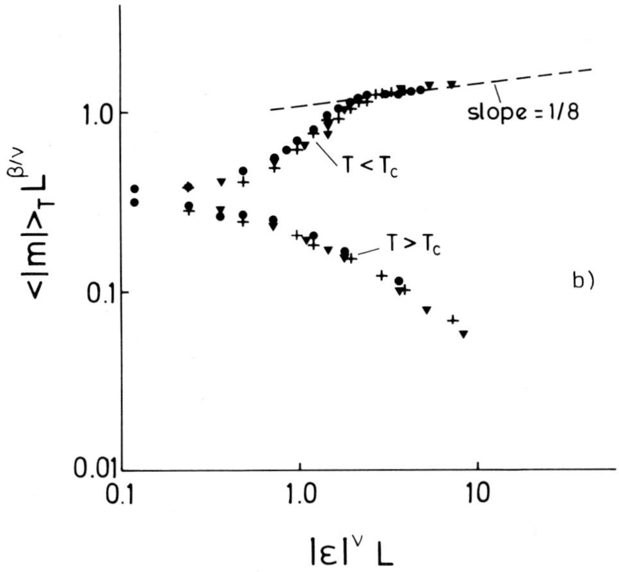

Fig. 27. Cont.

$\mathcal{H}_{\text{eff}}(\psi) \leq 1$ while the contributions from paths with $\mathcal{H}_{\text{eff}}(\{\psi\}) \gg 1$ can be neglected. Assuming a typical behavior of $\psi(x) = \psi_0 \exp(-x/\xi_{\|})$, the three parts of the integrand in (133) give rise to three contributions I, II, III of the integral,

$$\begin{aligned} &\mathcal{H}_{\text{eff}}(\psi) = I + II + III, \quad I \propto L_{\perp}^{d-1}\psi_0^2/\xi_{\|}, \\ &II \propto L_{\perp}^{d-1} t\psi_0^2\xi_{\|}, \quad III \sim L_{\perp}^{d-1} u_0\psi_0^4\xi_{\|}. \end{aligned} \tag{134}$$

For $t > 0$ terms I and II dominate. Requiring that both these terms are of order unity, we put

$$L_{\perp}^{d-1}\psi_0^2/\xi_{\|} = L_{\perp}^{d-1} t\psi_0^2\xi_{\|} = 1, \tag{135}$$

which yields the standard mean field results (using $k_BT\chi = L_{\perp}^{d-1}\int_0^{L_{\|}}\langle\psi(0)\psi(x)\rangle\, dx \approx L_{\perp}^{d-1}\psi_0^2\xi_{\|}$)

$$\xi_{\|} \propto t^{-1/2}, \quad \psi_0^2 \propto t^{-1/2}/L_{\perp}^{d-1}, \quad \chi \propto t^{-1}. \tag{136}$$

But a new situation emerges for $t = 0$, where II in (134) vanishes and I is balanced by III. Again we require that these terms are of order unity

$$L_{\perp}^{d-1}\psi_0^2/\xi_{\|} = L_{\perp}^{d-1} u_0\psi_0^4\xi_{\|} = 1. \tag{137}$$

Denoting the decay length $\xi_{\|}(t=0) = \xi_0$, we obtain from (137) $u_0\xi_{\|}^2\psi_0^2 = 1$ or

$$\xi_0 \propto u_0^{-1/3} L_{\perp}^{(d-1)/3}, \quad \psi_0^2 = u_0^{-1/3} L_{\perp}^{-2(d-1)/3}. \tag{138}$$

It is seen that $L_{\|}$ scales with a length $\xi_0 \propto L_{\perp}^{(d-1)/3} > L_{\perp}$ for $d > d^\star = 4$, which is the condition that Landau theory holds. Only right at the marginal dimension for

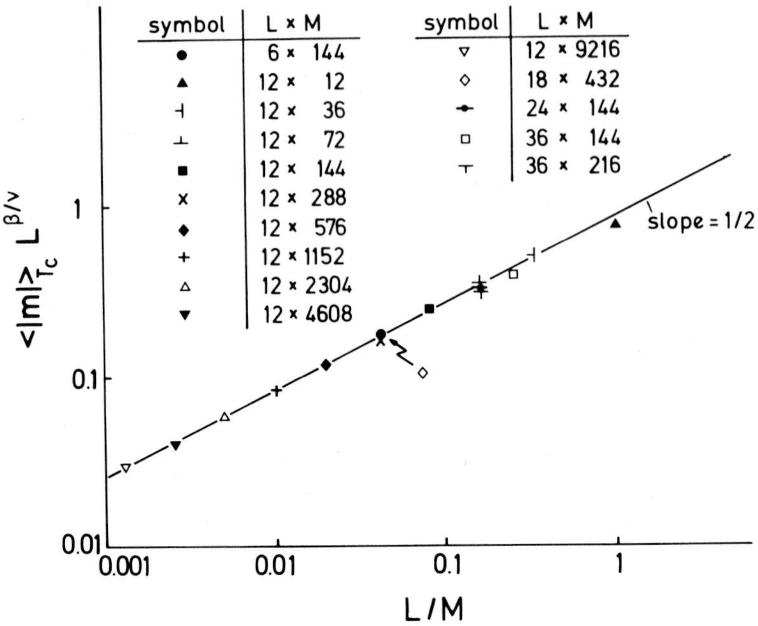

Fig. 28. Log-log plot of the scaled absolute value of the magnetization at T_c, $\langle |M| \rangle_{T_c} L_\perp^{\beta/\nu}$, versus the aspect ratio $L_\perp / L_\parallel$ ($L_\perp$ is denoted as L and $L_\parallel$ as M in this figure). Individual choices of these linear dimensions are indicated by different symbols as explained in the figure. Straight line is the predicted power law, (127). All data are for the nearest neighbor square lattice with two free boundaries of length $L_\parallel$ and periodic boundary conditions in the x-direction. From Albano et al. [149].

the validity of Landau theory, $d^\star = 4$, we have $\xi_0 \propto L_\perp$ and then $L_\parallel$ scales with $L_\perp$, that is, it is again simply the aspect ratio $L_\parallel / L_\perp$ that enters the description of system shape effects in finite size scaling, as in (121)-(127). Of course, the treatment outlined in (132)-(138) is a plausibility argument [48] only, but it is confirmed by more elaborate treatments [147]. Equations (122),(123) for $T = T_c$ are then replaced by

$$k_B T_c \chi(T_c) = L_\perp^{1/2} L_\perp^{(d-4)/6} \tilde{\chi} \left(\frac{L_\perp}{L_\parallel} L_\perp^{(d-4)/3} \right), \tag{139}$$

$$\langle |M| \rangle_{T_c} = L_\perp^{-(d-1)/6} L_\parallel^{-1/2} \tilde{M} \left(\frac{L_\perp}{L_\parallel} L_\perp^{(d-4)/3} \right). \tag{140}$$

While for the cubic geometry (all linear dimensions equal to L) considered in Sect.6 (103) implies for $T = T_c$ that $P_L(M) = L^{d/4} \tilde{P}'(0, M L^{d/2})$ and hence $k_B T_c \chi(T_c) = L^d \langle M^2 \rangle \propto L^{d/2}$ we have an enhancement here by a factor $L_\perp^{(d-4)/6}$ (for $L_\parallel \to \infty$). In contrast, for $d < d^\star$ both the cubic geometry and the present geometry yield (63),(126) $k_B T_c \chi(T_c) \propto L_\perp^{\gamma/\nu}$ or $L^{\gamma/\nu}$, respectively.

These considerations can again be carried over to systems with anisotropic critical behavior, such as uniaxial Lifshitz points [48]. In Landau theory, the uniaxial Lifshitz point arises if for one direction (for example, the z-direction) the coefficient of the $(\partial\psi/\partial x)^2$ term vanishes, and thus one must go to higher order in the gradient expansion [151]

$$\mathcal{H}_{\text{eff}}\{\psi\} = \int d\mathbf{r} \left\{ \frac{1}{2}\sum_{j=1}^{d-1}\left(\frac{\partial\psi}{\partial x_j}\right)^2 + \frac{1}{2}\left(\frac{\partial^2\psi}{\partial z^2}\right)^2 + \frac{1}{2}t\psi^2 + \frac{u_0}{4!}\psi^4 \right\}. \tag{141}$$

From this mean-field effective free energy functional one readily derives the wavevector-dependent susceptibility

$$\chi^{-1}(q) \propto t + q_\perp^2 + q_\parallel^4 = t(1 + q_\perp^2\xi_\perp^2 + q_\parallel^4\xi_\parallel^4), \tag{142}$$

and thus the exponents defined in (128) have the mean field values

$$\nu_\perp = \frac{1}{2}, \quad \nu_\parallel = \frac{1}{4}. \tag{143}$$

Together with the standard mean field results $\gamma = 1$, $\beta = 1/2$ (which still hold here) we see that hyperscaling hold for $d^\star = 4.5$ ($(d^\star - 1)\nu_\perp + \nu_\parallel = (d^\star - 1)\frac{1}{2} + \frac{1}{4} = \gamma + 2\beta = 2$).

We consider now a geometry where the ordering is inhomogeneous in the z-direction even in the mean field limit. Then (141) is replaced by

$$\mathcal{H}_{\text{eff}}(\psi) \approx L_\perp^{d-1}\int_0^{L_\parallel} dz \left\{ \frac{1}{2}\left(\frac{d^2\psi}{dz^2}\right) + \frac{1}{2}t\psi^2 + \frac{u_0}{4!}\psi^4 \right\} = I + II + III \tag{144}$$

Again we estimate the order of magnitude of the contributions I,II,III to the integral using $\psi = \psi_0 \exp(-z/\xi_\parallel)$ as a "trial function" in the path integral:

$$I \propto L_\perp^{d-1}\psi_0^2/\xi_\parallel^3, \quad II \propto L_\perp^{d-1}t\psi_0^2\xi_\parallel, \quad III \propto L_\perp^{d-1}u_0\psi_0^4\xi_\parallel. \tag{145}$$

For $t > 0$ and $L_\perp \to \infty$ we may ignore III and require

$$L_\perp^{d-1}\psi_0^2/\xi_\parallel^3 = L_\perp^{d-1}t\psi_0^2\xi_\parallel = 1. \tag{146}$$

This yields

$$\xi_\parallel \propto t^{-1/4}, \psi_0^2 \propto t^{-3/4}/L_\perp^{d-1}, \chi \propto \psi_0^2\xi_\parallel L_\perp^{d-1} \propto t^{-1}, \tag{147}$$

in agreement with (142), as expected (thus we check that our procedure makes sense!). But for $t = 0$ we again have to match I and III:

$$L_\perp^{d-1}\psi_0^2/\xi_\parallel^3 = L_\perp^{d-1}u_0\psi_0^4\xi_\parallel = 1, \tag{148}$$

and hence $\xi_\parallel(t = 0) = \xi_0$ becomes (remember $d^\star = 4.5$)

$$\xi_0 \propto u_0^{-1/7}L_\perp^{(d-1)/7} = u_0^{-1/7}L_\perp^{1/2}L_\perp^{(d-d^\star)/7}. \tag{149}$$

Now (122),(123) for $T = T_c$ are replaced by

$$k_B T_c \chi(T_c) = L_\perp^{d/2-1/4} L_\perp^{(d-d^\star)/14} \tilde{\chi} \left(\frac{L_\perp^{1/2}}{L_\parallel} L_\perp^{(d-d^\star)/7} \right), \tag{150}$$

$$\langle |M| \rangle_{T_c} = L_\perp^{-3/4} L_\parallel^{-1/2} L_\perp^{-3(d-d^\star)/14} \tilde{M} \left(\frac{L_\perp^{1/2}}{L_\parallel} L_\perp^{(d-d^\star)/7} \right). \tag{151}$$

Note that these forms match at $d = d^\star$ with the results which follow from (130),(131)

$$k_B T_c \chi(T_c) = L_\perp^{\gamma/\nu_\perp} \tilde{\chi}_c (L_\perp^{\nu_\parallel/\nu_\perp}/L_\parallel), \langle |M| \rangle_{T_c} = L_\perp^{-\beta/\nu_\perp} \tilde{M}_c (L_\perp^{\nu_\parallel/\nu_\perp}/L_\parallel) \tag{152}$$

if we insert the mean field exponents $\gamma = 1$, $\beta = 1/2$, $\nu_\perp = 1/2$, $\nu_\parallel = 1/4$ into (152). This fact corroborates once more the self-consistency of this simple treatment.

While for standard phase transitions it is believed that hyperscaling relations do hold for the physically accessible dimensionalities [8], this is not so for nonequilibrium phase transitions in driven systems [152,153]. The extension of finite size scaling concepts to such cases still is a subject of current discussion [48,154].

9 Crossover Scaling from One Universality Class to Another

There are many examples where one finds near a critical point crossover from one universality class [8] to another one: for example, adding a weak uniaxial anisotropy to the Hamiltonian $\mathcal{H} = -J \sum_{\langle i,j \rangle} \mathbf{S}_i \cdot \mathbf{S}_j$ of the isotropic n-vector model a crossover to the Ising universality class ($n = 1$) is induced. Systems close to a multicritical point, for example, a tricritical point [155] cross over from multicritical to ordinary critical behavior. Systems with large but finite range of the interaction cross over from Landau like critical behavior towards a nontrivial critical behavior close to T_c. This latter problem is practically relevant for the critical point of unmixing in polymer mixtures, where the degree of polymerization (chain length N) of the polymers plays a role similar to the interaction range in an Ising model or Landau theory [131,156].

This crossover behavior is explained schematically in Fig.29 [157]. One introduces scaling variables g (tangential to $T_c(p)$ in the point T_m, p_m) and t (normal to $T_c(p)$ at $p = p_m$). Both the critical line and the center of the crossover region can then be expressed in terms of the crossover exponent ϕ [158],

$$t_c = g^{1/\phi}/y_c, \quad t_{\text{cross}} = g^{1/\phi}/y_{\text{cross}}, \tag{153}$$

where we use the variable $y = g^{1/\phi}/t$ to characterize the crossover near the point $T_m = T_c(p_m)$ in the singular part of the free energy [158]:

$$F^{\text{sing}}(T, H, p) = t^{2-\alpha_m} \tilde{F}(H t^{-(\beta_m + \gamma_m)}, g^{1/\phi}/t). \tag{154}$$

Here α_m, β_m, γ_m are the standard critical exponents [7,8] in the point T_m, and $\tilde{F}(x, y)$ is a scaling function. The different critical exponents at $T_c(p)$ (or $y = y_c$, respectively) result from a singular behavior of $\tilde{F}(x, y)$ (for example, $\tilde{F}(0, y) \propto (y - y_c)^{2-\alpha}$).

Now one has to be careful if one carries out a finite size scaling analysis of numerical simulation data in such a situation: if finite size rounding sets in for $t \gg t_{\text{cross}}$ (or

$y \ll y_{\text{cross}}$, respectively), one will only see critical behavior associated with the point T_m (with exponents α_m, β_m, γ_m, ν_m, and so on), but not the true critical behavior. Only for very large systems the region $t_c < t < t_{\text{cross}}$ can be reached, and then one expects a singular dependence of finite size scaling quantities on the coordinate g that measures the distance from the multicritical point (T_m, p_m) in Fig. 29 [157].

In order to cope with such a situation, we have to include $g^{1/\phi}/t$ as an additional variable in the finite size scaling description. This yields for the probability distribution of the order parameter $P_L(M)$ [157], cf. (75)

$$P_L(M) = L^{\beta_m/\nu_m} \tilde{P}(L^{1/\nu_m} t, L^{\beta_m/\nu_m} M, L^{\phi/\nu_m} g). \tag{155}$$

Here we have rearranged the variables $L/\xi \sim L t^{\nu_m}$ and $y^\phi = g t^{-\phi}$ as $(L/\xi)^{\phi/\nu_m} y^\phi \propto (L^{\phi/\nu_m} t^\phi)(g t^{-\phi}) = L^{\phi/\nu_m} g$ to obtain the last argument of (155). This last argument has an obvious physical interpretation: along the crossover line in Fig. 29 the correlation length ξ is of order $\xi_{\text{cross}} \propto t_{\text{cross}}^{-\nu_m} = g^{-\nu_m/\phi} y_{\text{cross}}^{\nu_m}$. Thus we expect a variable $L/\xi_{\text{cross}} \propto L g^{\nu_m/\phi}$ or alternatively $L^{\phi/\nu_m} g$.

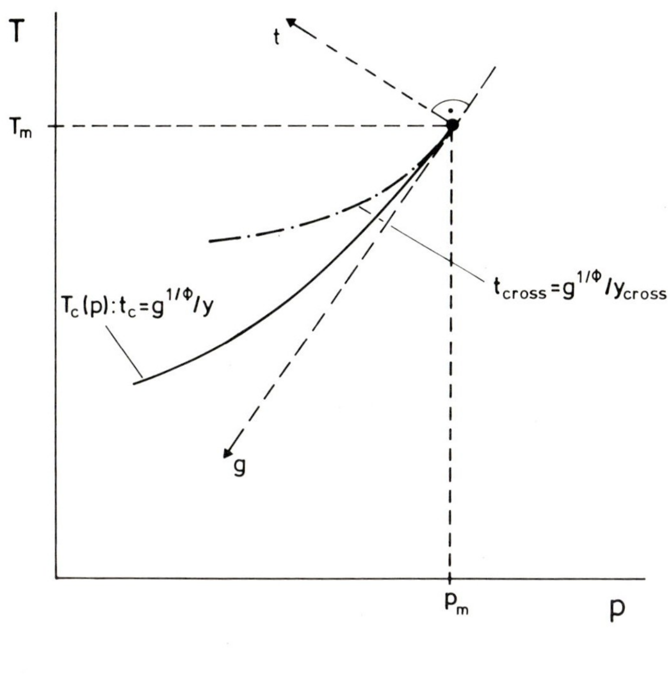

Fig. 29. Schematic phase diagram of a system exhibiting crossover between "ordinary" critical phenomena along the line $T_c(p)$ for $p < p_m$ and the special point $p = p_m$, $T_m = T_c(p_m)$. Considering the approach to the critical line (full curve) along an axis parallel to the t-axis, one will observe critical behavior associated with the special point, as long as one stays above the dash-dotted curve describing the center of the crossover region. Only in between this dash-dotted line and the critical line (which merge at $T_c(p_m)$) the correct asymptotic behavior for $p < p_m$ can be seen. From Binder and Deutsch [157].

From (155) it is straightforward to derive the quantities of interest, such as $\langle|M|\rangle$ and χ,

$$\langle|M|\rangle = L^{-\beta_m/\nu_m}\tilde{M}(L^{1/\nu_m}t, L^{\phi/\nu_m}g), \tag{156}$$
$$k_BT\chi = L^{-\gamma_m/\nu_m}\tilde{\chi}(L^{1/\nu_m}t, L^{\phi/\nu_m}g). \tag{157}$$

In the limit $L/\xi_{\text{cross}} \ll 1$ (or alternatively $L^{\phi/\nu_m}g \ll 1$) the last argument in (155)-(157) can be replaced by zero, and then these equations have their standard form with the exponents β_m, γ_m, ν_m of the special point (T_m, p_m). In the opposite limit, (155)-(157) are singular with respect to the last argument, $L^{\phi/\nu_m}g$. The standard form of finite size scaling with the exponents β, γ, ν is recovered by rearranging arguments of (155) as follows [157]

$$P_L(M) = L^{\beta/\nu}g^{-(\nu_m/\phi)[(\beta_m/\nu_m)-(\beta/\nu)]}\,\tilde{\tilde{P}}\,\{L^{1/\nu}g^{-(\nu_m/\phi)[(1/\nu_m)-1/\nu]}\tilde{t},$$
$$L^{\beta/\nu}g^{-(\nu_m/\phi)[(\beta_m/\nu_m)-(\beta/\nu)]}M\}, \quad \tilde{t} = t - t_c. \tag{158}$$

Here we have multiplied the prefactor and the two arguments of (155) with appropriate powers of the crossover variable $L^{\phi/\nu_m}g$ to obtain the L-dependence of standard finite size scaling, since this must be compatible with (155) as a special limiting case of that equation. One finds from (158) a singular g-dependence of $\langle|M|\rangle_{T_c(p)}$ and $k_BT_c\chi(T_c)$, namely

$$k_BT_c\chi(T_c) \propto L^{\gamma/\nu}g^{-(\nu_m/\phi)[(\gamma_m/\nu_m)-\gamma/\nu]}, \quad L^{\phi/\nu_m}g \gg 1, \tag{159}$$
$$\langle|M|\rangle_{T_c(p)} \propto L^{-\gamma/\nu}g^{(\nu_m/\phi)[(\beta_m/\nu_m)-\beta/\nu]}, \quad L^{\phi/\nu_m}g \gg 1. \tag{160}$$

For a tricritical point in $d = 3$ dimensions, $\nu_m = 1/2$, $\beta_m = 1/4$, $\gamma_m = 1$, $\phi = 1/2$ [155]. Using $\beta/\nu = (d-2+\eta)/2$ one finds $k_BT\chi_{\max} \propto L^{\gamma/\nu}g^{-\eta}$, $\langle|M|\rangle_{T_c(p)} \propto L^{-\beta/\nu}g^{-\eta/2}$. Since the exponent η is very small ($\eta \approx 0.03$ for the Ising model in $d = 3$ [8,66,71,72]), the g-dependence in (159),(160) is very weak.

This is not so when one considers the crossover to Landau-like mean field behavior in $d = 3$ dimensions, which occurs when $g = 1/R^d \to 0$ (R being the range of interaction), g hence is the inverse interaction volume, or inverse "coordination number"of interacting neighbors. However, one needs to be careful, since (155)-(160) imply the validity of hyperscaling, which is not true for mean field exponents in $d = 3$. However, from Sect.6 we know that formally finite size scaling carries over to this situation (in the L^d geometry) if the correlation length ξ_m is replaced by the thermodynamic length $\ell_m \propto (\chi/M^2)^{1/d} \propto t^{-(\gamma_m+2\beta_m)/d} = t^{-2/d}$ (101),(103). Then (155)-(160) get replaced by [157]

$$P_L(M) = L^{d/4}\tilde{P}(L^{d/2}t,\ L^{d/4}M, L^{d(4-d)/4}R^{-d}) \tag{161}$$
$$\langle|M|\rangle_{T_c} = L^{-d/4}\tilde{M}_c(L^{d(4-d)/4}R^{-d}),$$
$$k_BT_c\chi(T_c) = L^{d/2}\tilde{\chi}_c(L^{d(4-d)/4}R^{-d}). \tag{162}$$

Here the crossover exponent ϕ was simply taken from the Ginzburg criterion [159]: mean field behavior occurs for $R^d|t|^{(4-d)/2} \gg 1$, which implies $\phi = (4-d)/2$. Now the interpretation of the crossover scaling variable $L/R^{4/(4-d)}$ is that at the temperatures of the crossover $t_{\text{cross}}^{(4-d)/2} \propto R^{-d}$ the system size L has to be compared to the thermodynamic length $\ell \propto t^{-2/d}$, that is the length $\ell_{\text{cross}} \propto t_{\text{cross}}^{-2/d} \propto$

$(R^{-d})^{-4/[(4-d)d]} = R^{4/(4-d)}$. At this crossover line also the mean field correlation length $\xi \propto Rt^{-1/2} \propto R(R^{d/(4-d)}) = R^{4/(4-d)}$ is of the same order, as expected: In the classical mean field regime, for $t \gg t_{\rm cross}$, the length ξ is much larger than the thermal length ℓ, but which increases faster than ξ. As long as fluctuations can be neglected, L scales with ℓ, while in the fluctuation-dominated nonclassical critical regime, L scales with $\xi \propto t^{-\nu}$. In the crossover region, all these lengths are of the same order, and in this way it is possible to have a smooth crossover from one type of scaling (63),(75) to the other (103).

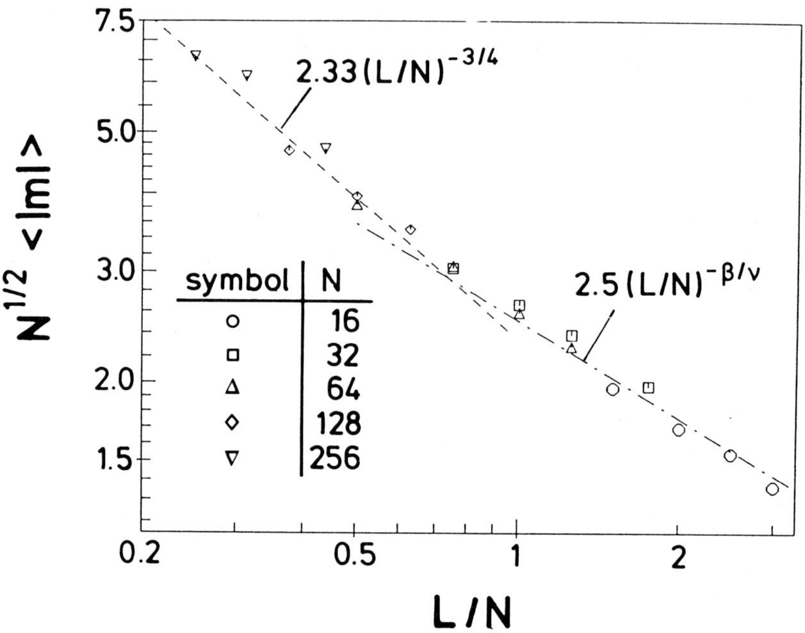

Fig. 30. Log-log plot of $N^{1/2}\langle|M|\rangle_{T_c}$ versus L/N, for a simple cubic lattice model of a symmetrical polymer mixture of chains consisting of N monomers, at a constant volume fraction of sites occupied by monomers, $\phi = 1/2$. Symbols indicate the different chain lengths used. The simulation was carried out in the grand-canonical ensemble using the bond fluctuation model of polymer chains. Dashed straight lines indicate the expected power laws, $(L/N)^{-\beta/\nu}$ in the Ising regime (with $\beta/\nu = (1+\eta)/2 \approx 0.515$) and $(L/N)^{-3/4}$ in the mean field regime. From Binder and Deutsch [157].

Applying this treatment to polymer mixtures in $d = 3$, one has to replace R^{-3} by $N^{-1/2}$ [13] and L^3 by the number of chains $n = L^3/N$ (changing R in an Ising lattice of linear dimension L does not change the number $n = L^d$ of Ising degrees of freedom, but changing N at fixed monomer density n changes. The number of degrees of freedom and not the volume itself is here the basic variable of interest). Then (162)

can be rewritten as

$$\langle |M| \rangle_{T_c} = L^{-3/4} N^{1/4} \tilde{M}_c(L^{3/4} N^{-1/4} N^{-1/2}) = N^{-1/2} \,\tilde{\tilde{M}}\, (L/N), \tag{163}$$

where in the last step we have redefined the scaling function in order to choose a simpler variable, L/N. Fig.30 shows a test of (163) [157]. It is seen that the expected scaling behavior is indeed obtained.

10 Concluding Remarks

In these lectures, some ideas and concepts about finite size effects at phase transitions have been interpreted by a phenomenological account of finite size scaling theory. The selection of material was biased from the point of view of Monte Carlo investigation of phase transitions of bulk systems. Very important other approaches (for example, the finite size scaling analysis of transfer matrix calculations [10,160,161], exact calculations in two dimensions using conformal invariance concepts [150,161], or finite size analysis of the distribution of zeroes of the partition function [162], exact calculations for spherical models [10,163] or in the framework of $\epsilon = 4-d$ expansions [147,164-166]) have hardly been mentioned. Even on a phenomenological level, our discussion is incomplete: there has been much recent work on universal finite size scaling amplitudes [167,168], finite size effects on interfacial critical phenomena [167,168], surface effects [139-141,169-172], and wetting phenomena [173-175], and so on. We also draw attention to work on finite size scaling of the ϕ^4 model [54,176,177] and Lennard-Jones fluids [67,178] as well as on studies of finite size effects in other statistical ensembles, such as the microcanonical ensemble [179], Gibbs ensemble [180], and so on. All such applications and extensions of finite size scaling theory have been out of scope here: but already the limited selection of material shows that this approach to critical phenomena is particularly illuminating and fruitful.

Acknowledgements

Part of the material presented here is due to a fruitful and pleasant collaboration with E. V. Albano, M. S. S. Challa, H.-P. Deutsch, A. Ferrenberg, J. D. Gunton, D. W. Heermann, K. Kaski, D. P. Landau, A. Milchev, M. Nauenberg, M. d'Onorio de Meo, W. Paul, V. Privman, A. Sadiq, J. S. Wang, and A. P. Young.

References

1. Binder, K., Heermann, D.W.: *Monte Carlo Simulation in Statistical Physics. An Introduction* (Springer, Berlin, Heidelberg) (1988)
2. Binder, K.: (ed) *Monte Carlo Methods in Statistical Physics*, 2nd edn (Springer, Berlin, Heidelberg) (1986)
3. Binder, K.: (ed) *Applications of the Monte Carlo Method in Statistical Physics*, 2nd edn (Springer, Berlin, Heidelberg) (1987)
4. Binder, K.: (ed) *The Monte Carlo Method in Condensed Matter Physics* (Springer, Berlin, Heidelberg) (1992)

5. Kalos, M.H., Whitlock, P.: *Monte Carlo Methods Vol 1* (Wiley, New York) (1986)
6. Privman, V.: (ed) *Finite Size Scaling and Numerical Simulation of Statistical Systems* (World Scientific, Singapore) (1990)
7. Stanley, H.E.: *An Introduction to Phase Transitions and Critical Phenomena* (Oxford University Press, Oxford) (1971)
8. Fisher, M.E.: Rev. Mod. Phys. **46** (1974) 597
9. Fisher, M.E.: In *Critical Phenomena, Proc. 1970 E. Fermi Int. School Physics*, ed by Green, M.S.: (Academic, New York) (1971) p 1
10. Barber, M.N: In *Phase Transitions and Critical Phenomena Vol 8* ed by Domb. C., Lebowitz, J.L.: (Academic, New York) (1983) p 145
11. Binder, K.: Ferroelectrics **73** (1987) 43
12. Cardy, J.L.: (ed) *Finite Size Scaling* (North-Holland, Amsterdam) (1988)
13. Binder, K.: Ann. Rev. Phys. Chem. **43** (1992) (in press)
14. Stauffer, D.: *Introduction to Percolation Theory* (Taylor and Francis, London) (1985)
15. Stauffer, D.: Phys. Rep. **54** (1979) 1
16. Kirkpatrick, S.: In *Ill-Condensed Matter*, ed by Balian, R., Maynard, R., Toulouse, G., (North-Holland, Amsterdam) (1979) p 321
17. Heermann, D.W., Stauffer D.: Z. Phys. **B40** (1980) 133
18. Binder, K.: , Z. Physik **B43** (1981) 119
19. Stoll, E., Binder, K., Schneider, T.: Phys. Rev. **B8** (1973) 3266
20. Binder, K., Landau, D.P.: Phys. Rev. **B30** (1984) 1477
21. Binder, K., Rauch, H.: Z. Phys. **219** (1969) 201
22. Stauffer, D., Binder, K.: Z. Phys. **B30** (1978) 313; **B41** (1981) 237
23. Müller-Krumbhaar, H., Binder, K.: Z. Phys. **254** (1972) 269
24. Kasteleyn, P.W., Fortuin, C.M.: J. Phys. Soc. Japan **26**, (Suppl.) (1969) 11; Fortuin, C.M., Kasteleyn, P.W.: Physica **57** (1972) 536
25. Potts, R.B.: Proc. Cambridge Philos. Soc. **48** (1952) 106
26. Wu, F.Y.: Rev. Mod. Phys. **54** (1982) 235
27. See also Fisher, M.E.: Physics **3** (1967) 267, and Refs. 28-30 for background on the concept of "physical clusters".
28. Binder, K.: Ann. Phys. **98** (1976) 390
29. Coniglio, A., Klein, W.: J. Phys. **A13** (1980) 2775
30. Hu, C.K.: Phys. Rev. **B29** (1984) 5103
31. Swendsen, R.H., Wang, J.S.: Phys. Rev. Lett. **58** (1987) 86
32. For a review, see: Swendsen, R.H., Wang, J.S., Ferrenberg, A.M.: in Ref. 4, Chap.4, and Ref. 33.
33. Wolff, U.: in this volume
34. Wolff, U.: Phys. Rev. Lett. **60** (1988) 1461; Nucl. Phys. **B300** [FS22] (1988) 501
35. Edward, R.G., Sokal, A.D.: Phys. Rev. **D38** (1988) 2009; **D40** (1989) 1374
36. Wolff, U.: Phys. Rev. Lett. **62** (1989) 361; Phys. Lett. **B228** (1989) 379; Nucl. Phys. **B322** (1989) 759
37. Hu, C.K., Mak, S.K.: Phys. Rev. **B40** (1989) 5007
38. D'Onorio, De Meo M., Heermann, D.W., Binder, K.: J. Stat. Phys. **60** (1990) 585
39. Hasenbusch, M.: Nucl. Phys. **B333** (1990) 581;
Hasenbusch, M., Meyer, S., Phys. Lett. **B241** (1990) 238; Phys. Rev. Lett. **66** (1991) 530

40. Wang, J.S., Selke, W., Dotsenko, V.S., Andreichenko, V.B.: Europhys. Lett. **11** (1990) 30; Physica **A164** (1990) 221
41. See Hohenberg, P.C., Halperin, B.I.: Rev. Mod. Phys. **49** (1977) 435, for a general review of critical slowing down.
42. Heermann, D.W., Burkitt, A.N.: Physica **A162** (1990) 210
43. Li, X.J., Sokal, A.D.: Phys. Rev. Lett. **63** (1989) 827; **67** (1991) 1482
44. Wang, J.S.: Physica **A164** (1991) 240
45. Margolina, A., Herrmann, H.J.: Phys. Lett. **104A** (1984) 295
46. Onsager, L.: Phys. Rev. **65** (1944) 117
47. Binder, K., Nauenberg, M., Privman, V., Young, A.P.: Phys. Rev. **B31** (1985) 1498
48. Binder, K., Wang, J.S.: J. Stat. Phys. **55** (1989) 87;
Binder, K.: in Ref 6, p 173
49. Binder, K.: Rept. Progr. Phys. **50** (1987) 783
50. Challa, M.S.S., Landau, D.P., Binder, K.: Phys. Rev. **B34** (1986) 1841
51. Landau, L.D., Lifshitz, E.M.: *Statistical Physics* (Pergamon, Oxford) (1958)
52. Binder, K.: Phys. Rev. **A25** (1982) 1699
53. Kaski, K., Binder, K., Gunton, J.D.: Phys. Rev. **B29** (1984) 3996
54. Milchev, A., Heermann, D.W., Binder, K.: J. Stat. Phys. **44** (1986) 749
55. Landau, D.P.: J. Magn. Magn. Mater. **31-34** (1983) 11
56. Murty, C.S.S., Landau, D.P.: J. Appl. Phys. **55** (1984) 2429; Challa, M.S.S., Landau, D.P.: Phys. Rev. **B33** (1986) 437
57. Landau, D.P., Binder, K.: Phys. Rev. **B31** (1985) 5946
58. Binder, K.: Z. Phys. **B61** (1985) 13
59. Bhatt, R.N., Young, A.P.: Phys. Rev. Lett. **54** (1985) 925; Phys. Rev. **B37** (1988) 5606
60. Scheucher, M.: Reger J D, preprint
61. Dünweg, B., Binder, K.: Phys. Rev. **B36** (1987) 6935
62. Schmid, F., Binder, K.: (in press) J. Phys. Condensed Matter
63. Sariban, A., Binder, K.: J. Chem. Phys. **86** (1987) 5859; Macromolecules **21** (1988) 711
64. Deutsch, H.P., Binder, K.: Europhys. Lett., in press; Macromolecules, submitted; Deutsch, H.P.: J. Stat. Phys., submitted
65. Peczak, P., Ferrenberg, A.M., Landau, D.P.: Phys. Rev. **B43** (1991) 6087
66. Ferrenberg, A.M., Landau, D.P.: Phys. Rev. **B44** (1991) 5081
67. Rovere, M., Heermann, D.W., Binder, K.: J. Phys. Condensed Matter **2** (1990) 7009
68. Novotny, R.M., Binder, K.: Z. Phys. **B77** (1989) 287
69. Wansleben, S.: Comp. Phys. Commun. **43** (1987) 9
70. Ferrenberg, A.M., Swendsen, R.H.: Phys. Rev. Lett. **61** (1988) 2635; **63** (1989) 1195
71. Pawley, G.S., Swendsen, R.H., Wallace, D.J., Wilson, K.G.: Phys. Rev. **B29** (1984) 4030
72. Blöte, H.W.J., De Bruin, J., Compagner, A., Croolkewit, J.H., Fonk, Y.T.J.C., Heringa, J.R., Hoogland, A., Van Willigen. A.L.: Europhys. Lett. **10** (1989) 105
73. Adler, J.: J. Phys. **A16** (1983) 3585
74. Liu, A.J., Fisher, M.E.: Physica **A165** (1989) 35

75. Landau, D.P.: Phys. Rev. **B13** (1976) 2997; Phys. Rev. **B14** (1976) 255
76. Binder, K.: Thin Solid Films **20** (1974) 367
77. Binder, K., Landau, D.P.: Phys. Rev. **B21** (1980) 1941
78. Landau, D.P.: Phys. Rev. **B27** (1983) 5604
79. Krinsky, S., Mukamel, D.: Phys. Rev. **B16** (1977) 2313
80. Swendsen, H., Krinsky, S.: Phys. Rev. Lett. **43** (1979) 177
81. Oitmaa, J.: J. Phys. **A14** (1981) 1159
82. Nightingale, M.P.: Phys. Lett. **A59** (1977) 486
83. Nauenberg, M.P., Nienhuis, B.: Phys. Rev. Lett. **33** (1974) 944
84. Le Guillou, J.C., Zinn-Justin, J.: Phys. Rev. Lett. **39** (1977) 95; Phys. Rev. **B21** (1980) 2976
85. Metropolis, N., Rosenbluth, A.W., Rosenbluth, M.N., Teller, A.H., Teller, E.: J. Chem. Phys. **21** (1953) 1087
86. Müller-Krumbhaar, H., Binder, K.: J. Stat. Phys. **8** (1973) 1
87. Binder, K.: In *Phase Transitions and Critical Phenomena*, ed by Domb, C., Green, M.S.: Vol 5b (Academic, New York) (1976) p 1
88. Glauber, R.J.: J. Math. Phys. **4** (1963) 294
89. Hohenberg, P.C., Halperin, B.I.: Rev. Mod. Phys. **49** (1977) 435
90. Wansleben, S., Landau, D.P.: Phys. Rev. **B43** (1991) 6006
91. Wansleben, S., Landau, D.P.: J. Appl. Phys. **61** (1987) 3968
92. Ferrenberg, A.M., Landau, D.P., Binder, K.: J. Stat. Phys. **63** (1991) 867
93. Alexander, H.W.:*Elements of Mathematical Statistics* (Wiley, New York) (1961)
94. Miyashita, S., Takano, H.: Progr. Theor. Phys. **73** (1985) 1122
95. Racz, Z.: Phys. Rev. **B13** (1976) 263;
Fisher, M.E., Rac, Z.: Phys. Rev. **B13** (1976) 5039
96. Kretschmer, R., Binder, K., Stauffer, D.: J. Stat. Phys. **15** (1976) 267
97. For a review, see Binder, K., Heermann, D.W., Milchev, A., Sadiq, A.: In *Heidelberg Colloquium on Glassy Dynamics*, ed by van Hemmen, L., Morgenstern, I.: Lect. Notes Phys., Vol 275 (Springer, Berlin, Heidelberg) (1987) p 154
98. Sadiq, A., Binder, K.: J. Stat. Phys. **35** (1984) 517
99. Stauffer, D., Binder, K.: Z. Physik **B30** (1978) 313; **41** (1981) 237
100. Binder, K.: Z. Physik **B45** (1981) 61
101. Binder, K.: Z. Physik **B61** (1985) 13
102. Imry, Y.: Phys. Rev. **B21** (1980) 2042
103. Fisher, M.E., Berker, A.N.: Phys. Rev. **B26** (1982) 2507
104. Blöte, H.W.J., Nightingale, M.P.: Physica **A112** (1982) 405
105. Cardy, J.L., Nightingale, M.P.: Phys. Rev. **B27** (1983) 4256
106. Privman V, Fisher M E J. Stat. Phys. **33** (1983) 285
107. Fisher, M.E., Privman, V.: Phys. Rev. **B32** (1985) 447
108. Privman, V., Rudnick, J.: J. Stat. Phys. **60** (1990) 551
109. Borgs, C., Kotecky, R.: J. Stat. Phys. **61** (1990) 79
110. Borgs, C., Kotecky, R., Miracle-Sole, S.: J. Stat. Phys. **62** (1991) 529
111. Lee, J., Kosterlitz, J.M.: Phys. Rev. **B43** (1991) 3265
112. Lee, J., Kosterlitz, J.M.: Phys. Rev. Lett. **65** (1990) 137
113. Fukugita, M., Mino, M., Okawa, M., Ukawa, A.: J. Stat. Phys. **59** (1990) 1397; J. Phys. **A23** (1990) L561
114. Peczak, P., Landau, D.P.: Phys. Rev. **B39** (1990) 11932

115. Challa, M.S.S., Hetherington, J.H.: Phys. Rev. Lett. **60** (1988) 77; Phys. Rev. **A38** (1988) 6324
116. Billoire, A., Gupta, S., Irbäck, A., Lacaze, R., Morel, A., Peterson, B.: Nucl. Phys. **B258** (1991) 231; Phys. Rev. **B42** (1990) 6743
117. Billoire, A., Lacaze, R., Morel, A.: preprint SPhT-91/122
118. Berg, B.A., Neuhaus, T.: preprints
119. Borgs, C., Janke, W.: preprint FUB-HEP 6/91
120. Cahn, J.W., Hilliard, J.E.: J. Chem. Phys. **28** (1959) 258; **31**, 688
121. Lebowitz, J.L., Penrose, O.: J. Math. Phys. **7** (1966) 98
122. Langer, J.S.: Ann. Phys. (N.Y.) **41** (1967) 108
123. Langer, J.S.: Ann. Phys. (N.Y.) **54** (1969) 258
124. Binder, K., Müller-Krumbhaar, H.: Phys. Rev. **B9** (1974) 2328
125. Binder, K., Stauffer, D.: Advances of Physics **25** (1976) 343
126. Binder, K.: J. Phys. (Paris) **41** , **C4**-51 (1980)
127. Heermann D W, Klein W, Stauffer D (1982) Phys. Rev. Lett. **49**, 1262
128. Heermann, D.W., Klein, W.: Phys. Rev. Lett. **50** (1983) 1962; Phys. Rev. **B27** (1983) 1732
129. Heermann, D.W., Coniglio, A.W., Klein, W., Stauffer, D.: J. Stat. Phys. **36** (1984) 447
130. Klein, W., Unger, C.: Phys. Rev. **B28** (1983) 445
131. Binder, K.: Phys. Rev. **A29** (1984) 341
132. Paul, W., Heermann, D.W., Binder, K.: J. Phys. **A22** (1989) 3325
133. Challa, M.S.S., Landau, D.P., Binder, K.: Phase Transitions **24-26** (1990) 343
134. Hüller, A. :Univ. of Erlangen preprint
135. Baumann, B.: Nucl. Phys. **B285** (1987) 391
136. Berg, B.A., Hansmann, U., Neuhaus, T.: preprint FSU-SCRI-91-125
137. Yang, C.N.: Phys. Rev. **85** (1951) 808
138. Binder, K., Landau, D.P.: Surface Science **108** (1981) 503;
Binder, K., Landau, D.P.: (1989) in *Molecule-Surface Interaction* Lawley, K.P., ed) p 91 (J. Wiley, New York)
139. Fisher, M.E., Nakanishi, H.: J. Chem. Phys. **75** (1981) 5857; **78** (1983) 3279
140. Albano, E.V., Binder, K., Heermann, D.W., Paul, W.: J. Chem. Phys. **91** (1989) 3700
141. Binder, K., Landau, D.P.: J. Chem. Phys. (1992) (in press)
142. Kihara, T., Midzuno, Y., Shizume, T.: J. Phys. Soc. Jpn. **9** (1954) 681;
Baxter, R.J.: J. Phys. **C6** (1973) L445
143. Fisher, M.E., Privman, V.: Commun. Math. Phys. **103** 527
144. Singh, S., Pathria, R.K.: Phys. Rev. **B37** (1988) 7806
145. Hasenfratz, P., Leutwyler, H.: Nucl. Phys. **B343** (1990) 241
146. Chen, K., Landau, D.P.: University of Georgia preprint
147. Zinn-Justin, J., Brézin, E.: Nucl. Phys. **B257** [FS14] (1985) 867
148. Albano, E.V., Binder, K., Heermann, D.W., Paul, W.: Surface Sci. **223** (1989) 151
149. Albano, E.V., Binder, K., Heermann, D.W., Paul, W.: Z. Physik **B77** (1989) 445
150. Cardy, J.L.: in *Phase Transitions and Critical Phenomena, Vol11* (Domb, C., Lebowitz, J.L., eds) p.55 (1987) (Academic Press, New York)
151. Hornreich, R.M., Luban, M., Shtrikman, S.: Phys. Rev. Lett. **35** (1975) 1678

152. Janssen, H.K., Schmittmann, B.: Z. Phys. **B64** (1986) 503
153. Leung, K.T., Cardy, J.: J. Stat. Phys. **44** (1986) 567
154. Leung, K.T.: Phys. Rev. Lett. **66** (1991) 453
155. Lawrie, I.D., Sarbach, S.: In *Phase Transitions and Critical Phenomena, Vol. 9* (1984) (ed by Domb, C. and Lebowitz, J.L.) p 1 (Academic Press, London)
156. de Gennes, P.G.: *Scaling Concepts in Polymer Physics* (Cornell University Press, Ithaca) (1979)
157. Binder, K., Deutsch, H.P.: Univ. of Mainz preprint
158. Riedel, E., Wegner, F.J.: Z. Phys. **225** (1969) 195
159. Ginzburg, V.L.: Sov.Phys. Solid State **2** (1960) 1824
160. For a review, see Nightingale, M.P.: in Ref. 6, p.287, and Ref.161
161. Henkel, M.: in Ref 6, p 353
162. Bhanot, G.V.: in Ref 6, p 435
163. Rudnick, J.: in Ref 6, p 141
164. Jasnow, D.: in Ref 6, p 99
165. Eisenriegler, E.: Z. Phys. **B61** (1985) 299;
Nemirowsky, A.M., Freed, K.F.: J. Phys. **A19** (1985) L319; J. Phys. **A19** (1985) 586;
Eisenriegler, E., Tomaschitz, R.: Phys. Rev. **B35** (1987) 4876
166. Dohm, V.: Z. Phys. **B75** (1989) 109
167. For a review, see Mon, K.K.: in Ref 6, p 261, and Ref 166
168. Privman, V.: in Ref 6, p 1
169. Krech, M., Dietrich, S.: Phys. Rev. Lett. **66** (1991) 345
170. Privman, V.: Phys. Rev. **B38** (1988) 9261
171. Fisher, M.E., de Gennes, P.G.: C. R. Acad. Sci (Paris) Ser. **B287** (1978) 207
172. Nightingale, M.P., Blöte, H.W.J.: Phys. Rev. Lett. **60** (1988) 1562
173. Albano, E.V., Binder, K., Heermann, D.W., Paul, W.: J. Stat. Phys. **61** (1990) 161
174. Parry, A.O., Evans, R.: Phys. Rev. Lett. **64** (1990) 439
175. Kroll, D.M., Gompper, G.: Phys. Rev. **B39** (1989) 433
176. Nicolaides, D., Bruce, A.D.: J. Phys. **A21** (1988) 233
177. Bruce, A.D.: J. Phys. **A18** (1985) L873
178. Welding, N.B., Bruce, A.D.: Univ. of Edinburgh preprint
179. Desai, R.C., Heermann, D.W., Binder, K.: Phys. Rev. Lett. **49** (1982) 708
180. Mon, K.K., Binder, L.: Univ. of Mainz preprint

This article was processed using the LaTeX macro package with LMAMULT style

High Precision Simulations With Fast Algorithms

Ulli Wolff

CERN, Theory Division, CH-1211 Genève 23, Switzerland

Abstract. A brief non-rigorous introduction to the Monte Carlo method for the simulation of lattice quantum field theoretical models is given. The associated problem of critical slowing down in the continuum limit is addressed together with its partial solution for some classes of systems. The version of overrelaxation mostly used in practice, with alternating microcanonical and standard updates, is analyzed exactly for free fields and described in numerical applications to spin and gauge systems. Cluster algorithms are briefly introduced, and applications to the two dimensional $O(3)$ σ–model are described in some detail. They include the exploration of scaling behavior, a numerical estimate of scattering phases and the demonstration and test of a general technique to compute running renormalized coupling constants in asymptotically free theories. We conclude with remarks on the present status of algorithm development.

1 Introduction

Quantum field theories for the description of elementary particles and their interactions are only formally defined by specifying fields and an action with certain symmetries. The actual derivation of predictions requires a limiting procedure where one first regularizes the theory under study at short distance with an ultraviolet cutoff. This ad hoc modification is unphysical and presumably has to do with the end of the domain of validity of the field theoretical model. Only the cutoff independent long distance features of renormalizable theories can be hoped to approximate the underlying physics at present energies with all reference to shorter distances embodied in some phenomenological parameters.

One possible cutoff procedure is to discretize space time by a lattice with spacing a. It has the virtue of manifestly reducing the number of degrees of freedom per volume to become finite and makes thus a numerical computation straightforward in principle. Apart from the spacing one has the system size L and physical scales like masses m as dimensionful quantities. Finite size effects characterized by the dimensionless combination mL, although also artificial, are often not too difficult to control and sometimes even a useful way to probe the system. On the other hand $\xi = (ma)^{-1}$, which has to be extrapolated to infinity to remove the cutoff, is more problematic. It corresponds to a correlation length in units of lattice spacings, and when ξ gets large the problem of critical slowing down arises in Monte Carlo simulations. Depending

on how quantities of interest approach their asymptotic value — proportional to ξ^{-2} or ξ^{-1} or only $1/\ln \xi$ — it becomes difficult to verify convergence and to extrapolate. It is in this area that significant progress has been made over the last five years, some of which will be reported in these lectures[1].

In some interesting two dimensional models (nonlinear σ–models) correlation lengths of order $\xi = 100$ can be simulated accurately now. Some appropriate physical quantities get close enough to the cutoff limit then to reliably extrapolate them to zero lattice spacing. The progress up to now has been less pronounced in four dimensional gauge theories. Cluster and multigrid techniques, which led to the most spectacular breakthrough for σ–models, are not (yet) fully developed, and it is not yet possible to have as large systems as in two dimensions. However, with massively parallel computers and further algorithmic progress we may well arrive at a similar computing ability for four dimensions.

The following notes are divided into four parts. After this introduction we start with a simple non-rigorous introduction to Monte Carlo simulation and define parameters that allow us to judge the quality of an algorithm. In section 3 overrelaxation algorithms are discussed which, although outperformed in some cases, are at present the most universally applicable tool (including gauge theory). In section 4 the cluster method is reviewed. While the description of the algorithm is kept brief, most space is devoted to physical results in two dimensions that have been obtained. We conclude with an outlook on the present situation with regard to algorithm development.

2 Monte Carlo Evaluation of Path Integrals

2.1 Free fields on the lattice

In this subsection we develop some notation by briefly discussing free fields on the lattice. They will furthermore be useful as an example, where also the dynamical properties of Monte Carlo algorithms can be derived analytically.

We consider a D–dimensional hypercubic lattice with sites

$$x = (x^0, x^1, \ldots, x^{D-1}), \quad x^\mu = n^\mu a, \tag{1}$$

where the n^μ are integer and a is the dimensionful lattice spacing. We will mostly use lattice units, $a = 1$, but reintroduce a on some special occasions to emphasize its rôle for the continuum limit. A real scalar field $\varphi(x)$ is defined on each site, and, with periodic boundary conditions of equal period L (L/a integer) in all directions, it obeys

$$\varphi(x) = \varphi(x + L\hat{\mu}) \tag{2}$$

with $\hat{\mu}, \mid \hat{\mu} \mid = 1$, a unit vector along any of the axes. We may thus take $(L/a)^D$ field values as independent, for instance those with

$$\{\varphi(x) \mid 0 \leq x^\mu < L\}. \tag{3}$$

[1] As the field is under very active study, this overview cannot be complete. The choice of examples presented here is biased toward the author's contributions. In particular, no small-step methods, as they are used in connection with fermions, are discussed.

We further need the lattice derivative

$$\partial_\mu \varphi(x) = \frac{\varphi(x + \hat{\mu}a) - \varphi(x)}{a}, \tag{4}$$

and may now write the free field action

$$S(\varphi) = \frac{1}{2}\sum_x a^D \left\{ \sum_\mu (\partial_\mu \varphi(x))^2 + m^2 \varphi(x)^2 \right\} \tag{5}$$

which, for smooth classical fields sampled on a grid, tends to the integral $\frac{1}{2}\int \mathrm{d}^D x$ $\{(\partial_\mu \varphi)^2 + m^2 \varphi^2\}$ in the limit $a \to 0$ at fixed L.

For particle physics we wish to integrate over all sampled fields and extract the physics from (euclidean) Green functions of the type

$$\langle \varphi(x_1) \cdots \varphi(x_n) \rangle = \frac{1}{Z} \int \prod_x \mathrm{d}\varphi(x)\, \mathrm{e}^{-S(\varphi)}\, \varphi(x_1) \cdots \varphi(x_n), \tag{6}$$

where Z is the normalizing partition function. We would like to emphasize a few points about this formulation of a quantum field theory:

- It is approximated by a high but finite dimensional integral, which is well defined for $m^2 > 0$ in (5) and can in principle be computed numerically.
- This is still true when an interaction term like $\lambda\varphi^4$ is added to (5); the definition is independent of perturbation theory and Feynman diagrams.
- These nice properties are due to the presence of both an ultraviolet cutoff $a \neq 0$ and an infrared cutoff $L < \infty$, whose rôles have to be kept track of, for instance by extrapolating.

A simple and important observable allows measurement of the mass gap. The standard procedure is first to define fields at zero spatial momentum

$$\phi_0(t) = \sum_{\mathbf{x}} \varphi(x^0 = t, \mathbf{x}) \tag{7}$$

that depend on the coordinate $x^0 = t$ by convention singled out as euclidean time. The gap $\hat{m}$ follows from the observed temporal decay

$$\langle \phi_0(t)\phi_0(0) \rangle = \text{const}\left(\mathrm{e}^{-t\hat{m}} + \mathrm{e}^{-\hat{m}(L-t)}\right), \tag{8}$$

where the second term takes into account periodicity in time. In the free case the energy $\hat{m}$ is related to m by

$$m = \frac{2}{a}\sinh(\hat{m}a/2),\ \hat{m} = m(1 - \frac{1}{24}(ma)^2 + \ldots). \tag{9}$$

Here we have an instance of a (UV–)cutoff correction or lattice artifact. The expected relation $\hat{m} = m$ emerges only in the cutoff or continuum limit $a \to 0$ at fixed m, where the correlation length in lattice units $\xi = (\hat{m}a)^{-1}$ diverges. In other words, we tune the system to a critical point. The renormalized mass in the presence of interactions would still be extracted by fitting with (8) at large separation. Similar procedures allow us to derive renormalized couplings and other physical quantities from Green functions.

2.2 Monte Carlo simulation of lattice path integrals

The clue to a numerical evaluation of (6) is the observation that it is a structure known from equilibrium statistical mechanics. In

$$\langle A(\varphi)\rangle = \frac{1}{Z}\int \prod_x \mathrm{d}\varphi(x)\,\mathrm{e}^{-S(\varphi)}\,A(\varphi) \tag{10}$$

the exponential Boltzmann factor controls the support of the integral in phase space with the observable $A(\varphi)$ as a slowly varying modification. This is true at least for certain classes of observables like polynomial ones, and for some of them the Monte Carlo techniques can be hoped to be practical.

We generate a sequence of field configurations $\varphi^{(n)}$ in a computer memory,

$$\varphi^{(0)} \to \varphi^{(1)} \to \varphi^{(2)} \to \ldots, \tag{11}$$

that have the Boltzmann part "built in", i.e. for late members in the sequence (large enough n) configurations have a probability of appearance

$$P(\varphi^{(n)} = \varphi) = \frac{1}{Z}\exp(-S(\varphi)). \tag{12}$$

Then an approximation to (10) is given by the sample average

$$\langle A(\varphi)\rangle \approx \frac{1}{N}\sum_{i=n+1}^{i=n+N} A(\varphi^{(i)}). \tag{13}$$

In (11) the recursion starts from some initial $\varphi^{(0)}$, and $\varphi^{(i)} \to \varphi^{(i+1)}$ means that for given $\varphi^{(i)}$ we choose $\varphi^{(i+1)}$ at random with a probability $T(\varphi^{(i)}, \varphi^{(i+1)})$. The set of probabilities T together with a constructive prescription on how to realize the transition make up a Monte Carlo algorithm. There are clearly constraints from (12), and furthermore we have to address the questions of how large n has to be and what "$\approx$" means in (13).

For some arguments to follow now it is helpful to think of an ensemble of independent simulations of the kind just discussed. They lead to distributions $R^{(i)}(\varphi)$ for the i'th configuration in the sequences, including an arbitrary initial distribution $R^{(0)}(\varphi)$. The recursion

$$R^{(i+1)}(\varphi') = \int \mathrm{D}\varphi\, R^{(i)}(\varphi)\, T(\varphi,\varphi') \tag{14}$$

gives in an obvious shorthand notation

$$R^{(i)} = R^{(0)}T^{i}. \tag{15}$$

To converge to the desired distribution (12) we clearly have to obey the stability equation

$$PT = P. \tag{16}$$

We also need ergodicity in the sense of $T^k(\varphi,\varphi') > 0$ for any pair of arguments φ, φ' and k larger than some k_0. This implies that our algorithm can take us from any configuration to any other one in some number of steps. For most applications in field theory this is not a problem, in principle. One can always blend in some ergodic but not necessarily efficient (see below) update steps. In practice some moves called for in implementing (12) may take a forbiddingly large number of iterations. Naturally, this leads to problems to be discussed in the following sections. With stability and ergodicity and T being a probability ($T(\varphi,\varphi') \geq 0$, $\int \mathrm{D}\varphi'\, T(\varphi,\varphi') = 1$) there are theorems [1] proving that beside the eigenvalue one with eigenvector P, the remaining ones are smaller than one in modulus. This guarantees convergence to P for an arbitrary start of the simulation(s).

Most of the time one enforces (16) by the sufficient but not necessary condition of detailed balance

$$P(\varphi)T(\varphi,\varphi') = P(\varphi')T(\varphi',\varphi). \tag{17}$$

A typical example is a local heatbath algorithm. One sweeps the lattice in some order with elementary transitions,

$$T = \prod_x H_x \qquad \text{(some order)}, \tag{18}$$

and takes for the local steps

$$H_x(\varphi,\varphi') = \text{const} \prod_{y \neq x} \delta(\varphi(y) - \varphi'(y))\, P(\varphi') \tag{19}$$

with the constant determined by normalization. In words: in each local step all but one of the degrees of freedom are frozen, and the remaining one is resampled with the equilibrium distribution. Each H_x fulfills detailed balance, and this implies stability (16) for the product T.

2.3 Speed of an algorithm, critical slowing down

From the spectral properties of the update transition probabilities T one can show [1] that the distributions of configurations approach the Boltzmann factor monotonically in some suitably defined norm with a bound given by

$$\| R^{(n+1)} - P \| \leq \mathrm{e}^{-1/\tau_{\mathrm{exp}}} \| R^{(n)} - P \| \,. \tag{20}$$

The exponential autocorrelation time τ_{exp} defined here is the longest time scale in the problem. It controls after how many iterations an algorithm "forgets" the arbitrary initial configuration and approaches equilibrium. Thus, to start the estimation of observables, one has to have $n \gg \tau_{\mathrm{exp}}$ with systematic errors being of the order $\exp(-n/\tau_{\mathrm{exp}})$.

The question about statistical accuracy of estimation at equilibrium is a slightly different one. We have to investigate the expected discrepancy between estimate and true mean,

$$\sigma_A^2 = \left\langle \left(\frac{1}{N} \sum_{i=n+1}^{i=n+N} A(\varphi^{(i)}) - \langle A(\varphi) \rangle \right)^2 \right\rangle. \tag{21}$$

With the autocorrelation function

$$\Gamma_A(i,j) = \left\langle \left(A(\varphi^{(i)}) - \langle A \rangle \right) \left(A(\varphi^{(j)}) - \langle A \rangle \right) \right\rangle \tag{22}$$

and the equilibrium relation $\Gamma(i,j) \approx \Gamma(| i-j |)$ this is rewritten for samples much larger than τ_{exp} as

$$\sigma_A^2 = \frac{1}{N^2} \sum_{i,j=1}^{N} \Gamma_A(i,j) \approx \frac{1}{N} \sum_{t=-\infty}^{\infty} \Gamma(| t |) =: \frac{\Gamma_A(0)}{N} 2\tau_{\mathrm{int},A}. \tag{23}$$

We defined the integrated autocorrelation time

$$\tau_{\mathrm{int},A} = \frac{1}{2} + \sum_{t=1}^{\infty} \Gamma_A(t)/\Gamma_A(0). \tag{24}$$

As $\Gamma_A(0)$ is the variance of A, we see from (23) that $2\tau_{\mathrm{int},A}$ is the number of iterations necessary to produce an effectively independent estimate for A. It is the scale relative to which the length of a certain simulation has to be judged when a calculation of A is the goal and equilibration has been achieved. If Γ is one pure exponential then $\tau_{\mathrm{int},A}$ approximately equals that scale. In general, however, Γ has many components decaying on various time scales up to the one given by τ_{exp}. The integral quantity $\tau_{\mathrm{int},A}$ is usually smaller than, but of a similar order of magnitude to τ_{exp}, and $\tau_{\mathrm{int},A}$ is normally easier to determine for physical observables.

Critical slowing down amounts to the fact that typically all the autocorrelation times explode as one approaches the continuum limit of a field theoretical model. This is described by dynamical critical exponents z and z_A,

$$\tau_{\mathrm{exp}} \propto \xi^{z} \tag{25}$$

$$\tau_{\mathrm{int},A} \propto \xi^{z_A}. \tag{26}$$

Standard local Monte Carlo algorithms have $z, z_A \approx 2$. One interprets this as being due to the growing mismatch of the physical scale ξa of (static) correlations and the lattice scale $1a$ over which the field values of neighbors influence an elementary updating move.

A natural unit for the various τ are complete passes through the lattice with a given algorithm, or sweeps. They have a computational complexity of const$\times(L/a)^D$ operations per sweep. If we estimate $\langle A \rangle$ taking the continuum limit at fixed size in physical units L, then the cost grows as

$$\text{cost} \propto (L/a)^{D+z_A}. \tag{27}$$

It should be clear now that reducing z_A for interesting quantities to one or even zero, as has become possible for some systems recently, allows us to remove the artificial cutoff much further from the physical scale and/or enhances the accuracy of our results.

3 Overrelaxation

3.1 Gaussian case

To discuss local updating in more detail we now bring out the dependence of the free field action (5) on one specific field value $\varphi(x_0)$,

$$S(\varphi) = \frac{1}{2\sigma^2}\left(\varphi(x_0) - \sigma^2 \sum_{|x-x_0|=1} \varphi(x)\right)^2 + \text{terms independent of } \varphi(x_0). \tag{28}$$

The width is given by

$$\sigma^2 = \frac{1}{m^2 + 2D}. \tag{29}$$

The local heatbath procedure in this case amounts to replacing $\varphi(x_0)$ by a new value $\varphi'(x_0)$,

$$\varphi(x_0) \to \varphi'(x_0) = \sigma^2 \sum_{|x-x_0|=1} \varphi(x) + \sigma\eta, \tag{30}$$

with a unit width Gaussian random number η. Adler proposed a generalization of heatbath for the Gaussian case [2] that had previously been known in the context of solving difference equations. He observed that

$$\varphi'(x_0) = (1-\omega)\varphi(x_0) + \omega\sigma^2 \sum_{|x-x_0|=1} \varphi(x) + \sqrt{\omega(2-\omega)}\sigma\eta, \tag{31}$$

leads to equilibrium for the free field for all values $0 < \omega < 2$. While $\omega = 1$ corresponds to heatbath, the range of overrelaxation (OR), $1 < \omega < 2$ will be of special interest. The limiting case $\omega = 2$ is called microcanonical. Random numbers no longer enter, and, since each local scalar field is just reflected through the average field of its neighbors, the deterministic discrete evolution conserves energy. In this case the stability property (16) is valid, but ergodicity does not hold.

Both for theoretical analysis [3] and for programming on vector or parallel machines it is advantageous to choose an order for local updates (in (18) for instance) such that a maximal number of independent ones (commuting H_x in (18)) are grouped together. This is achieved by checkerboard ordering. According to a parity function

$$(-1)^x = (-1)^{x^0+x^1+\ldots+x^{D-1}} \tag{32}$$

we divide all sites into even and odd (assuming L/a even). For nearest neighbor coupled theories local updates on each of the sublattices are independent of each other. As we Fourier expand the fields,

$$\varphi(x) = L^{-D} \sum_p \tilde{\varphi}(p) e^{ip\cdot x} \tag{33}$$

the components

$$\tilde{\varphi}_\pm(p) = \frac{1}{2}\left[\tilde{\varphi}(p) \pm \tilde{\varphi}(p + (\pi, \pi, \ldots, \pi))\right] \tag{34}$$

support fields on the respective sublattices. Momenta range over

$$p_\mu = \frac{2\pi}{L} n_\mu, \quad -\pi \le p_\mu < \pi. \tag{35}$$

A full checkerboard update sweep leads to an evolution that couples only the two momentum modes in $\tilde{\varphi}_\pm$,

$$\tilde{\varphi}'_\pm(p) = (1-\omega)\tilde{\varphi}_\pm(p) + \omega\lambda(p)\tilde{\varphi}_\mp(p) + \tilde{r}_\pm. \tag{36}$$

Here $\tilde{r}_\pm$ are the Fourier transformed random numbers and

$$\lambda(p) = \frac{2D - \hat{p}^2}{2D + m^2} \simeq 1 - \frac{m^2 + \hat{p}^2}{2D} + \ldots \tag{37}$$

will be important for the rest. It depends on the lattice momenta

$$\hat{p}_\mu = 2\sin(p_\mu/2) \hat{=} \frac{2}{a}\sin(ap_\mu/2), \quad -2 \le \hat{p}_\mu < 2, \tag{38}$$

and extends over an interval such that $\mid \lambda(p) \mid \le \lambda(0) < 1$. For

$$\chi(p) = \begin{pmatrix} \tilde{\varphi}_+(p) \\ \tilde{\varphi}_-(p) \end{pmatrix} \tag{39}$$

an update on the even sublattice followed by one on the odd, i. e. a full sweep, reads in two-component notation

$$\chi \to \chi' = M(\lambda,\omega)\chi + A(\lambda,\omega)r \tag{40}$$

with

$$M = \begin{pmatrix} 1-\omega & \omega\lambda \\ (1-\omega)\omega\lambda & 1-\omega+(\omega\lambda)^2 \end{pmatrix} \tag{41}$$

and some matrix A that we will not need in detail. Upon iteration we have for the n'th configuration

$$\chi^{(n)} = M^n \chi^{(0)} + \sum_{i=1}^{n} M^{n-i} A r^{(i)}, \tag{42}$$

where $r^{(i)}$ are the random numbers associated with the transition from $n-1$ to n. For large n the first term in (42) will be seen to be negligible because of equilibration. Taking into account that successive Gaussian random numbers are independent,

$$\langle r^{(n)} r^{(m)} \rangle \propto \delta_{n,m}\,, \tag{43}$$

we find the decay of correlations between members of the sequence controlled by M as

$$\langle \chi^{(n+t)} \chi^{(n)} \rangle \propto \mid M \mid^t =: \exp(-t/\tau(p)). \tag{44}$$

By $\mid M \mid$ we mean the largest eigenvalue of M, which, for a converging simulation algorithm, has to come out smaller than one, of course. Component indices of χ are suppressed here. As different pairs of momenta in χ do not couple — this is of course a pathology of free fields — we have an autocorrelation time for each such pair.

It now is an easy matter to specialize to heatbath, $\omega = 1$. Eigenvalues of

$$M = \begin{pmatrix} 0 & \lambda \\ 0 & \lambda^2 \end{pmatrix}$$

are 0 and λ^2, and this gives

$$\tau(p) = \frac{-1}{2\ln|\lambda|} \simeq \begin{cases} \frac{D}{\hat{p}^2+m^2} & \text{for} \quad m^2, \hat{p}^2 \ll 1 \\ O(1) & \text{for} \quad \hat{p}^2 \not\approx 0, 4D \end{cases} . \tag{45}$$

For small momenta, $p = 0$ for example, τ grows like $(ma)^{-2} = \xi^2$ (we reintroduced a that had been put to $a = 1$ during the calculation), and we see that $z = 2$. The same is true for the z_A of observables, that couple strongly enough to the low momentum modes.

Next we are ready to demonstrate how OR does better than heatbath. For general ω the eigenvalues of M are

$$\rho_\pm = 1 - \omega + (\omega\lambda)^2/2 \pm \sqrt{[1 - \omega + (\omega\lambda)^2/2]^2 - (1-\omega)^2}. \tag{46}$$

It is now possible to choose ω just below 2 such that the argument under the square root becomes negative for all $\lambda(p)$, and then we have

$$|\rho_\pm| = |1 - \omega| \quad \text{independent of } p. \tag{47}$$

It is optimal to just start fulfilling this condition for $p = 0$ with

$$4\frac{\omega - 1}{\omega^2} = \lambda(0)^2 \tag{48}$$

which implies close to the scaling limit $m \to 0$

$$\omega \simeq 2\left(1 - \frac{m}{\sqrt{D}}\right) + O(m^2) \tag{49}$$

and

$$\tau(p) \simeq \frac{\sqrt{D}}{2m}. \tag{50}$$

The linear appearance of m in this formula means that we reduced z from two to one. As we just change a parameter in the local procedure, this is in practice an enormous improvement, and it comes essentially free of cost. In $D = 4$ the gain for zero-momentum observables is a factor 4ξ.

For the precise OR implementation just discussed with a tunable parameter ω not many workable extensions to non-Gaussian models have been found. A different variant, which I would like to call hybrid overrelaxation, is however applicable to many models. It is in fact at present the optimal known method for pure Yang Mills theory on the lattice. Here we only need an analogue of the microcanonical $\omega = 2$ steps. We will then update with N such steps followed by one ordinary sweep of heatbath, for instance. It will turn out that making N large in the continuum limit will produce results similar to the tuning of ω in (49). To the author's knowledge no analysis of this variant of OR in the exactly solvable Gaussian case has been presented. This gap is presently filled [4], and some results will be shown here.

The place of $M(\lambda,\omega)$ in the iteration (42) is now taken by a matrix

$$K(\lambda, N) = M(\lambda, 1)M(\lambda, 2)^N. \tag{51}$$

The eigenvalues of K are computed by first carrying out a similarity transformation diagonalizing $M(\lambda, 2)$, whose eigenvalues are a conjugate pair of phases $\exp(\pm i\theta)$ with

$$\mid \lambda(p) \mid = \cos(\theta(p)/2). \tag{52}$$

Now the eigenvalues of K are

$$\kappa_1 = 0, \quad \kappa_2 = \mid \lambda \mid \cos((N+1/2)\theta). \tag{53}$$

If we now keep $N\theta(0)$ fixed as $m \to 0$ such that κ_2 is close to 1/2, for example, we effectively have $z = 1$ for the zero momentum mode. This is because the autocorrelations between configurations separated by our compound updates do not grow (for that mode), but one such iteration requires a number of sweeps that grows linearly with the correlation length. In this case we include a factor $N + 1$ in the conversion from eigenvalues to times τ, which hence always refer to single sweeps. While we focus on one mode, there are however many closely spaced values of $\theta(p)$ in a large system, and for some of them the cosine in (53) will be close to one in magnitude. That mode is then slowed down with $\tau \approx -(N+1)/\ln \mid \lambda \mid = O(m^{-3})$. For a representative choice of parameters this is shown in Fig.1. The figure suggests that on average the τ are of order N with only isolated neighborhoods in p-space being slower. Their position depends sensitively on N. If, instead of a fixed number of microcanonical steps, we choose this number at random with average N — for instance between 1 and $2N - 1$ with equal likelihood — the picture changes. The eigenvalues of the associated matrix $\bar{K}(\lambda, N)$ are

$$\bar{\kappa_1} = 0, \quad \bar{\kappa_2} = \mid \lambda \mid \cos((N+1/2)\theta)\frac{\sin((N-1/2)\theta)}{(2N-1)\sin(\theta/2)}. \tag{54}$$

The corresponding plot for the same parameters as before is shown in Fig.2. A similar situation has been found [5] with hybrid Monte Carlo. In both cases the phenomenon has to do with the large number of cyclic modes in free field theory. In [5] it was concluded that randomizing the microcanonical (trajectory) length remains advantageous also after inclusion of a φ^4 interaction, which was very weak, however. The author tested the randomization of N in a hybrid OR simulation of the asymptotically free CP^3 model [6] and found absolutely no difference compared to fixed N.

3.2 Overrelaxation for a compact spin system

In many interesting field theoretical models fields do not take values in linear spaces, but in compact manifolds like real spheres S^n or Lie groups like $SU(n)$. A typical family of such systems are the $O(n)$ invariant nonlinear σ–models. The partition function with the standard lattice discretization is given by

$$Z = \int \prod_x \mathrm{d}^n s(x)\, \delta(s(x)^2 - 1)\, \mathrm{e}^{\beta \sum_{x\mu} s(x)\cdot s(x+\hat{\mu})}, \tag{55}$$

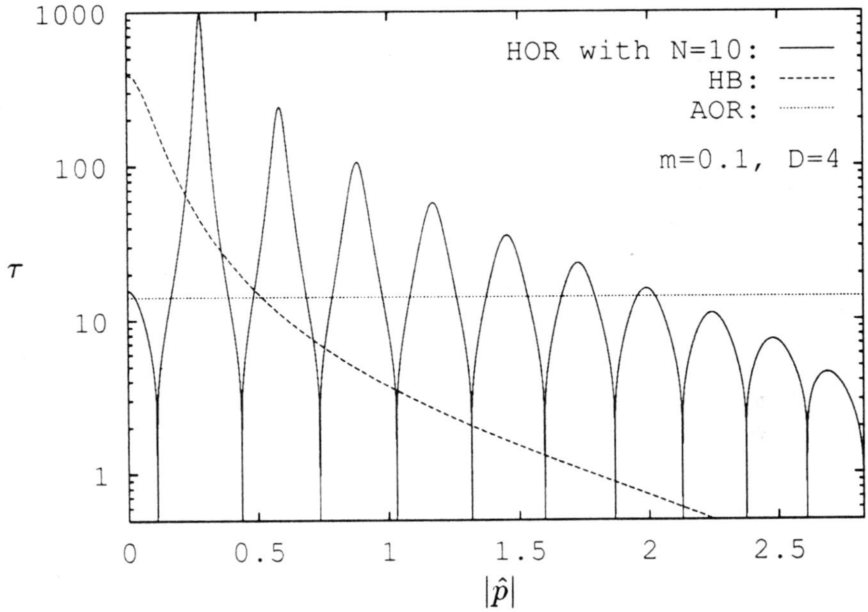

Fig. 1. Momentum dependent autocorrelation times in units of sweeps for heatbath and hybrid overrelaxation.

where $s(x) \in S^{n-1}$ has n components. For $n = 2$ this is the x-y model, which in two dimensions is expected to show a Kosterlitz Thouless transition controlled by vortices. In the same space time dimension the models with $n \geq 3$ are believed to be asymptotically free. While we will investigate the physics of the latter case in great detail in the next section, we now want to introduce hybrid OR for the $O(n)$ models.

For a local update we again focus our attention on one given site x_0. The Boltzmann factor depends on $s(x_0)$ through

$$\exp\{s(x_0) \cdot h(x_0) + \text{terms independent of } s(x_0)\}$$

with a local magnetic field

$$h(x_0) = \beta \sum_{|x-x_0|=1} s(x). \tag{56}$$

In standard local updates this distribution of $s(x_0)$ is generated with the Metropolis or (in closed form for n=3) heatbath procedure. For the additional microcanonical steps in hybrid OR we may reflect the spin through a plane orthogonal to the local field,

$$s \to s' = -s + 2\frac{h(h \cdot s)}{h^2}. \tag{57}$$

This update may be characterized as follows:

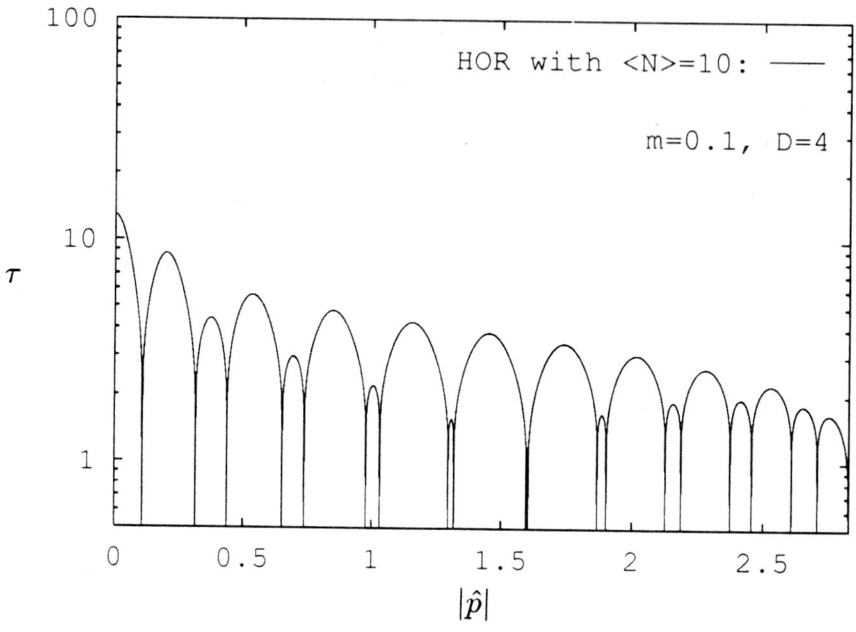

Fig. 2. As Fig.1 for a fluctuating number of microcanonical sweeps with $N = 10$ on average.

- The energy is unchanged, $h \cdot s = h \cdot s'$.
- The measure in (55) is invariant, since the reflection belongs to $O(n)$.
- Under the constraint of conserved energy, s' is as far from s as possible in a notion of distance suggested by the model, i.e. $s \cdot s'$ is minimal.

These (in the present case admittedly rather trivial) observations are useful for other cases, as one rephrases in a model independent way what we do in the $O(n)$ system.

The authors of [7] have applied hybrid OR to the $O(3)$ model in rather large scale simulations. In Fig.3 we reproduce a plot from their paper. It shows essentially the integrated autocorrelation time of the total magnetization versus N_{or}/ξ with N_{or} being the number of microcanonical steps that we called N previously. The tiny additional powers of ξ in the quantities plotted are not essential for our purpose. The fact that we see an approximately universal curve of data at widely differing correlation lengths demonstrates that z is not far from one. For instance, the cluster of points around $N_{or}\xi^{-1.1} \approx 0.8$ contains $\xi \approx 11 \ldots 225$. In the paper the estimate is $z = 1.1$. The authors actually tried various combinations of microcanonical and normal updates and conclude from their data as a general strategy that a z close to the optimal one is achieved by raising N_{or} with ξ such that measurements taken in between remain with constant correlation. This coincides with our analytical findings in the free field model.

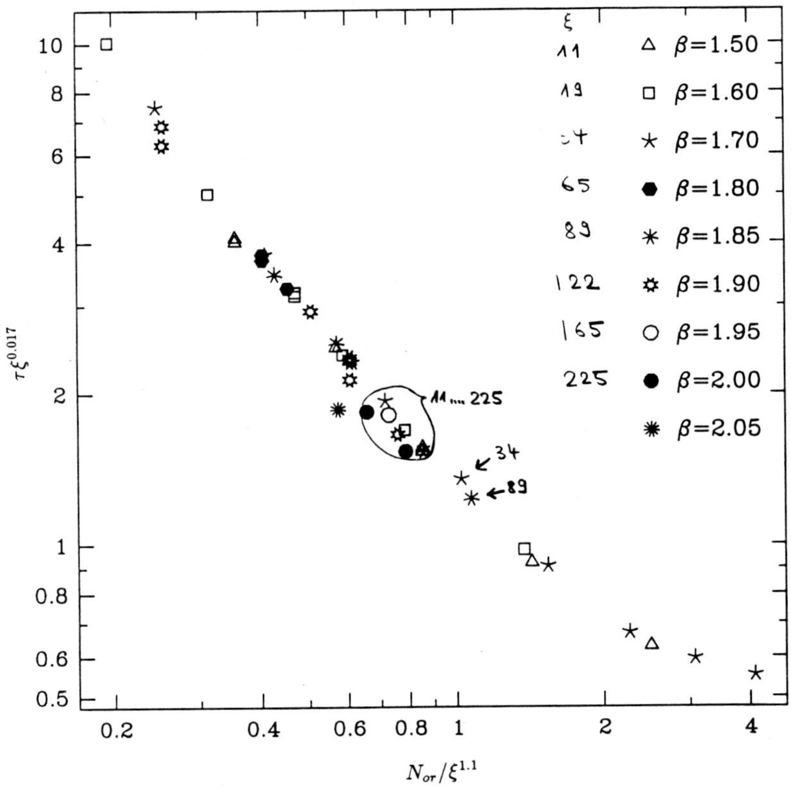

Fig. 3. Plot of autocorrelation times for hybrid overrelaxation in the $O(3)$ model reproduced from ref.[7].

3.3 Pure gauge theory, $SU(2)$ and $SU(3)$

We now briefly consider lattice gauge theory where field variables $U_\mu(x) \in SU(n)$ reside on the links of our lattice. The standard Wilson action

$$S(U_\mu(x)) = \frac{\beta}{n} \sum_p \operatorname{Re}\operatorname{tr}(1 - U_p) \tag{58}$$

is formulated in terms of gauge invariant traces over plaquette variables

$$U_p = U_\mu(x) U_\nu(x + \hat{\mu}) U_\mu(x + \hat{\nu})^{-1} U_\nu(x)^{-1}, \tag{59}$$

and the sum is over all plaquettes. It approaches the continuum Yang Mills action

$$\frac{1}{4g^2} \int \mathrm{d}^4x F^i_{\mu\nu}(x) F^i_{\mu\nu}(x)$$

for smooth fields as the lattice spacing a goes to zero. The partition function

$$Z = \int_{SU(n)} \prod_{x\mu} \mathrm{d}U_\mu(x)\,\mathrm{e}^{-S(U)} \tag{60}$$

involves invariant integrations with the group measure on every link, and expectation values of observables are defined in the usual way. In updating a given link $x_0\mu_0$ the relevant part of the action now is

$$S(U) = -\frac{\beta}{n}\mathrm{Re}\,\mathrm{tr}(U_{\mu_0}(x_0)H^\dagger_{\mu_0}(x_0)) + \text{terms independent of } U_{\mu_0}(x_0). \tag{61}$$

In this case the environment enters through a sum over three link products ("staples") of incomplete plaquettes after factoring out $U_{\mu_0}(x_0)$, symbolically given as

$$H^\dagger = \sum_{\text{staples}} UUU \tag{62}$$

with $2(D-1)$ terms from all directions along the axes orthogonal to the link under consideration.

The construction of microcanonical steps is now different for $n=2$ and $n=3$. The first case is easy if one observes the correspondence between $SU(2)$ and S^3. We expand H in terms of Pauli matrices $\boldsymbol{\tau}$,

$$H = h_0 + i\mathbf{h}\cdot\boldsymbol{\tau}, \quad h_0^2 + \mathbf{h}^2 = 1 \tag{63}$$

and likewise for $U_{\mu_0}(x_0)$. Now the trace in (61) goes over to a scalar product in the component vectors like $(h_0, \mathbf{h})$, and the reflection (57) can immediately be generalized. Rather than spelling this out in detail we now give the translation back into $SU(2)$ language. A special feature of $SU(2)$ related to the S^3 correspondence is the fact that sums of $SU(2)$ matrices like H can always be written as

$$H = \rho V, \quad \rho \in R,\ \rho \geq 0,\ V \in SU(2). \tag{64}$$

With this polar decomposition we now have the "flip" operation [8]

$$U \to U' = VU^\dagger V, \tag{65}$$

that leaves the energy invariant, as a trivial calculation shows. The above algorithm has been tested in [9] on 32^4 lattices at two different β values. Although systematic studies of autocorrelation times are very demanding and do not exist, the autocorrelation functions of large Wilson loops in Fig.4 for both heatbath and hybrid OR with $N=4$ very clearly show the gain that OR achieves.

For $SU(3)$ the situation is more complicated, as an analogue of (64) does not exist. A possible solution is however to still associate an $SU(3)$ element with H, namely a V, such that, when U runs through $SU(3)$,

$$\mathrm{Re}\,\mathrm{tr}(UH^\dagger)|_{U=V} = \text{maximal} \tag{66}$$

holds. With this group element constructed numerically in some approximation we then propose the substitution (65). In $SU(3)$ this in general results in an energy

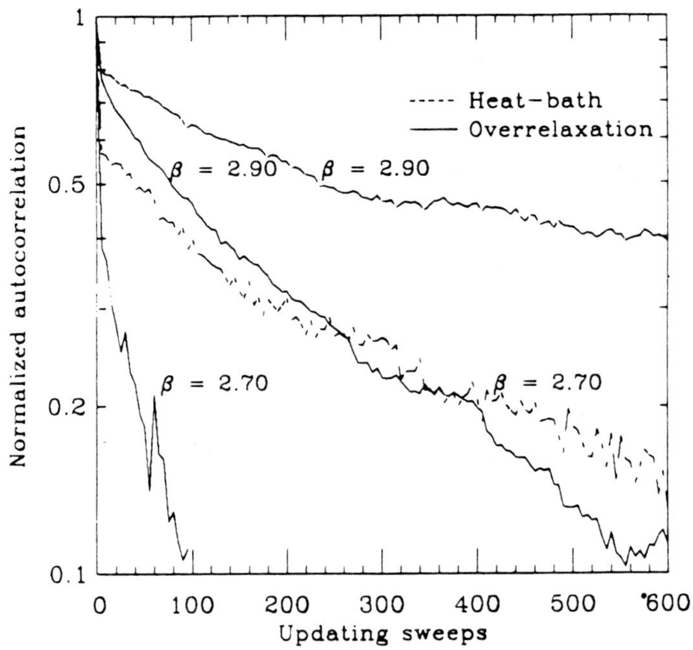

Fig. 4. Autocorrelation function for $SU(2)$ gauge theory on 32^4 with heatbath vs. OR with N=4 from ref.[9].

change, and a nontrivial accept/reject step has to follow as with Metropolis updating. This procedure coincides with the previous one for $SU(2)$. If the "sum over staples" is nearly proportional to a group element, as one naively expects for large β, the acceptance should be very high and the update almost microcanonical. Note that an error in solving (66) does not lead to systematic errors in the simulation but just influences the acceptance. In ref.[10] an acceptance of more than 99% was indeed found for physically interesting β in $SU(3)$ lattice gauge theory. Compared with a number of algorithms including pseudo-heatbath, Metropolis and hybrid Monte Carlo, OR was found to be most efficient. Again, however, the study was not systematic but just compared a few quantities on even rather small lattices. Also the implementation was somewhat different here: one almost microcanonical step of the type just described was followed *locally* with one (presumably not too large) Metropolis step. (Note that many hits or any other good approximation to heatbath would actually "undo" the OR step.) A different possibility from the one described here would be microcanonical updates in $SU(2)$ or even $U(1)$ subgroups that one embeds locally into the $SU(3)$ links.

To summarize, we think that OR is "state of the art" at present for simulations of 4–D pure gauge theory. The recently emerged tuning strategy of choosing N proportional to the scale should be employed. With OR and rapidly developing parallel computing facilities, for which OR is ideally suited, we may soon see results which are not so far away from what can be done in two dimensions. A first result of this kind is an impressive plot of the interquark force in $SU(2)$ from ref. [11] reproduced in Fig.5. The calculation is on a $48^3 56$ lattice, and the scale associated with the string tension in lattice units is about 15.

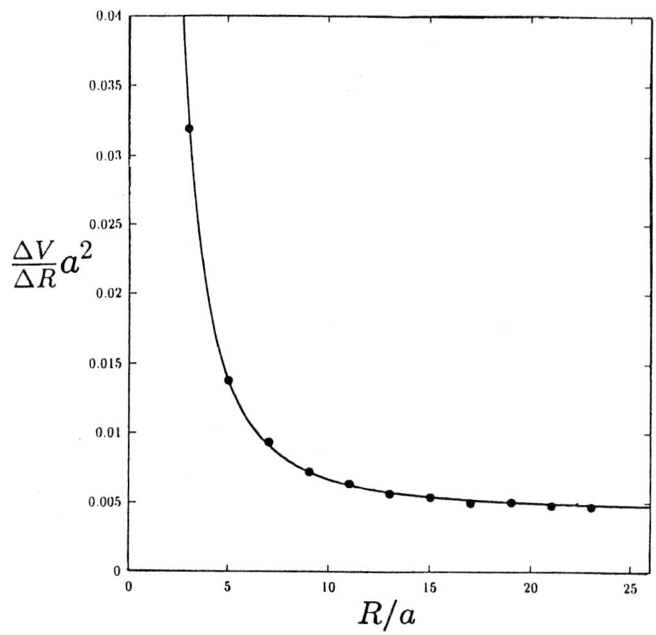

Fig. 5. Interquark force for $SU(2)$ gauge theory from ref.[11].

4 Cluster Algorithm and $O(3)$ σ–Model Results

4.1 Ising model, Swendsen Wang algorithm

Cluster algorithms were first developed[12] for the Ising model

$$Z = \sum_{\{\sigma(x)=\pm 1\}} e^{\beta \sum_{x\mu} \sigma(x)\sigma(x+\hat{\mu})} . \tag{67}$$

As this algorithm appears as one component in the updates for σ–models that we want to study here, we introduce it first. Due to the simple nature of the Ising field variables, the only possible moves are sign-flips. Therefore a general nonlocal algorithm has to construct domains C, such that

$$\sigma(x) \to -\sigma(x)\Big|_{\text{all } x \in C} \tag{68}$$

are appropriate collective moves for (67). Swendsen and Wang achieve this with the help of a percolation process. We derive it by introducing additional variables into (67) without changing it, apart from overall factors that we do not keep track of for simplicity (they cancel in expectation values).

As a first step note that

$$Z = \sum_{\{\sigma(x)\}} \prod_{x\mu} \left[(1-p) + p\delta_{\sigma(x),\sigma(x+\hat{\mu})}\right] \tag{69}$$

has the same relative weight for configurations as in (67), provided we put

$$p = 1 - \mathrm{e}^{-2\beta}. \tag{70}$$

We then introduce a field $k_\mu(x) = 0, 1$ to enumerate the two terms on every link

$$Z = \sum_{\{\sigma(x), k_\mu(x)\}} \left(\frac{p}{1-p}\right)^{\sum_{x\mu} k_\mu(x)} \Delta(k \mid \sigma), \tag{71}$$

where all the δ symbols from (69) are combined in

$$\Delta(k \mid \sigma) = \begin{cases} 0 \text{ if on any link } \sigma(x)\sigma(x+\hat{\mu})k_\mu(x) = -1 \\ 1 \text{ else} \end{cases}. \tag{72}$$

It enforces a constraint that bonds can only be "active" ($k_\mu(x) = 1$) between parallel spins and vice versa that spins connected by such bonds have to be aligned. The algorithm now consists of alternating updates of spins and of bonds. With given spins, the bonds can be chosen independently of their previous values to be

$$k_\mu(x) = \begin{cases} 1 \text{ with probability } p\delta_{\sigma(x),\sigma(x+\hat{\mu})} \\ 0 \text{ else} \end{cases}. \tag{73}$$

This corresponds to (perfect) independent sampling for this subset of the variables. For given bonds, one decomposes the sites into a maximal number of clusters such that no pair of sites in two different clusters is connected by an active bond. This step also appears in the study of bond percolation [13], and it can be carried out with a number of operations proportional to the volume. Then (71) instructs us to assign to each cluster as a whole (i.e. to all its spins in parallel) one of the two possible spin orientations with equal probability. In this step a nonlocal rearrangement of the spins takes place, to which the good efficiency ($z \approx 0.3$), that has been found numerically for Swendsen Wang updating in the Ising model, has to be attributed.

One can moreover prove an identity [14] , that exhibits the connection between the clusters and spin physics,

$$\langle\sigma(x)\sigma(y)\rangle = \left\langle \sum_{\text{clusters } c} \Theta_c(x)\Theta_c(y) \right\rangle \tag{74}$$

with $\Theta_c(x)$ the characteristic function of a cluster. The relation demonstrates that in a well defined sense the clusters have a size coupled to the correlation length. In particular, they percolate in the magnetized phase of the Ising model, and the threshold coincides with the critical point. The relation also has practical relevance, since it allows us to estimate correlations with a reduced variance.

A closely related technique has been proposed in [14] in the form of the single cluster algorithm. There one picks a random site x_0 (the "seed") and grows a cluster out of that site. Only those bonds are examined that one meets in this growth process, and they are activated as before. When the growth stops, all spins on the one cluster formed are flipped. This easily programmable version has been found more efficient, in particular in more than two dimensions. In $D = 4$ it seems to give $z = 0$ whereas the original many cluster technique has $z = 1$ [15]. Results in three dimensions are $z_E = 0.28$ and $z_\chi = 0.14$ for energy and squared total magnetization respectively [16].

4.2 Embedded Ising variables

Sometime after the discovery of the cluster algorithm for Ising and Potts models by Swendsen and Wang it was realized that it can also be applied to systems with continuous spin manifolds through embedding. A simple example is φ^4 with action (5) augmented by a quartic local interaction term. There one may introduce the sign of the field as an explicit degree of freedom [17]. Starting from a given configuration of the simulation one writes

$$\varphi(x) = \mid \varphi(x) \mid \sigma(x). \tag{75}$$

The cluster update of $\{\sigma(x)\}$ proceeds with an effective action

$$S_{\text{eff.}}(\sigma) = -\sum_{x\mu} \mid \varphi(x) \mid\mid \varphi(x+\hat{\mu}) \mid \sigma(x)\sigma(x+\hat{\mu}) + \text{terms independent of } \sigma\,. \tag{76}$$

This represents a slightly generalized Ising system (at fixed $\mid \varphi \mid$) with bond dependent ferromagnetic couplings. Cluster algorithms can be applied to it by simply substituting these coupling values into (70). The cluster update for the sign of φ has to be supplemented with other steps that can also change the magnitudes to make the compound ergodic. It is plausible, however, that at large quartic coupling, where φ^4 in the critical region has a double well structure, the behavior is Ising-like and the sign fluctuations are most important. In this region it is indeed found in ref.[17] that the cluster steps significantly accelerate the simulation.

Also the $O(n)$ nonlinear σ–model (55) may be simulated with Ising embedding [14]. In this case we first pick a random spin direction $r \in R^n$, $\mid r \mid = 1$, independently of the spin field $s(x)$ and decompose

$$s(x) = \sigma(x) \mid r \cdot s(x) \mid r + s_\perp(x)\,. \tag{77}$$

Ising flips are now mapped to reflections of $O(n)$ spins through the plane orthogonal to r. As any two n-component unit vectors can be connected by a reflection through some plane, these updates alone are even ergodic, if all directions can appear for r. It is clearly natural to choose r with the same uniform measure on the sphere as the spins themselves. The effective embedded Ising action has again ferromagnetic random bond strengths,

$$S_{\text{eff.}}(\sigma) = -\sum_{x\mu} | r\cdot s(x) || r\cdot s(x+\hat{\mu}) | \sigma(x)\sigma(x+\hat{\mu}) + \text{terms independent of } \sigma\,. \quad (78)$$

All applications (both single and many cluster) of this algorithm have yielded results which are consistent with the complete absence of critical slowing down, $z = 0$ [18]. Typical bounds quoted are $z \leq 0.1$ [19]. The autocorrelation times for typical observables, in particular with single cluster updating, just stay constant and are typically even smaller than one. This means that building and reflecting a number of clusters such that the computational effort corresponds to one pass through the lattice, achieves complete decorrelation. Moreover, variance reduction — as discussed in the Ising case — is also available for the σ–models. In this very favorable situation it is possible to get accurate results close to the continuum limit, with relatively modest computer time, as will be described in the following.

4.3 S-matrix of $O(n)$ σ–models

For the $O(n)$ σ–models an exact S-matrix has been proposed by Zamolodchikov in ref.[20]. Before we turn to numerical methods for computing scattering phases, we now briefly sketch the arguments that lead to the exact S-matrix. One starts with the standard scenario for the two dimensional model: asymptotic freedom and mass generation with an $O(n)$ vector multiplet of massive fundamental particles. They are associated with one particle states $| p, a\rangle$, to which the fundamental spin $s(x)$ field couples,

$$\langle 0 | s^a(x) | p, b\rangle = \delta^{ab} Z^{1/2} e^{ip\cdot x}, \quad (79)$$

and a, b are $O(n)$ indices here, and p is the (on shell) momentum. This standard scenario is supported by the perturbative renormalization group and by the $1/n$ expansion. It has also been criticized in a series of papers by Patrascioiu and Seiler, see for instance [21]. It is clearly of interest to see how far the standard picture is consistent with numerical results, although a real proof will never result from this.

For the S-matrix it is of particular importance that beyond the features just mentioned, the $O(n)$ models possess an infinite number of conserved (nonlocal) charges. This leads to the absence of particle production, and the set of outgoing momenta is restricted to be identical to the incoming ones. Moreover, there are identities that allow to express arbitrary matrix elements in terms of 2–particle scattering. In [20] this was combined with general principles like unitarity, crossing and analyticity, and it turned out that then the S-matrix for a massive n-plet is almost fixed. The simplest of the remaining choices agrees with perturbation and $1/n$ expansion and is proposed as the exact S-matrix.

The Zamolodchikov 2–particle S-matrix can be decomposed into isospin channels $I = 0$ (scalar), $I = 1$ (antisymmetric tensor) and $I = 2$ (symmetric tensor) [22]. For each of them there is a real phase δ_I. Their dependence on the momentum transfer is conveniently given in terms of the rapidity difference $\theta = \eta - \chi$, where the momentum p of one incoming particle is

$$p = \begin{pmatrix} m \cosh \eta \\ m \sinh \eta \end{pmatrix}, \tag{80}$$

and there is a corresponding expression for the other momentum in terms of χ. The phases for the case $n = 3$ are given by

$$e^{2i\delta_0} = \frac{\theta + 2\pi i}{\theta - 2\pi i} \tag{81}$$

$$e^{2i\delta_2} = \frac{\theta - \pi i}{\theta + \pi i} \tag{82}$$

$$e^{2i\delta_1} = e^{2i(\delta_0 + \delta_2)}. \tag{83}$$

This momentum dependence of δ_I is shown in Fig.6. All phases vanish logarithmically for large momenta in accordance with asymptotic freedom.

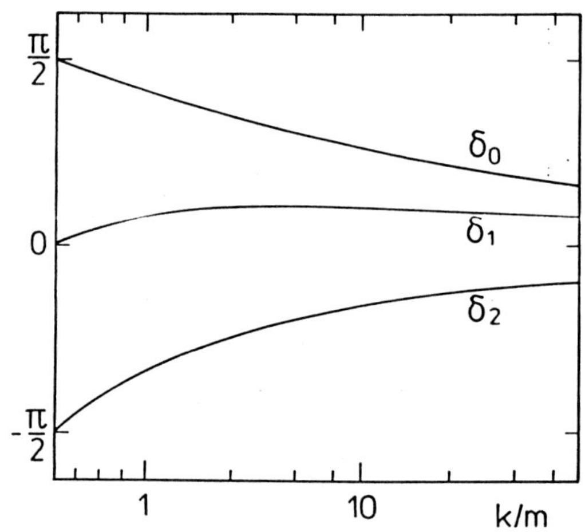

Fig. 6. Exact scattering phases in the 2–particle isospin channels of the $O(3)$ model.

There is no known practical way to simulate real scattering in Minkowski space. Lüscher noticed however [23] interesting quantitative relations between the S-matrix and leading finite volume effects in euclidean space for $mL < \infty$. This is intuitively plausible in the naive sense that wave packets associated with outgoing particles in scattering states in infinite volume have some probability to overlap again and again on a torus. For the case at hand, Lüscher's method requires information about the spectrum of 2–particle states (the lowest states even under charge conjugation $s^a(x) \to -s^a(x)$) in the various isospin channels at zero total momentum. Where there would be a continuum in infinite volume, there are densely spaced discrete levels for finite L. The extension T in euclidean time should be effectively infinite compared to L to have zero temperature and to be able to resolve the spectrum of interest by analyzing time decay properties. For each such energy W found, one has to compute k by putting

$$W = 2\sqrt{m^2 + k^2}. \tag{84}$$

Then a point of the graph $(k, \delta_I(k))$ is given by

$$\mathrm{e}^{2i\delta_I(k)} = \mathrm{e}^{-ikL}. \tag{85}$$

In free theory with vanishing phases (85) would just be the usual periodicity condition for the equal and opposite momenta of the two particles. For the modification in the interacting case to hold, it is important that polarization effects due to selfinteractions of the particles around the torus have to be negligible. This can be arranged for short range interactions. For the σ–model these interactions are known to be exponentially small in mL [24]. Formula (85) can be made particularly transparent in the nonrelativistic limit. We refer the reader to [22] for such a discussion.

We now come to numerical results using the finite size method for scattering phase shifts. A typical run with the single cluster method in ref.[22] was at parameters $L = 128, T = 256, \xi = (ma)^{-1} = 13.632(6)$. For a first orientation we checked the influence of cutoff effects on the energy momentum dispersion relation in the 1–particle sector of the spectrum. For spatial momentum $p^1 = 2\pi\nu/L$ we determined the corresponding energy $\hat{p}^0$ from decay in euclidean time and compared with $p^0 = \sqrt{m^2 + (p^1)^2}$. Some results are in the following table:

ν	p^1/m	$\hat{p}^0/p^0$
1	0.67	0.9992(3)
2	1.34	0.9993(3)
⋮	⋮	⋮
8	5.35	0.9904(4)

The discrepancy visible in the last column can be fitted as $1 - 0.06(p^1 a)^2$, a typical small cutoff effect of the type that is expected in physical observables [25]. It grows with momentum, as waves with larger wave numbers feel the discretization more and more. But we see that for a rather relativistic momentum at $\nu = 8$ we still only have effects at the percent level.

To access levels in the 2–particle sectors we measured the matrix of correlations

$$C^I_{pp'}(t) = \langle O^{ab}_I(0,p) O^{ab}_I(t,p') \rangle, \tag{86}$$

with composite fields

$$O^{ab}_I(t,p) = P_I(ab \mid cd) \sum_{xy} \mathrm{e}^{ip(x-y)} s^c(t,x) s^d(t,y). \tag{87}$$

In these formulas p, p', x, y are purely spatial, a, b, c, d are $O(n)$ indices, and we use isospin projectors P_I, which for $I = 2$ for instance, reads

$$P_2(ab \mid cd) = \frac{1}{2}(\delta_{ac}\delta_{bd} + \delta_{ad}\delta_{bc}).$$

A somewhat involved procedure, for whose details we refer to [22], now allows us to derive a number of levels from the correlation matrix C. They are converted to phases by (84) and (85) and shown together with the exact curves (81),(82),(83) in Fig.7. The various symbols refer to simulations at different parameters. The filled symbols stem from the run mentioned before with $L = 128$ and $(ma)^{-1} \approx 13.6$. The open symbols refer to an approximately two-fold larger lattice spacing at the same physical size mL. As expected they have a tendency to deviate more from the continuum curves at larger momenta. For the simple error bars we used a physically smaller lattice to verify that our results are insensitive to the polarization effects mentioned before.

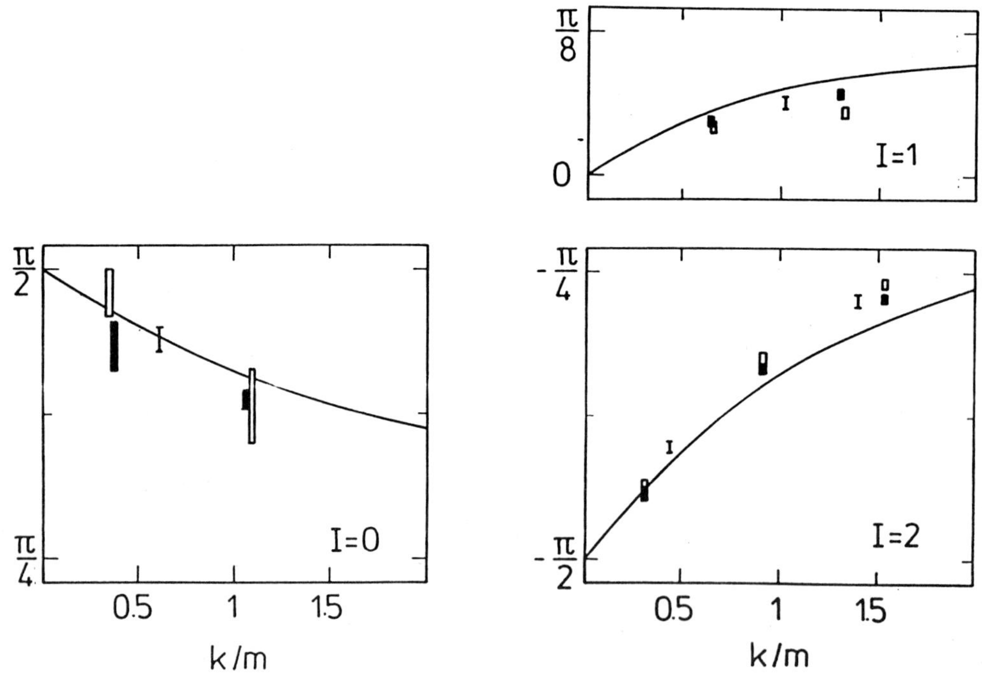

Fig. 7. Numerically determined scattering phases in the 2–particle isospin channels $I = 0, 1, 2$ of the $O(3)$ model together with curves predicted by the Zamolodchikov S-matrix (see text for an explanation of the symbols).

We see a close coincidence between theoretical prediction and numerical results with small discrepancies characteristic of finite lattice spacing. In principle, with more data than those shown here, one could attempt a systematic extrapolation to the continuum for this model. We feel, however, that the existing data already leave little doubt that the method works, and that the Zamolodchikov S-matrix indeed describes the $O(3)$ model. The latter is not entirely trivial, as the derivation, as we mentioned, involves various assumptions including one mainly based on simplicity. The numerical calculation required rather precise data on 4–point functions, which could be obtained with reasonable CPU–effort by the cluster technique.

4.4 Some properties of asymptotically free field theories

In this section we summarize some properties of asymptotically free theories like $O(n)$ invariant spin models in two and pure gauge theories in four dimensions. Although well known to particle physicists, they are decisive for a discussion of scaling and running coupling "constants" that follows. All the models — whether defined formally in the continuum or on a lattice — have at first sight only one free parameter, the bare coupling g_0^2. On the lattice it is often given by the analogous inverse temperature, $\beta = 1/g_0^2$ for the spin models or $\beta = 2n/g_0^2$ for $SU(n)$ gauge theory. As an implicit scale the cutoff has to be added, and we will restrict ourselves here to the lattice with spacing a. In the regularized theory one computes physical scales that we generically call m_i, and they appear as dimensionless ratios

$$m_i a = f_i(g_0^2) = 1/\xi_i. \tag{88}$$

The continuum limit is reached at a critical coupling, $g_0 \to g_{0c}$, where all correlation lengths ξ_i diverge. We interpret this as the spacing a vanishing with some mass, say m_0, staying finite. This is then the only parameter in the continuum theory. This kind of limit is called dimensional transmutation in particle physics. The critical bare coupling is $g_{0c} = 0$ according to the standard picture based on perturbation theory. It should be clear that both the lattice scale a and the distinguished mass m_0 are only (convenient) bookkeeping devices. In principle one could restrict oneself to just talking about dimensionless universal combinations like

$$\frac{m_i}{m_j}\Big|_{\text{continuum}} = \lim_{g_0\to 0} \frac{f_i}{f_j} = \lim_{g_0\to 0} \frac{\xi_j}{\xi_i}. \tag{89}$$

Apart from direct physical scales like masses, there is a class of quantities in asymptotically free models that have an interpretation as renormalized coupling constants. A typical example is some scattering cross section depending on a euclidean momentum transfer Q^2. After removing kinematical factors such that the perturbation expansion starts with the bare coupling, one has in the regularized (lattice) theory

$$g^2(Q) = g_0^2 + \sum_{k=2}^{\infty} c_k(Qa) g_0^{2k}. \tag{90}$$

We can also employ such a quantity to fix and parameterize the continuum limit by keeping $g(Q_0)$ at a certain numerical value and sending $a \to 0$ by tuning g_0

critical. Now Q_0 takes the rôle of a continuum scale in place of a mass like m_0. It should be emphasized that many such quantities are possible. One may use various cross sections, and, as we shall see later, there are also very different scale dependent ("running") couplings having to do with the volume dependence on the torus. What they all have in common is the existence of a formal expansion like (90) and their dependence on a physical scale staying finite (relative to all other physical scales) in the cutoff limit. If we have two such couplings $g^{(1)}(Q)$ and $g^{(2)}(Q)$ one may reorganize (90) to an expansion yielding in the cutoff limit

$$\left(g^{(1)}(Q)\right)^2 = \left(g^{(2)}(\tilde{Q})\right)^2 + \sum_{k=2}^{\infty} d_k^{(1,2)}(Q/\tilde{Q}) \left(g^{(2)}(\tilde{Q})\right)^{2k}. \tag{91}$$

Renormalizability guarantees all the coefficients $d_k^{(1,2)}$ to exist and remain finite in the continuum limit. Asymptotic freedom implies that when $Q, \tilde{Q} \gg m_0$ — at physically high energy or short distance — the renormalized perturbation expansion (91) can be meaningful (for not too large $|\ln(Q/\tilde{Q})|$) since both couplings are small. Therefore it is in principle sufficient to know one running coupling at high energy, because one can then relate it to others analytically by perturbation theory. There is thus a whole perturbative sector of the theory. Its connection to nonperturbative scales like the massgap m_0 is a typical problem for numerical calculations. There one could, for instance, seek the answer to a question like: at which ratio Q/m_0 does a certain coupling $g^2(Q)$ reach a certain value like 0.1? With one such information the scale of the whole perturbative sector is set relative to a nonperturbative scale.

Among the perturbatively computable quantities is the change of g with Q at high energy,

$$Q\frac{\mathrm{d}g^2}{\mathrm{d}Q} = \beta(g^2). \tag{92}$$

It can be considered as an infinitesimal version of (91) with both couplings identical and infinitesimally differing Q and $\tilde{Q}$. The perturbative renormalization group β–function is given as

$$\beta(u) = -u^2 \sum_{l=0}^{\infty} b_l u^l \tag{93}$$

with b_0 and b_1 universal for differently defined couplings. These two coefficients are usually known, while the first nonuniversal one b_2 is only known for a few models and couplings. The negative sign of β (asymptotic freedom) for small arguments implies that any renormalized coupling, once perturbative evolution is applicable at high energy, vanishes asymptotically as

$$g^2(Q) \simeq \frac{b_0}{\ln(Q)} \quad \text{as } Q \to \infty. \tag{94}$$

Defining the continuum limit by keeping some renormalized coupling fixed also determines the rate at which the bare coupling vanishes with the lattice spacing. It is given by the bare β-function $\bar{\beta}$ in

$$a\frac{\mathrm{d}g_0^2}{\mathrm{d}a} = -\bar{\beta}(g_0^2), \tag{95}$$

and it again has an expansion like in (93) with the universal b_0, b_1 and some $\bar{b}_2, \ldots$ This simply follows from the constancy of the renormalized coupling combined with (90), (92) and simple dimensional analysis. The differential equation (95) fixes all functions f_i in (88) up to a multiplicative constant which is *beyond* perturbation theory,

$$f_i(g_0^2) \propto (b_0 g_0^2)^{-b_1/b_0^2} \exp(-1/b_0 g_0^2)(1 + O(g_0^2)). \tag{96}$$

From this asymptotic behavior of all masses in lattice units one can define an artificial perturbative scale Λ_L which in the two loop approximation is given by

$$\Lambda_L^{(2)} = \frac{1}{a}(b_0 g_0^2)^{-b_1/b_0^2} \exp(-1/b_0 g_0^2). \tag{97}$$

Masses can now be referred to this scale by

$$\frac{m}{\Lambda_L}\Big|_{\text{continuum}} = \lim_{g_0^2 \to 0} \frac{m(g_0^2)}{\Lambda_L^{(2)}} \tag{98}$$

One speaks of asymptotic scaling when the right hand side looks constant to some accuracy and can hence be extrapolated to $g_0^2 = 0$. Asymptotic scaling is the traditional way to try to establish the connection between nonperturbative and perturbative continuum physics. The convergence rate is proportional to g_0^2 or a higher power of it, if Λ_L can be approximated by more loops. Because of the exponential growth of correlation lengths with vanishing g_0 implied by (96) it is hardly possible to vary $g_0^2 \propto 1/\ln \xi$ by significant amounts. This led to asymptotic scaling being an essentially untested hypothesis in many older studies with large errors. The new techniques for two dimensional spin models allow now some variation of $\ln \xi$ with much more accurate results. It turned out that, rather than allowing a reliable extrapolation $g_0^2 \to 0$, the variation is just enough to show that this is numerically hopeless in those models. A demonstration for the $O(3)$ model [26] is given in Fig.8. There we also see as a horizontal line what is proposed to be the exact value in [27]. No smooth asymptotic approach to a constant value is visible, and any extrapolation of the data is clearly impossible.

The situation in the $SU(3) \times SU(3)$ and CP^3 spin models, that could recently be studied in detail due to progress with multigrid methods [40], is just the same as $O(3)$. To conclude, let us emphasize the all important difference between asymptotic scaling and scaling. The latter means convergence proportional to a^2 or, in less favorable cases, a (inverse powers of ξ) rather than logarithmic. An example was the relativistic dispersion relation and the scattering phases presented earlier.

4.5 Numerical estimation of physical running couplings

We now outline a method to compute physical running coupling constants at high energy including the normalization of energy relative to the nonperturbative sector. It has been carried out for the $O(3)$ model [28] and relies only on the verification of scaling with convergence proportional to powers of the lattice spacing. The development of techniques for such a computation is attractive to ultimately compare with experimentally measured values of the strong QCD coupling constant $\alpha_s(M_Z) \approx 0.11$. Here the mass M_Z of the Z-particle is, from the point of view of QCD, just some high

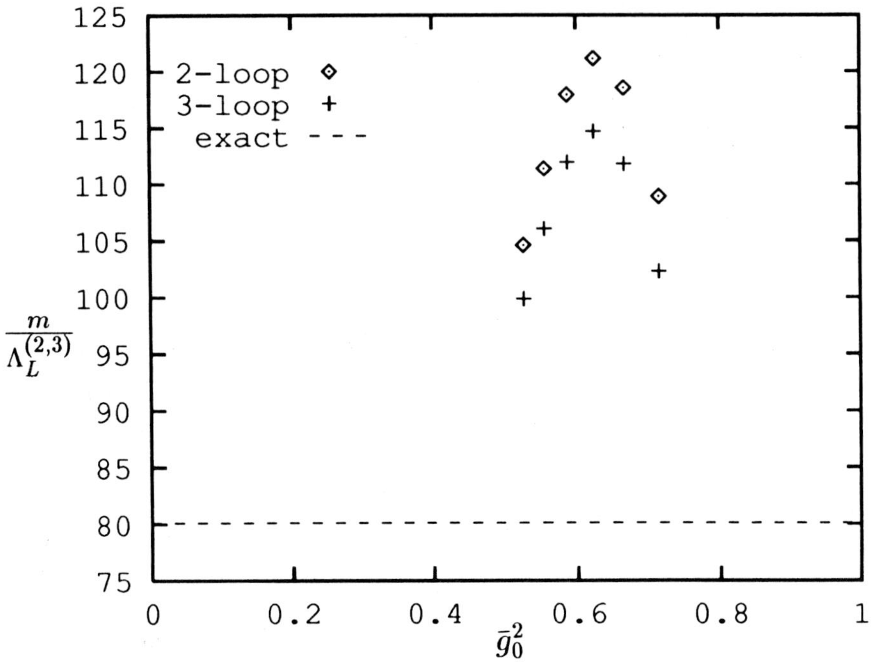

Fig. 8. Ratio of massgap to perturbative Λ_L-scale in the two and three loop approximation. Asymptotic scaling would correspond to a constant ratio. The data shown range over correlation lengths from $\xi = 6.9$ to $\xi = 122$.

energy scale, about 200 times that given by the string tension or 118 times the ρ-mass. From what was presented in the last section it is clear that such a number — in principle — is calculable without free parameters in asymptotically free theories.

As different physical couplings can be connected by perturbation theory at high energy, we choose a coupling for our computation guided by technical convenience. The decisive point is to take the finite spatial size L as the (only) free scale parameter, on which the coupling depends. Then universal continuum results only require $a \ll L$ to hold, and one does not have to *simultaneously* arrange $a \ll Q^{-1} \ll L$ for another scale. The finite size coupling used for the $O(3)$ model is

$$\bar{g}^2(L) = m(L)L, \tag{99}$$

where $m(L)$ is the finite volume massgap [29]. It is defined from the temporal decay of spatial 0–momentum Green functions as before. While $\bar{g}^2$ is indeed a perturbative coupling with an expansion (90) set up for small systems [29], $m(\infty)L \ll 1$, it is well defined for all L and diverges like

$$\bar{g}^2 \simeq m(\infty)L \quad \text{as } L \to \infty \tag{100}$$

in the nonperturbative regime.

With a convenient coupling chosen, the next important point is to break the problem of connecting low and high energy into several steps. This is done by introducing an analogue of β-functions for finite scale factors, the step scaling function

$$\sigma(s,u) = \bar{g}^2(sL)\Big|_{u = \bar{g}^2(L)} \cdot \tag{101}$$

Its relation to the β–function is

$$\ln s = -\int_u^{\sigma(s,u)} \frac{dv}{\beta(v)}. \tag{102}$$

We emphasize here that $\sigma(s,u)$ is a concept on the renormalized level independent of the lattice. Let us imagine we knew the exact step scaling function, say for scale factor $s = 2$. We could then start at a scale which we expect to be perturbative, for instance L_1 where $\bar{g}^2(L_1) = g_1^2 = 0.9176$, to use numerical examples from [28]. Now, using σ, we iterate in the following way:

$$\begin{aligned}
\sigma(2,g_1^2) &= \bar{g}^2(2L_1) && = g_2^2 > g_1^2 \\
\sigma(2,g_2^2) &= \bar{g}^2(4L_1) && = g_3^2 > g_2^2 \\
\vdots \qquad & \qquad \vdots && \qquad \vdots \\
\sigma(2,g_k^2) &= \bar{g}^2(2^k L_1) && = g_{k+1}^2 > g_k^2 \\
&= m(2^k L_1) 2^k L_1 && \\
&\approx m(\infty) 2^k L_1 &&
\end{aligned}$$

Here the first few steps can be used to check if we really started in the perturbative range. If true, this evolution should agree with a perturbative approximation of the step scaling function in (102). After some number of steps one arrives at long distance (bottom of the iteration scheme). Here we can make contact with a nonperturbative scale. In the present example this is particularly easy as it appears in the coupling $\bar{g}^2$ itself. We finally learn what L_1 we started with in units of the inverse massgap, namely

$$L_1 = g_{k+1}^2 2^{-k} m(\infty)^{-1}. \tag{103}$$

At this point the goal of knowing a running coupling at high energy with an absolutely normalized scale relative to a mass has been achieved. It remains now to describe how we are able to compute the step scaling function nonperturbatively.

At finite lattice spacing there is an approximate step scaling function Σ with

$$\sigma(s,u) = \lim_{a\to 0} \Sigma(s,u,a/L). \tag{104}$$

A sequence of $\Sigma(2,u,a/L)$ is now constructed with fixed u and shrinking a/L. We start by tuning the bare coupling g_0 on some lattice of, for instance, $L = 10a$, such that $\bar{g}^2 = u$ is met. Then with the *same* g_0, a simulation on $L = 20a$ gives a renormalized charge $\Sigma(2,u,1/10)$. If this is repeated for many pairs of lattices, for example $L/a = (7,14),(8,16),(10,20),\ldots$, then the resulting values for Σ can be extrapolated to the continuum. The convergence rate expected for the coupling (99) is $(a/L)^2$ (neglecting the variation of logarithms). This is shown in an example from [28] in Fig.9.

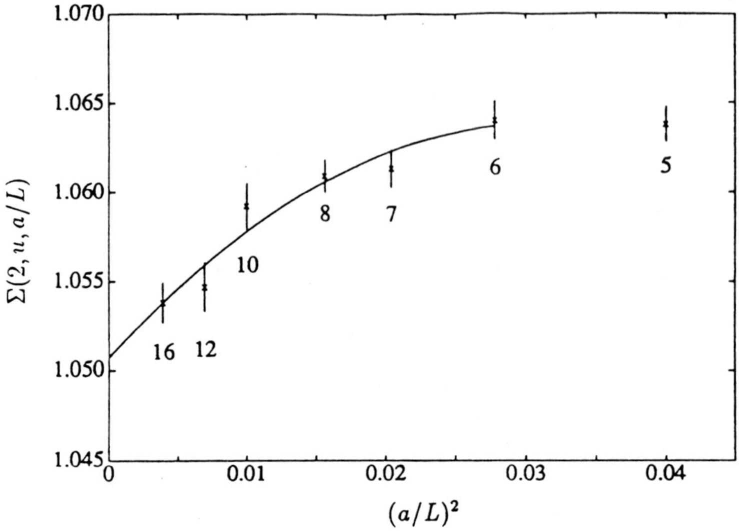

Fig. 9. Extrapolation of the step scaling function to the continuum.

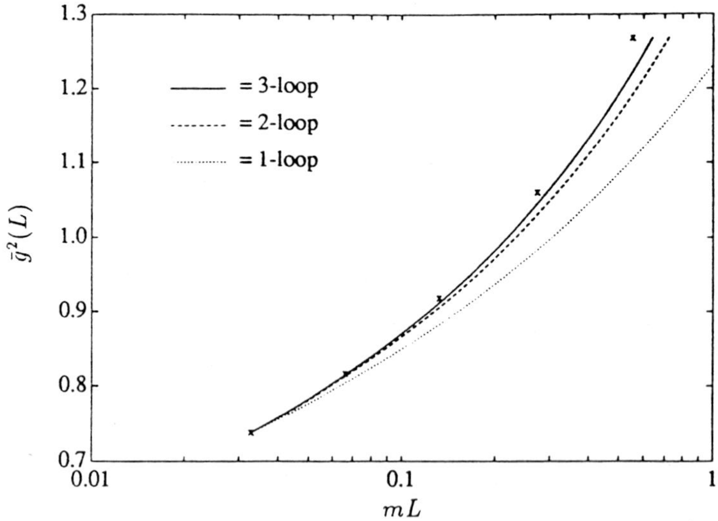

Fig. 10. Running of the coupling constant $\bar{g}^2(L)$ with the scale in units of the infinite volume massgap. The data points have been obtained numerically by extrapolation to the continuum.

In Fig.10 five data points constructed by the above procedure are shown together with the perturbative evolution starting from the point at shortest distance. We see that the evolution rapidly breaks away from the perturbative curves only for scales L approaching $m(\infty)^{-1}$.

While each value of a running coupling like $\bar{g}^2(L)$ implicitly defines a scale, one can also associate a finite Λ–parameter scale with the known short distance asymptotics. Similarly to the bare charge on the lattice we define

$$\Lambda^{(2)}(L) = \frac{1}{L}(b_0\bar{g}^2)^{-b_1/b_0^2} \exp(-1/b_0\bar{g}^2) \tag{105}$$

and

$$\Lambda = \lim_{L\to 0} \Lambda^{(2)}(L). \tag{106}$$

As is well known, Λ parameters belonging to different couplings and also those of minimal subtraction and of the bare lattice charge can be exactly related by 1–loop calculations. This is plausible, as they are obtained from the behavior of infinitesimal couplings at infinite energy. Now the convergence rate from higher loop corrections in (105) is proportional to powers of $\bar{g}^2(L)$. The situation for the scale connected with our charge (99) is shown in Fig.11, and the dashed line is again from [27].

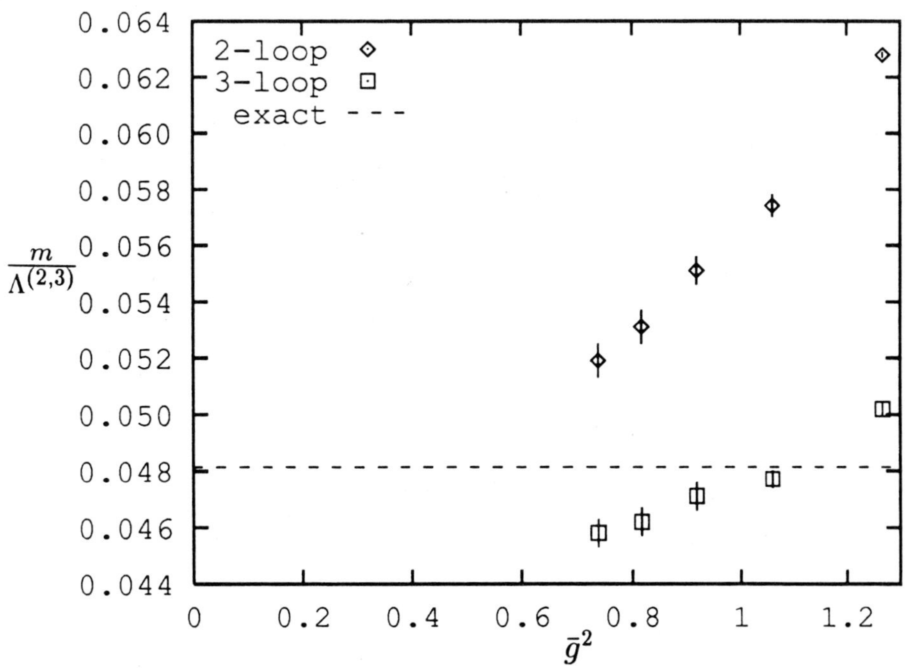

Fig. 11. Ratio of massgap to the Λ–scale associated with a renormalized coupling in the two and three loop approximations.

Also here a reliable extrapolation to $\bar{g}^2 = 0$ is not possible. On the other hand it is clear from Fig.10 that the coupling at $L = 100m^{-1}$ has been very well determined. This shows, that Λ-parameters, both for bare and renormalized couplings, are not very convenient parametrizations for numerical work. It seems better, with methods like the one presented, to avoid them and to compute directly quantities more closely related to experimental numbers like $\alpha_s(M_Z)$. In [28] also the step of converting from $\bar{g}^2$ to the coupling defined by minimal subtraction (independent of a finite volume) has been carried out for the $O(3)$ model. The result is

$$\bar{g}^2_{MS}(\mu = 1/L) = \bar{g}^2(L) + 0.156\,\bar{g}^4(L) - 0.035\,\bar{g}^6(L), \tag{107}$$

which looks reasonably stable for the smallest coupling $\bar{g}^2 = 0.74$ that was used in our rescaling steps.

At present a similar method is developed for four dimensional pure gauge theory [30]. Since small volume massgaps are not so easily measurable in this case, a different finite volume physical quantity has been proposed there. It is an adaptation of background field methods to finite volume [31]. The physical method behind it is essentially an estimate of the change in free energy from varying a dimensionless parameter associated with Dirichlet boundary conditions in one direction. First numerical tests in $SU(2)$ look positive and will hopefully be reported on in the near future.

5 Outlook

Considerable effort has been devoted to the generalization of cluster algorithms beyond ferromagnetic Ising and Potts models and $O(n)$ invariant spin systems. Some nice successes have been scored for models of interest to condensed matter physics like SOS [32], interface roughening [33], fully frustrated Ising [34] and antiferromagnetic Potts [35] models. Interesting cluster simulations of gauge invariant models have so far been restricted to those with Ising variables on links. An elegant method [36] was proposed and tested in $D = 3$, and another variant [37], although less efficient for three dimensions, has the virtue of being generalizable to $D = 4$ and to the inclusion of Higgs fields [38]. Embedded subsystems in $U(1)$ [39] or $SU(n)$ gauge theories, for which the $Z(2)$ algorithms could be used, have not been found so far.

Also the more modest goal of extending cluster techniques to other nonlinear σ–models like CP^n or chiral $SU(n) \times SU(n)$ systems has not been met. Just recently, however, Hasenbusch, Mack and Meyer have reported on a multigrid algorithm [40], that was shown to have dynamical exponents $z \approx 0.2$ at least effectively for correlation lengths up to about 100 in $O(3)$, CP^3 and $SU(3) \times SU(3)$ spin models. They propose block moves of a smooth, for instance pyramidal, shape in a hierarchical fashion on all scales. The smoothness, as opposed to step function like moves, is absolutely crucial to get acceptance for physically relevant updates. The implementation of their multigrid updates proceeds in the unigrid fashion, where one goes back to the original fine lattice from every level of updates to compute the change in energy. This has the slight disadvantage of costing an extra $\log(L/a)$ factor in the operations count, but it is relatively easy to implement. Thus, multigrid techniques seem to be a more versatile tool than cluster algorithms, and a lot of research is devoted to their generalizations to gauge theory. One should bear in mind, however, that the crossover in efficiency between multigrid and the simple hybrid overrelaxation algorithms in spin systems is only reached at correlation length 20 or so. Although this is clearly model as well as program dependent, OR may be the method of choice for pure gauge theory for quite a while.

Finally we would like to mention a different area, where multigrid is developed vigorously, namely for the inversion of the lattice Dirac operator in a given gauge field. This problem consumes the bulk of computing power in computations involving quark fields, and there is slowing down, too. Again, there are optimized more conventional methods like conjugate gradient, which should eventually, for very large scale applications on parallel machines, be outperformed by nonlocal methods like multigrid.

Acknowledgements

I would like to thank my co-workers in the presented and ongoing calculations, Martin Lüscher, Rajamani Narayanan, Rainer Sommer and Peter Weisz, for the enjoyable and exciting collaboration.

References

1. Sokal, A.D.: *Monte Carlo Methods in Statistical Physics: Foundations and New Algorithms*, Lecture Notes of the Cours de Troisième Cycle de la Physique en Suisse Romande, Lausanne 1989
2. Adler, S.: Phys. Rev. D23 (1981) 2901
3. Neuberger, H.: Phys. Rev. Lett. 59 (1987) 1877
4. Wolff, U.: *Dynamics of hybrid overrelaxation in the Gaussian model*, preprint CERN-TH 6408/92
5. Mackenzie, P.: Phys. Lett. 226B (1989) 369
6. Wolff, U.: *Scaling topological charge in the CP^3 spin model*, preprint CERN 6407/92
7. Apostolakis, J., Fox, G.C., Baillie, C.F.: Nucl.Phys. B (Proc. Suppl.) 20 (1991) 678
8. Creutz, M.: Phys. Rev. D36 (1987) 515;
   Brown, F.R., Woch, T.J.: Phys. Rev. Lett. 58 (1987) 2394
9. Decker, K., de Forcrand, Ph.: Nucl.Phys. B (Proc. Suppl.) 17 (1990) 567
10. Gupta, R., Kilcup, G.W., Patel, A., Sharpe, S.R., de Forcrand, Ph.: Mod. Phys. Lett. A3 (1988) 1367
11. Booth, S.P., et al.: (UKQCD Collaboration), *Towards the continuum limit of SU(2) lattice gauge theory*, Liverpool preprint LTH 271 (1991)
12. Swendsen, R.H., Wang, J.-S.: Phys. Rev. Lett. 58 (1987) 86
13. For a review, see, e.g., Stauffer, D.: Phys. Rep. 54 (1979) 1
14. Wolff, U.: Phys. Rev. Lett. 62 (1989) 361
15. Tamayo, P., Brower, R.C., Klein, P.W.: J. Stat. Phys. 58 (1990) 1083
16. Wolff, U.: Phys. Lett. 228B (1989) 379.
17. Brower, R.C., Tamayo, P.: Phys. Rev. Lett. 62 (1989) 1087
18. Frick, Ch., Jansen, K., Seuferling, P.: Phys. Rev. Lett. 63 (1989) 2613
19. Edwards, R., Sokal, A.: Phys. Rev. D40 (1989) 1374
20. Zamolodchikov, A.B., Zamolodchikov, Al.B.: Nucl. Phys. B133 (1978) 525

21. Patrascioiu, A., Seiler, E.: *The Difference between Abelian and NonAbelian models: Fact and fancy*, München preprint MPI-PH-91-87 (1991)
22. Lüscher, M., Wolff, U.: Nucl. Phys. B339 (1990) 222
23. Lüscher, M.: Commun. Math. Phys. 105 (1986) 153
24. Lüscher, M.: *On a relation between finite size effects and elastic scattering processes*, lecture given at Cargèse (1983), *in* Progress in gauge field theory, eds. Hooft, G.'t: et al. (Plenum, New York, 1984);
Commun. Math. Phys. 104 (1986) 77
25. Symanzik, K.: Nucl. Phys. B226 (1983) 187, 205;
and *Cutoff dependence in lattice ϕ_4^4 theory* in Recent developments in gauge theories (Cargèse 1979) eds. Hooft, G.'t: et al. (Plenum, New York, 1980)
26. Wolff, U.: Nucl. Phys. B334 (1990) 581; Phys. Lett. 222B (1989) 473
27. Hasenfratz, P., Maggiore, M., Niedermayer, F.: Phys. Lett. 245B (1990) 522;
Hasenfratz, P., Niedermayer, F.: Phys. Lett. 245B (1990) 529
28. Lüscher, M., Weisz, P., Wolff, U.: Nucl. Phys. B359 (1991) 221
29. Lüscher, M.: Phys. Lett. 118B (1982) 391
30. Lüscher, M., Narayanan, R., Weisz, P., Wolff, U.: *The Schrödinger Functional — a Renormalizable Probe for Non-Abelian Gauge Theories*, in preparation
31. Wolff, U.: Nucl. Phys. B265 [FS15] (1986) 506 and 537
32. Evertz, H.G., Hasenbusch, M., Marcu, M., Pinn, K., Solomon, S.: Nucl.Phys. B (Proc. Suppl.) 20 (1991) 80
33. Hasenbusch, M., Meyer, S.: Phys. Rev. Lett. 66 (1991) 530
34. Kandel, D., Ben-Av, R., Domany, E.: Phys. Rev. Lett. 65 (1990) 941
35. Wang, J.-S., Swendsen, R.H., Kotecký: Phys. Rev. Lett. 63 (1989) 109
36. Ben-Av, R., Kandel, D., Katznelson, E., Lauwers, P.G.: J. Stat. Phys. 58 (1990) 125
37. Brower, R.C., Huang, S.: Phys. Rev. D62 (1990) 708
38. Bunk, B.:*An Efficient Algorithm for Cluster Updates in $Z(2)$ Lattice Gauge Theory*, in the Proceedings of the Workshop on Fermion Algorithms, HLRZ, Jülich 1991, eds. Herrmann, H.J., Karsch, K.: (World Scientific 1991)
39. For a new approach for $U(1)$ see:
Adler, S.L., Bhanot, G.: Phys. Rev. Lett. 66 (1991) 1807;
Adler, S.L., Bhanot, G., Lippert, T., Schilling K., Ueberholz, P.: *Defeating Critical Slowing Down for Abelian Gauge Dynamics*, Princeton preprint IASSNS-HEP-91/39
40. Hasenbusch, M., Meyer, S, Mack, G: Nucl.Phys. B (Proc. Suppl.) 20 (1991) 110;
Hasenbusch, M., Meyer, S.: Phys. Rev. Lett. 68 (1992) 435;
Testing Accelerated Algorithms in the Lattice CP^3 Model, Kaiserslautern preprint Th 91-23 (1991)

This article was processed using the LaTeX macro package with LMAMULT style

The Present and Future of Lattice QCD

T. DeGrand

Department of Physics, University of Colorado, Boulder, Colorado 80309

Abstract. These lectures provide an introduction to lattice methods for nonperturbative studies of Quantum Chromodynamics. I give a brief introduction to the subject, then discuss areas of current interest: spectroscopy, matrix elements, and the physics of high temperature QCD.

1 Introduction

Lattice methods have transformed strong interaction physics. We are (apparently) on the verge of being able to compute the masses and matrix elements of hadrons from first principles using lattice-based numerical simulations of QCD. These lectures are intended to be an introduction to the lattice approach to QCD. I am going to focus on physics issues rather than computational issues.

I will discuss four topics. Chapter 2 is about the basics: why and how one defines a gauge theory on the lattice, then a description of confinement signals, the continuum limit, and a few words about lattice fermions. In Chapter 3 I will describe how spectroscopy calculations are done, from basics to the construction and reconstruction of hadronic wave functions. I will describe recent result for the spectroscopy of ground and excited states. I will talk about the physics of QCD at high temperature as revealed by lattice simulations in Chapter 4. Chapter 5 will be concerned with lattice methods for calculating matrix elements, and I will describe as a case study the calculation of heavy meson decay constants. Finally, I am supposed to describe the "future" of lattice calculations, and I will make a few oracular (ill-considered?) remarks.

2 Beginnings

2.1 Basics

The lattice is a mathematical trick. It provides a cutoff removing the ultraviolet divergences of quantum field theories. As with any regulator, it must be removed after renormalization. Contact with experiment only exists in the continuum limit, when the lattice spacing is taken to zero.

We are drawn to lattice methods by our desire to study nonperturbative phenomena. Older regularization schemes are tied closely to perturbative expansions: one calculates a process to some order in a coupling constant; divergences are removed order by order in perturbation theory. The lattice, however, is a nonperturbative cutoff. Before a calculation begins, all wavelengths less than a lattice spacing are removed.

On the lattice, a field theory becomes mathematically well defined. This fact is not useful merely to mathematicians. It means that one can formulate problems in quantum field theory in such a way that they can be studied using a combination of analytical and numerical methods. Generally one cannot carry out analytical studies of a field theory for physically interesting parameter values. However, lattice techniques lend themselves naturally to implementation on digital computers, and one can perform more-or-less realistic simulations of quantum field theories, revealing their nonperturbative structure, on present day computers. I think it is fair to say that most of the qualitative insight into, and all of the quantitative results about, nonperturbative QCD which have been obtained in the last decade, could not have been obtained without the use of numerical methods.

It is very easy to define scalar fields on the lattice. One just replaces the space-time coordinate x_μ by a set of integers n_μ ($x_\mu = an_\mu$, where a is the lattice spacing). Field variables $\phi(x)$ are defined on sites $\phi(x_n) \equiv \phi_n$, The action, an integral over the Lagrangian, is replaced by a sum over sites

$$S = \int d^D x \mathcal{L} \to a^4 \sum_n \mathcal{L}(\phi_n). \tag{1}$$

and the generating functional for Euclidean Green's functions is replaced by an ordinary integral over the lattice fields

$$Z = \int (\prod_n d\phi_n) e^{-S}. \tag{2}$$

Gauge fields are a little more complicated. Gauge fields carry a space-time index μ in addition to an internal symmetry index a ($A_\mu^a(x)$) and are associated with a path in space $x_\mu(s)$: a particle traversing a contour in space picks up a phase factor

$$\begin{aligned} \psi &\to P(\exp\, ig \int_s dx_\mu A_\mu)\psi \\ &= U(s)\psi(x). \end{aligned} \tag{3}$$

P is a path-ordering factor analogous to the time-ordering operator in ordinary quantum mechanics. Under a gauge transformation g, $U(s)$ is rotated at each end:

$$U(s) \to g^{-1}(x_\mu(s))U(s)g(x_\mu(0)). \tag{4}$$

These considerations led Wilson [1] to formulate gauge fields on a space-time lattice, as follows:

The fundamental variables are elements of the gauge group G which live on the links of a D-dimensional lattice, connecting x and $x+\mu$: $U_\mu(x)$, with $U_\mu(x+\mu)^\dagger = U_\mu(x)$

$$U_\mu(n) = \exp(igaT^a A_\mu^a(n)) \tag{5}$$

for $SU(N)$ (g is the coupling, a the lattice spacing, A_μ the vector potential, and T^a is a group generator). An action is specified by recalling that the classical Yang-Mills action involves the curl of A_μ, $F_{\mu\nu}$. Thus a lattice action ought to involve a product of U_μ's around some closed contour. The simplest such contour has a perimeter of four links

$$\beta S = \frac{1}{g^2} \sum_n \sum_{\mu>\nu} \text{Re Tr } U_\mu(n) U_\nu(n+\hat{\mu}) U_\mu^\dagger(n+\hat{\nu}) U_\nu^\dagger(n). \tag{6}$$

This action is called the Wilson action after its inventor.

Let us see how this action reduces to the standard action. Specializing to U(1),

$$S = \frac{1}{g^2} \sum_n \sum_{\mu>\nu} \text{Re } (\exp(iga[A_\mu(n) + A_\nu(n+\hat{\mu}) - A_\mu(x+\hat{\nu}) - A_\nu(n)])). \tag{7}$$

The naive continuum limit is taken by assuming that the lattice spacing a is small, and Taylor expanding

$$A_\mu(n+\hat{\nu}) = A_\mu(n) + a\partial_\nu A_\mu(n) + \ldots \tag{8}$$

so the action becomes (dropping the constant)

$$\begin{aligned} S &= \frac{1}{g^2} \sum_n \sum_{\mu>\nu} \text{Re } (\exp(iga[a(\partial_\nu A_\mu - \partial_\mu A_\nu) + O(a^2)])) \\ &= -\frac{1}{4g^2} a^4 \sum_n \sum_{\mu\nu} g^2 F_{\mu\nu}^2 + \ldots \\ &= -\frac{1}{4} \int d^4x F_{\mu\nu}^2 \end{aligned} \tag{9}$$

transforming the sum on sites back to an integral.

On the lattice we sacrifice Lorentz invariance but preserve full gauge invariance. This preservation is important for nonperturbative physics. For example, gauge invariance is a property of the continuum theory which is nonperturbative, so maintaining it as we pass to the lattice means that all of its consequences (including current conservation and renormalizability) will be preserved.

One of these consequences is Elitzur's theorem [2]: only gauge invariant quantities have nonzero expectation values.

Under a gauge transformation link variables transform as

$$U_\mu(x) \to V(x) U_\mu(x) V(x+\hat{\mu})^\dagger \tag{10}$$

and site variables as

$$\psi(x) \to V(x)\psi(x) \tag{11}$$

so the only gauge invariant operators we can use as order parameters are closed loops of U's (Fig. 1a)

$$\text{Tr} \ldots U_\mu(x) U_\mu(x+\hat{\mu}) \ldots \to \text{Tr} \ldots U_\mu(x) V^\dagger(x+\hat{\mu}) V(x+\hat{\mu}) U_\mu(x+\hat{\mu}) \ldots \tag{12}$$

or matter fields connected by "strings" of U's (Fig. 1b)

$$\bar{\psi}(x_1) U_\mu(x) U_\mu(x+\hat{\mu}) \ldots \psi(x_2) \tag{13}$$

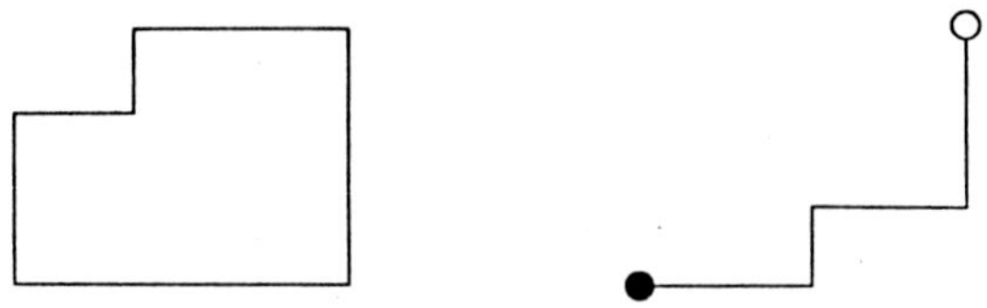

(a) (b)

Fig. 1. Gauge invariant observables are either (a) closed loops of link variables or (b) ordered chains ("strings") of links connecting quarks and antiquarks.

The best way people have found to check for quark confinement is to ask the following question: In a world in which there are no light quarks, what is the potential $V(r)$ between a heavy $q\bar{q}$ pair? If the limit as R goes to infinity of $V(R)$ is infinite, we have confinement, if not, quarks are not confined. $V(R)$ can be computed by considering the partition function in Euclidean space for gauge fields in the presence of a classical current distribution

$$Z_J = \int [dU] \exp(-\beta S + i \int J_\mu A_\mu d^4x). \tag{14}$$

(for a non-Abelian gauge group insert color sums $J_\mu^a A_\mu^a$.) If J_μ represents a point particle moving along a world line, it is a δ-function on that world line (parameterized by ℓ_μ)

$$i \int J_\mu A_\mu d^4x = i \oint A_\mu dl_\mu. \tag{15}$$

Gauge invariance, which implies current conservation, says that the current line cannot end, and so the loop is a closed loop. Let's make it rectangular for simplicity, extending a distance R in some spatial direction and a distance T in the Euclidean time direction. This is the famous Wilson loop (See Fig. 2a.).

Z_J describes the loop immersed in a sea of gluons - we can think of it as the partition function for a $q\bar{q}$ pair which is created at t=0, pulled apart a distance R, and then allowed to annihilate at a time T later.

To find the pair's energy, differentiate $\ln(Z)$ with respect to T (physically, we are measuring the response of the system as we stretch the loop a bit) [3]. But this is not quite right. This calculation will include the vacuum energy of the gluons, which diverges. What we really want is the change in energy of the system when we include the pair and when it is not present

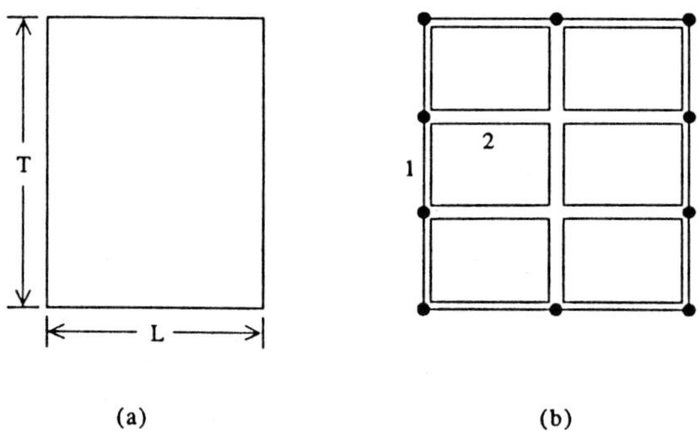

Fig. 2. An $L \times T$ Wilson loop (a) and the pattern of tiling which occurs when it is evaluated in strong coupling (b).

$$\begin{aligned} E_{q\bar{q}} &= -\frac{\partial}{\partial T}[\ln Z_J - \ln Z_{J=0}] \\ &= -\frac{\partial}{\partial T}\ln(Z_J/Z_{J=0}) \\ &= -\frac{\partial}{\partial T}\ln\frac{\int[dU]e^{\beta S}e^{i\int A\cdot dl}}{\int[dU]e^{\beta S}} \\ &= -\frac{\partial}{\partial T}\ln\langle W\rangle_{J=0} \end{aligned} \tag{16}$$

where $\langle W\rangle$ is the expectation value of the Wilson loop in the background gluon field. Thus the behavior of $\langle W\rangle$ will tell us whether we have confinement or not.

For example, if $\langle W\rangle \simeq \exp(-m(2R+2T))$ (a so-called "perimeter law") then $E = 2m$, and m is just the quark mass. The energy of the pair is independent of R so that the quarks can separate arbitrarily far apart. However, if $\langle W\rangle \simeq \exp(-\sigma RT)$ (an "area law"), $E = \sigma R$, and quarks are confined with a linear confinement potential. The parameter σ is called the "string tension". In general, $\langle W\rangle$ is complicated [4]: $-\ln\langle W\rangle = \sigma RT + 2m(R+T) + \text{constant} + T/R + \ldots$ Nevertheless, any area law term will dominate at large R and T and the theory will confine.

Now let us give an explicit demonstration of confinement for U(1) gauge theory [5]. The Haar measure for $U(1)$ is ($\theta_\mu = gaA_\mu$)

$$\int dU = \int_{-\pi}^{\pi}\frac{d\theta}{2\pi} \tag{17}$$

and

$$Z = \int [dU]\exp\beta S \tag{18}$$

where S is given by (6).

Strong coupling corresponds to small β. Now we write down two trivial (but useful) results:

$$\int_{-\pi}^{\pi} \frac{d\theta}{2\pi} e^{i(\theta+\phi)} = 0$$
$$\int_{-\pi}^{\pi} \frac{d\theta}{2\pi} e^{i(\theta+\phi_1)} e^{-i(\theta+\phi_2)} = e^{i(\phi_1-\phi_2)}. \tag{19}$$

For small β we may expand the exponential in Z as a power series in β:

$$Z = \int \prod_{links} \frac{d\theta}{2\pi} \prod_{plaquettes} (1 + \beta(\mathrm{Tr}U_p + \mathrm{h.\ c.}) + \ldots)$$
$$= 1 + O(\beta^2) \tag{20}$$

since all linear terms vanish.

Now we compute the expectation value of an L $\times$ T Wilson loop (Fig. 2a): There are LT plaquettes enclosed by it.

$$\langle W(L,T)\rangle = \frac{1}{Z} \int \prod \frac{d\theta}{2\pi} \exp(i \sum_{WL} \theta) \prod_{plaquettes} (1 + \beta \cos(\sum \theta) + \ldots) \tag{21}$$

It's easy to convince one's self that $\langle W\rangle$is proportional to β^{LT}. Pick one link on the boundary to integrate. We need a term in exp (βS) to contribute and cancel the phase of the boundary link appearing in the Wilson loop (link 1 of Fig. 2b). Now we need another O(β) term to cancel the phases of link 2 (of Fig. 2b). This term contributes an amount proportional to β. Only if we "tile" the whole loop as shown do we get a nonzero expectation value; each plaquette contributes a factor β so

$$\langle W(L,T)\rangle \sim \beta^{LT}$$
$$E_{q\bar{q}} = \sigma L \tag{22}$$
$$\sigma = -\ln \beta$$

This is confinement by linear potential. The calculation can be generalized to all gauge groups. This calculation was first done by Wilson [1] and was the first demonstration of quark confinement in a field theory in more than two space-time dimensions, albeit in an unphysical limit.

2.2 Relativistic Fermions on the Lattice

Defining fermions on the lattice presents a new problem: doubling. The naive procedure of discretizing the continuum fermion action results in a lattice model with many more low energy modes than one originally anticipated. Let's illustrate this with free field theory.

The free Euclidean fermion action in the continuum (in four dimensions) is

$$S = \int d^4x [\bar{\psi}(x)\gamma_\mu \partial_\mu \psi(x) + m\bar{\psi}(x)\psi(x)]. \tag{23}$$

One obtains the so-called naive lattice formulation by replacing the derivatives by symmetric differences: we explicitly introduce the lattice spacing a in the denominator and write

$$S_L^{naive} = \sum_{n,\mu} \bar{\psi}_n \frac{\gamma_\mu}{2a}(\psi_{n+\mu} - \psi_{n-\mu}) + m \sum_n \bar{\psi}_n \psi_n. \tag{24}$$

Let's write down the propagator:

$$G(p) = (i\gamma_\mu \sin p_\mu a + ma)^{-1} = \frac{-i\gamma_\mu \sin p_\mu a + ma}{\sum_\mu \sin^2 p_\mu a + m^2 a^2} \tag{25}$$

and identify poles in $G(p)$ with masses of particles. Let's suppose $m = 0$. Then, besides the pole at $p = (0,0,0,0)$ there are fifteen other poles, at $p = (\pi, 0, 0, 0)$, $(0, \pi, 0, 0,)$, $\ldots (\pi, \pi, \pi, \pi)$. This is a model for sixteen massless fermions, not one.

(a) Wilson Fermions. There are two ways to deal with the doublers. The first way is to alter the dispersion relation so that it has only one low energy solution. The other solutions are forced to $E \simeq 1/a$ and become very heavy as a is taken to zero. The simplest version of this solution (and almost the only one seen in the literature) is due to Wilson: add a second derivative-like term

$$S^W = -\frac{r}{2a} \sum_{n,\mu} \bar{\psi}_n (\psi_{n+\mu} - 2\psi_n + \psi_{n-\mu}) \tag{26}$$

to S^{naive}. The parameter r must lie between 0 and 1; $r = 1$ is almost always used and "$r = 1$" is implied when one speaks of using "Wilson fermions." The propagator is

$$G(p) = \frac{-i\gamma_\mu \sin p_\mu a + ma - r\sum_\mu (\cos p_\mu a - 1)}{\sum_\mu \sin^2 p_\mu a + (ma - r\sum_\mu (\cos p_\mu a - 1))^2}. \tag{27}$$

It has one pole at $p_\mu = (im, 0, 0, 0)$, plus other poles at $p \simeq r/a$. In the continuum these states become infinitely massive and decouple (although this is not a trivial statement).

With Wilson fermions it is conventional not to use not the mass but the "hopping parameter" $\kappa = \frac{1}{2}(ma + 4r)^{-1}$, and to rescale the fields $\psi \to \sqrt{2\kappa}\psi$. The action for an interacting theory is conventionally written

$$S = \sum_n \bar{\psi}_n \psi_n - \kappa \sum_{n\mu} (\bar{\psi}_n (r - \gamma_\mu) U_\mu(n) \psi(n+\mu) + \bar{\psi}_n (r + \gamma_\mu) U_\mu^\dagger \psi(n - \mu). \tag{28}$$

Wilson fermions are closest to the continuum formulation– there is a four component spinor on every lattice site for every color and/or flavor of quark. Constructing currents and states is just like in the continuum.

Unfortunately, the Wilson term explicitly breaks chiral symmetry. This has the consequence that the zero bare quark mass limit is not respected by interactions; the quark mass is additively renormalized. The value of κ_c, the value of the (bare) hopping parameter at which the pion mass vanishes, is not known a priori before beginning a simulation; it must be computed. This is done in a simulation involving Wilson fermions by varying κ and watching the pion mass extrapolate quadratically to zero as $m_\pi^2 \simeq \kappa_c - \kappa$ ($\kappa_c - \kappa$ is proportional to the quark mass for small m_q).

(b) Staggered Fermions. In this formulation one reduces the number of fermion flavors by using one component "staggered" fermion fields rather than four component

Dirac spinors. The Dirac spinors are constructed by combining staggered fields on different lattice sites.

Let us call the staggered field on site n χ_n and define the matrix field

$$\psi_{\alpha k}(n) = \sum_b (\gamma^{n+b})_{\alpha k} \chi_{n+b} \tag{29}$$

where

$$\gamma^{n+b} = \gamma_1^{n_1+b_1} \gamma_2^{n_2+b_2} \gamma_3^{n_3+b_3} \gamma_4^{n_4+b_4} \tag{30}$$

and b runs over the sites of a hypercube $b = (0,0,0,0)$, (1,0,0,0), ... (1,1,1,1). $\bar{\psi}_{\alpha k}$ is defined analogously,

$$\bar{\psi}_{\alpha k}(n) = \sum_b (\gamma^{n+b})^\dagger_{\alpha k} \bar{\chi}_{n+b}. \tag{31}$$

One interprets the α indices as spin (= 1 to 4) and k as flavor (= 1 to 4). The free action (with $m_q = 0$) is

$$S = \frac{1}{128} \sum \bar{\psi}_n \gamma_\mu (\psi_{n+\mu} - \psi_{n-\mu}) \tag{32}$$

$$= \frac{1}{2} \sum \bar{\chi}_n \eta_{n,\mu} (\chi_{n-\mu} - \chi_{n+\mu}) \tag{33}$$

where $\eta_{n,\mu} = (-1)^{n_1+n_2+n_3+n_4}$. Gauge fields can be added in the standard way, and explicit mass terms give an extra $m_q \sum \bar{\chi}\chi$ term.

Staggered fermions preserve an explicit chiral symmetry as $m_q \to 0$ even for finite lattice spacing, as long as all four flavors are degenerate. They are preferred over Wilson fermions in situations in which the chiral properties of the fermions dominate the dynamics–for example, in studying the chiral restoration/deconfinement transition at high temperature. They also present a computationally less intense situation from the point of view of numerics than Wilson fermions, for the trivial reason that there are less variables. However, flavor symmetry and translational symmetry are all mixed together. Construction of meson and baryon states (especially the Δ) is more complicated than for Wilson fermions [6].

2.3 Taking the Continuum Limit

When we define a theory on a lattice we implicitly assume that a is an ultraviolet cutoff and $g(a)$ is the bare coupling defined with the cutoff. When we take a to zero we must also specify how $g(a)$ behaves. The proper continuum limit comes when we take a to zero holding physical quantities fixed, not when we take a to zero holding g fixed.

On the lattice, if all quark masses are set to zero, the only dimensionful parameter is the lattice spacing, so all excitations have masses

$$m = \frac{1}{a} f(g(a)) \tag{34}$$

where $f(g(a))$ is dimensionless. We must tune $g(a)$ so

$$\lim_{a \to 0} \frac{1}{a} f(g(a)) \to \text{constant}. \tag{35}$$

¿From the point of view of the lattice theory, we must tune g so that correlation lengths $1/ma$ diverge. This will occur only at the locations of second (or higher) order phase transitions in the lattice theory. Recall that the β-function is defined by

$$\beta(g) = a\frac{dg}{da}g(a) = \frac{dg(a)}{d\ln(1/\Lambda a)}. \tag{36}$$

(Λ is a dimensional parameter introduced to make the argument of the logarithm dimensionless.) At a critical point $\beta(g_c) = 0$. Thus the continuum limit is the limit

$$\lim_{a\to 0} g(a) \to g_c. \tag{37}$$

In QCD $g_c = 0$ and the two-loop β-function is prescription independent,

$$\beta(g) = -b_1 g^3 + b_2 g^5, \tag{38}$$

and so we want to observe

$$a\Lambda = \left(\frac{1}{g^2(a)}\right)^{b_2/(2b_1^2)} \exp\left(-\frac{1}{b_1 g^2(a)}\right), \tag{39}$$

or equivalently ma showing the same dependence as a goes to zero, in order to be in an asymptotic scaling regime.

Thus to compute a physically interesting (i.e. continuum) result in lattice gauge theory one must do three things: (1) Compute masses in lattice units at several g's (measure ma vs. $g(a)$). (2) Observe ma scaling according to the β-function as g is varied. (3) Use one dimensionful parameter to set the overall mass scale and predict all other dimensionful parameters in its units.

There is a less stringent form of scaling which is often discussed in the literature. The idea is that since mass ratios show less g dependence than the individual masses, one ought to require mass ratios which are g-independent even though individual masses do not scale correctly. That is,

$$m_1(a)/m_2(a) = m_1(0)/m_2(0)(1 + O(a^2/m^2)) \tag{40}$$

or

$$m_1(a)/m_2(a) = m_1(0)/m_2(0)(1 + O(\exp(-1/(b_1 g(a)^2)))) \tag{41}$$

This is called "scaling" as opposed to "asymptotic scaling" where the perturbative $f(g)$ is seen.

This program is presently feasible only for glueballs in quenched QCD. Simulations with fermions have another dimensionful parameter besides the lattice spacing, namely the quark mass. In principle, one must tune the bare quark mass along with the lattice spacing (or coupling) so that the physical quark mass is held fixed. I have only seen this done for quenched calculations after an extrapolation to $m_q = 0$. Instead, one commonly presents figures involving mass ratios. Examples of such plots are the so-called Edinburgh plot (m_p/m_ρ vs. m_π/m_ρ) and the Rome plot (m_p/m_ρ vs. $(m_\pi/m_\rho)^2$). Here one displays mass ratios for a variety of quark masses, all heavier than the physical quark masses, and tries to tell whether or not an extrapolation to physical quark mass will yield correct spectroscopy. Of course, one must have a model for the extrapolation. It may be hard to disentangle the variation in hadron mass with bare quark mass from variation due to the coupling constant.

2.4 All the Hard Work

A "generic" Monte Carlo simulation in QCD breaks up naturally into two parts. In the "configuration generation" phase one constructs an ensemble of states with the appropriate Boltzmann weighting. One complication for QCD which spin models don't have is fermions. The fermion path integral is not a number and a computer can't simulate fermions directly. However, one can formally integrate out the fermion fields. For n_f degenerate flavors of staggered fermions

$$Z = \int [dU][d\psi][d\bar{\psi}] \exp(-\beta S(U) + \sum_{i=1}^{n_f} \bar{\psi} M \psi)$$

$$= \int [dU](\det M)^{n_f} \exp(-\beta S(U)). \tag{42}$$

One can make the determinant positive-definite by writing $\det(M^\dagger M)^{n_f/2}$. The determinant introduces a nonlocal interaction among the U's:

$$Z = \int [dU] \exp(-\beta S(U) - \frac{n_f}{2} \mathrm{Tr} \ln(\mathrm{M}^\dagger \mathrm{M})). \tag{43}$$

All large scale dynamical fermion simulations today generate configurations using some variation of the microcanonical ensemble. That is, they introduce momentum variables P conjugate to the U's and integrate Hamilton's equations through a simulation time t

$$\dot{U} = iPU \tag{44a}$$

$$\dot{P} = -\frac{\partial S_{eff}}{\partial U}. \tag{44b}$$

The integration is done numerically by introducing a timestep Δt. The momenta are repeatedly refreshed by bringing them in contact with a heat bath and the method is thus called Refreshed or Hybrid Molecular Dynamics [7].

For special values of n_f (multiples of 2 for Wilson fermions or of 4 for staggered fermions) the equations of motion can be derived from a local Hamiltonian and in that case Δt systematics in the integration can be removed by an extra Metropolis accept/reject step. This method is called Hybrid Monte Carlo [8].

The reason for the use of these small timestep algorithms is that for any change in any of the U's, $(M^\dagger M)^{-1}$ must be recomputed. When (44b) is integrated all of the U's in the lattice are updated simultaneously, and only one matrix inversion is needed per timestep.

The major computational problem dynamical fermion simulations face is inverting the fermion matrix M. It has eigenvalues with a very large range – from 2π down to $m_q a$ – and in the physically interesting limit of small m_q the matrix becomes ill-conditioned. At present it is necessary to compute at unphysically heavy values of the quark mass and to extrapolate to $m_q = 0$. The standard inversion algorithm today is conjugate gradient or one of its variants. Algorithms like Gauss-Seidel just don't converge at small quark masses. Nobody has invented a multiscale method that works well at the parameter values used by large scale simulations. We need better methods to invert the fermion matrix.

Calculations which do not include the fermion determinant in the updating (the so-called "quenched approximation") are computationally simpler. They typically also use local updating. Probably the best algorithm at present is a microcanonical over-relaxation [9] algorithm mixed with some stochastic update like Metropolis or heat bath to insure ergodicity.

Our algorithms certainly look very crude compared to those discussed by Wolff. I think that there are two reasons for this. First, the fundamental interactions (gauge fields in a plaquette, to say nothing of the nonlocal effective interactions induced by fermions) are complicated and hard to visualize. Second, a phenomenologically interesting calculation must have a lattice spacing less than 0.1 fm and a simulation size greater than 1-2 fm (in four dimensions). This makes the calculations quite unwieldy. Today's QCD simulations range from $16^3 \times 32$ to 32^4 points and run from hundreds (quenched) to thousands (full QCD) of hours on the fastest supercomputers in the world. Even then, the simulations run only for a few hundred autocorrelation times (if we are lucky) and in fact we are only at the stage where we can begin to understand our error bars. There is a continual need for new ideas about algorithms.

Anyhow, now we have generated some configurations. What can we do with them?

3 Lattice Spectroscopy

3.1 Correlation Functions

Masses are computed in lattice simulations from the asymptotic behavior of Euclidean-time correlation functions. A typical (diagonal) correlator can be written as

$$C(t) = \langle 0|O(t)O(0)|0\rangle. \tag{45}$$

Making the replacement

$$O(t) = e^{Ht}Oe^{-Ht} \tag{46}$$

and inserting a complete set of energy eigenstates, (45) becomes

$$C(t) = \sum_n |\langle 0|O|n\rangle|^2 e^{-E_n t}. \tag{47}$$

At large separation the correlation function is approximately

$$C(t) \simeq |\langle 0|O|1\rangle|^2 e^{-E_1 t} \tag{48}$$

where E_1 is the energy of the lightest state which the operator O can create from the vacuum. In some cases that state is the vacuum itself, in which $E_1 = 0$. If the operator does not couple to the vacuum, then in the limit of large t one hopes to to find the mass E_1 by measuring the leading exponential falloff of the correlation function, and most lattice simulations begin with that measurement. If the operator O has poor overlap with the lightest state, a reliable value for the mass can be extracted only at a large time t. This makes the actual calculation of the energy E_1 more difficult.

This is the basic way hadronic masses are found in lattice gauge theory. The many calculations differ in important specific details of choosing the operators $O(t)$. However, I would like to make a few general remarks before continuing further.

In Monte Carlo we calculate any observable Γ by importance sampling: we do N measurements using the field variables $\phi^{(i)}$ appropriate to the sample [10].

$$\langle \Gamma \rangle \simeq \bar{\Gamma} \equiv \frac{1}{N} \sum_{i=1}^{N} \Gamma[\phi^{(i)}] \tag{49}$$

As the number of measurements N becomes large the quantity $\bar{\Gamma}$ will become a Gaussian distribution about a mean value. Its standard deviation is roughly

$$\sigma_{\bar{\Gamma}}^2 = \frac{1}{N}\left(\frac{1}{N} \sum_{i=1}^{N} |\Gamma[\phi^{(i)}]|^2 - \bar{\Gamma}^2\right). \tag{50}$$

We can also calculate this quantity in field theory: it is just

$$\sigma_{\bar{\Gamma}}^2 = \frac{1}{N}(\langle |\Gamma|^2 \rangle - \langle \Gamma \rangle^2). \tag{51}$$

This formula is very useful in analyzing the uncertainty in one's measurements.

There are several types of behavior a correlation function can display which are relevant to a spectroscopy calculation. If the operator O has nonzero vacuum expectation value, then the correlation function approaches a constant at large t:

$$\lim_{t \to \infty} C(t) \to |\langle 0|O|\mathbf{p} = 0\rangle|^2 \exp(-mt) + |\langle 0|O|0\rangle|^2. \tag{52}$$

The statistical fluctuations on $C(t)$ are given by (50) and we find after a short calculation that

$$\sigma \to \frac{C(0)}{\sqrt{N}}. \tag{53}$$

Thus the signal to noise ratio collapses at large t like $\sqrt{N}\exp(-mt)$. This situation occurs in the calculation of the mass of the 0^{++} glueball (or of any particle with vacuum quantum numbers).

Now we assume that the operator O has no vacuum expectation value. Then $C(t)$ shows a pure exponential behavior at large t. The fluctuations in $C(t)$ are given by (51). The second term of (51) falls off like $\exp(-2mt)$ and will give a constant signal-to-noise ratio at large separation. The first term of (51), however, is itself a correlation function of the operator $|O(t)|^2$ and will die away with a mass characteristic of the lightest particle $|O|^2$ can make from the vacuum. If O is a meson operator (a $\bar{q}q$ pair) O^2 will be a $q^2\bar{q}^2$ state, which will most likely couple to a $\pi\pi$ pair. Its correlator will fall like $\exp(-2m_\pi t)$. In the baryon sector this operator will make a $q^3\bar{q}^3$ state, and the lightest such state is three pions. Thus we expect a signal to noise ratio to be at best a constant (for the pion) or a growing function of t:

$$\sigma / C_H(t) \simeq \exp(m_H - m_\pi)t \tag{54}$$

for mesons, and

$$\sigma / C_H(t) \simeq \exp(m_H - 3/2 m_\pi)t \tag{55}$$

for baryons.

Fig. 3 shows an example of the behavior in the signal to noise ratio from a recent simulation [11] with two flavors of dynamical Wilson fermions with $\kappa = 0.1670$ and gauge coupling $\beta = 5.3$, with curves of (54) and (55) drawn "by eye" through the large-t points.

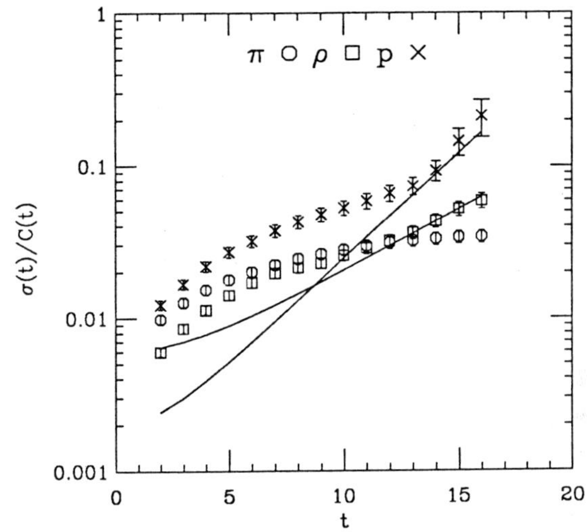

Fig. 3. Ratio of the error on a propagator $\sigma(t)$ to the value of a correlation function $C(t)$ for the pion (crosses), rho (circles) and proton (diamonds), from a dynamical Wilson fermion simulation at $\beta = 5.3$, $\kappa = 0.1670$. The lines are the parameterization of (54)-(55), with arbitrary overall normalization.

Finally, observe that (49) can also be used to find the correlation between fluctuations of $C(t)$ at different times:

$$N\sigma^2(t_1,t_2) = \langle (C(t_1) - \langle C(t_1)\rangle)(C(t_2) - \langle C(t_2)\rangle)\rangle \tag{56}$$

$$= \langle O(t_1)O(0)O(t_2)O(0)\rangle - \langle O(t_1)O(0)\rangle \langle O(t_2)O(0)\rangle \,. \tag{57}$$

In the limit when t_1 and t_2 become large, but the difference $t_1 - t_2$ remains small, the first term dominates the correlation and we expect

$$N\sigma^2(t_1,t_2) = \langle O(t_1)O(t_2)O(0)O(0)\rangle \tag{58}$$

or

$$N\sigma^2(t_1,t_2) = C(t_1-t_2)C(0) \tag{59}$$

so

$$N\sigma^2(t_1,t_2) \simeq C(0)^2 \exp(-m|t_1-t_2|). \tag{60}$$

This large positive correlation means that if $C(t)$ on a lattice at one t value is high, all the $C(t)$'s within a t interval of roughly $1/m$ will also be high. These correlations are unavoidable and must be taken into account when the data is fit.

This means that naive error analysis is at best misleading. It is an incorrect assumption to think that the "naive" χ^2

$$\chi^2 = \sum_i \frac{(y_i - y(x_i))^2}{\sigma_i^2} \tag{61}$$

is a good indicator of goodness of fit. The right approach is to minimize a "correlated χ^2"

$$\chi^2 = \sum_{ij} \frac{(y_i - y(x_i))}{\sigma_i} \frac{(y_j - y(x_j))}{\sigma_j} H_{ij}^{-1} \tag{62}$$

where the correlation matrix is given by the data itself:

$$H_{ij} = \sum_l \frac{(y_i^l - \langle y_i \rangle)}{\sigma_i} \frac{(y_j^l - \langle y_j \rangle)}{\sigma_j} \tag{63}$$

(I have normalized $H_{ii} = 1$). In Fig. 4 I display a contour plot of H_{ij} for a typical hadron correlator $C(t) = \langle O(t)O(0) \rangle$. Notice that H_{ij} is above 50 per cent for $\Delta t = 4$. The value of this χ^2 at the fit is a meaningful indicator of goodness of fit, in contrast to the value of the naive χ^2 ($H_{ij} = \delta_{ij}$).

A good reference to error analysis in the lattice context is Ref. [12].

It is very difficult to overwhelm a correlator by going to enormous separation. The only way to succeed in a spectroscopy calculation will be to find operators which couple very strongly to the desired state.

3.2 Good Trial Wave Functions for Quark Bound States

If a poor interpolating field O is chosen, a reliable value for the mass can be extracted only a large time t. In practice one must make the lattice larger in the time direction to compensate. Furthermore, since the signal to noise ratio can fall as the time separation of the measurement increases, higher statistics are needed. Both of these effects on lattice measurements of particle masses are expensive in terms of computer resources, so there is considerable incentive to find operators which couple very strongly to the ground state.

Let us focus on meson spectroscopy. Suppressing all color and spin indices, an operator which creates a meson is

$$O(t) = \sum_{x,y} c_q(x,t)^\dagger c_{\bar{q}}(y,t)^\dagger \Psi(x,y) \tag{64}$$

where $c_i(x,t)$ is an annihilation operator for a quark or an antiquark at space-time position (x,t) and Ψ is a c-number function, the wave function for the state. In the past, one chose Ψ to be a delta function in the relative coordinate $x - y$ so that O is a current which can create the state from the vacuum. Then $O(t) = \sum_x J(x)$, where $J = \bar{\psi}\Gamma\psi$, Γ is a Dirac matrix. However, especially for small quark masses or small lattice spacings, these pointlike wave functions perform very poorly as operators which generate hadronic states. This is not surprising, since the lightest physical hadrons are large objects (with radii on the order of 1 fermi) relative to the typical lattice spacings

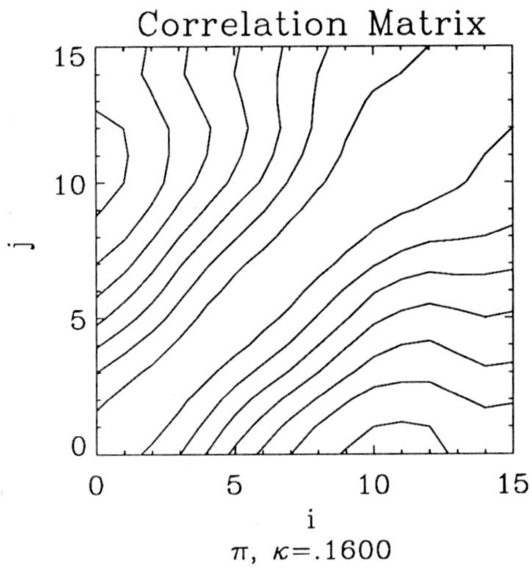

Fig. 4. Contours of constant autocorrelation in a "typical" hadronic correlation function (an HEMCGC Wilson pion). Contours run from .95 at the center, decreasing by 0.1 per contour.

used in lattice QCD (≤ 0.1 fermi). One therefore expects a trial wave function which has physical extent over several lattice spacings to be evolved more readily into the true ground state by the projection operator e^{-Ht}.

There are at least two classes of trial wave functions. The first are wave functions which depend explicitly and separably on the relative (r) and center of mass (R) coordinates of the quarks, which are called "bound state" wave functions: $\Psi(x,y) = \Psi(R-r/2, R+r/2)$. I call the second kind of wave function a "shell model" wave function. In these wave functions there is some externally imposed confinement volume; inside that volume the bound quarks and antiquarks move independently. For example, a "shell" meson wave function with equal mass quarks could be written as $\Psi(x,y) = \phi_R(x)\phi_R(y)$. The first kind of wave function is obviously more physical. However, because of the way the correlation function is actually computed, the "shell" wave function may be preferred in actual calculations. This can be seen if we write out a meson correlation function as

$$C_{ij}(t) = \langle 0| \sum_{x_1,y_1} \sum_{x_2,y_2} \Psi_i(x_1,y_1)^\dagger G_q(x_1,t;x_2,0) G_{\bar{q}}(y_1,t;y_2,0) \Psi_j(x_2,y_2)|0\rangle . \qquad (65)$$

The treatment of the two Ψ's in (65) is quite different. One of them will be used as the source for propagators while the other will be used to weight the propagators at different locations on the lattice. With the bound state wave function as a source it turns out that one needs at least two matrix inversions (per color and spin index) to construct $C(t)$.

However, when one uses a shell model wave function for the source the calculation of propagators factorizes: one solves $M(G\phi) = \phi$ for the individual propagator and then just reconstructs the meson as if one had a source which was a local operator.

One can use a shell model wave function at the sink, also. This gives a positive-definite correlation function and a variational bound on the ground state energy. A bound state wave function for the sink, perhaps even a point wave function (as is most commonly used) can be averaged over all spatial locations. This has the advantage that one is keeping only zero momentum states in the mode sum in (45). The disadvantage of this scheme is that one's bound on the lightest energy is not variational; one must be sure one is measuring an asymptotic exponential before quoting a mass value.

If the points x and y are not coincident the wave function is not gauge invariant and care must be taken if the expectation value of the correlator is not to vanish due to Elitzur's theorem [2]. Two ways to insure that $C(t)$ is nonvanishing are either to gauge fix or to explicitly include the gauge links in the wave function, so that color information is parallel transported from point x to point y. Gauge fixing is a minor overhead compared to the time spent generating propagators and, as it allows one to use arbitrary Ψ's, is my method of choice (but not everyone's) [13].

Several kinds of wave functions have been used for spectroscopy. The earliest of these calculations was carried out by the APE group [14]. They used shell model wave functions which are uniform cubes ($\phi = 1$ inside a cubic confinement volume and zero outside), with the gauge variables fixed to a smooth gauge. Another class of shell model wave functions was introduced by the Wuppertal group [15]. and more recently they have been used by Gupta, et. al. [16]. They take wave functions which are the solution to a three dimensional scalar field equation with a delta function source, i. e.,

$$(D^2 + m^2)\Psi(x-y) = \delta^3(x-y). \tag{66}$$

The mass m can be used as a variational parameter. The use of the covariant derivative D makes the correlation function $C(t)$ gauge invariant. Finally, several groups have used "wall" wave functions: the High Energy Monte Carlo Grand Challenge [17] (HEMCGC) and the Staggered Collaboration [18] have used shell model wave functions which are uniform over an entire timeslice of the lattice. They gauge fix to Coulomb gauge. These simple wave functions worked well for the parameters of these simulations and should do so as long as the lattice size is not too much bigger than a typical hadron diameter. A comparison of wall and point trial wave functions is shown in Fig. 5.

Trial wave functions for glueballs predate trial wave functions for quark states by several years. The physics is very similar [19]. The simplest operator which can make a scalar glueball is the lattice analog of $\sum_{ij} \mathrm{Tr} F_{ij}^2$, a single plaquette averaged over orientations, but it has terrible contributions from ultraviolet modes. The cure is once again to block or smear out the high momentum components of the interpolating field. Most of the successful methods for smearing do not involve gauge fixing. Rather, one defines new links which sample the local environment [20]. For example, one can define a variable $V_\mu(x)$ on the sites of the lattice as a "fat link" which spans two lattice sites and is a linear combination of a product of links along the direct path $U_\mu(x)U_\mu(x+\mu)$ plus offset paths like $\sum_\nu U_\nu(x)U_\mu(x+\nu)U_\mu(x+\mu+\nu)U_\nu^\dagger(x+2\mu)$. One then uses various closed paths of the V's as interpolating fields for glueballs. One can repeat the procedure to define "super V's" which span four sites–or more. The idea is to make a glueball operator which is a big fuzzy ball of string whose physical size is 1/2 to 1 fermi across.

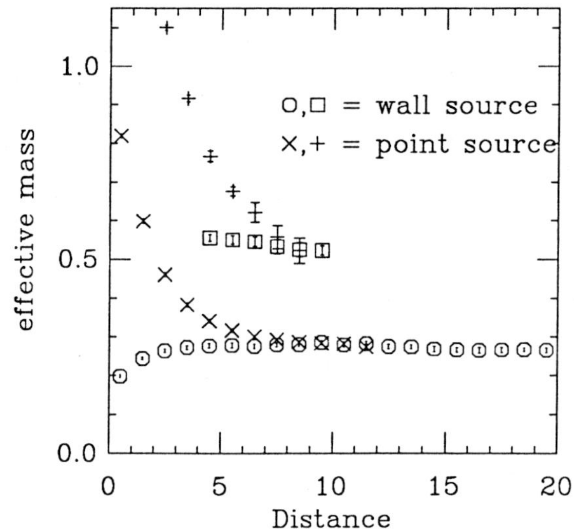

Fig. 5. Pion and rho effective mass from point and wall sources (HEMCGC data).

One can also reconstruct quark wave functions (distribution amplitudes) directly from Monte Carlo simulations. The wave function $\psi_G(r)$ of a meson H in a gauge G is defined as

$$\psi_G(r) = \sum_{\mathbf{x}} \langle H | c_q^\dagger(\mathbf{x}) c_{\bar{q}}^\dagger(\mathbf{x}+\mathbf{r}) | 0 \rangle \tag{67}$$

where $c_q^\dagger(\mathbf{x})$ and $c_{\bar{q}}^\dagger(\mathbf{y})$ are quantum mechanical operators which create a quark and an antiquark at locations $\mathbf{x}$ and $\mathbf{y}$. One computes ψ from convolutions of quark and antiquark propagators $G(x,y)$ with the quark spins coupled appropriately

$$C(\mathbf{r},t) = \sum_{\mathbf{x}} \langle 0 | \Psi(\mathbf{y}_1,\mathbf{y}_2) G_q(\mathbf{y}_1,0;\mathbf{x},t) G_{\bar{q}}(\mathbf{y}_2,0;\mathbf{x}+\mathbf{r},t) | 0 \rangle \tag{68}$$

where $\Psi(\mathbf{y}_1,\mathbf{y}_2)$ is some input trial wave function. At large t if the mass of the hadron is m_H, then $C(\mathbf{r},t) \simeq \exp(-m_H t)\psi_G(\mathbf{r})$ and so by plotting $C(\mathbf{r},t)$ as a function of $\mathbf{r}$ we can reconstruct the wave function up to an overall constant. The convolutions can be done using (fast) Fourier transforms.

In 1985 Velikson and Weingarten [21] studied meson wave functions in SU(2) and Gottlieb [22] carried out the first study of wave functions with SU(3) Wilson fermions. Recently, Chu, Lissia and Negele have looked at gauge invariant wave functions (with a product of links connecting the quarks) [23]. My student M. Hecht and I are currently carrying out an extensive study of wave functions for Wilson quarks.

The goal of these studies is two-fold: First, visualizing wave functions is a powerful diagnostic for lattice studies. One can see whether the wave function of a hadron is squeezed by the simulation volume; if it is, then a calculation of spectroscopy may be compromised. A picture of the wave function provides a hint for a good trial wave function for spectroscopy.

Second, these wave functions are really the minimal Fock space states of zero momentum hadrons in a particular gauge. The value of the wave function at the origin, $\psi(0)$, is measured in vector meson electromagnetic decay or in leptonic decays of pseudoscalars, and derivatives of the wave function at the origin are related to form factors (through the formalism of Brodsky and Lepage [24]). For heavy quark systems, overlaps of the wave functions themselves are related to electromagnetic transition rates.

Here are some examples. In Fig. 6 I show a snapshot of the Coulomb gauge pion's wave function and in Fig. 7 the wave function of the d quark in a proton, with the two u quarks offset by 0, 2, 4 and 6 lattice spacings, both in the plane $z = 0$, from quenched $\beta = 6$ simulations with Wilson fermions at $\kappa = 0.153$ ($m_\pi a = 0.42$). There is clearly a lot of qualitative science in these pictures.

3.3 Generic Features of Recent Calculations

The first lattice Monte Carlo studies of QCD spectroscopy were done in 1981, [25] but only since 1988 have simulations been done with large enough data sets and good enough algorithms and trial wave functions that the statistical errors can be made small enough to begin compare different groups' work.

Most results can be summarized in one plot, an Edinburgh plot of the ratio of particle energies (m_p/m_ρ vs. m_π/m_ρ). In the continuum limit and on an infinite lattice these mass ratios will be a smooth function of the quark mass. However, on a finite lattice that need not be the case. One should be very cautious in attempting to extrapolate lattice results from heavy to light quark mass.

For future reference, most collaborations schedule presentation of their results for the Lattice Conference, which is usually held in late fall. You should consult the latest Proceedings of that series for the most up to date results. The newest data I present here are through Autumn 1991.

All QCD simulations, quenched or full, with staggered or Wilson quarks, have shown the expected $m_\pi^2 \simeq m_q$ behavior at small quark mass. (Of course, for Wilson fermions the relation between bare hopping parameter and renormalized mass is not known, but one can see the square of the pion mass interpolating linearly to zero at some critical value of the hopping parameter.) An example of this behavior from recent data is shown in Fig. 8.

Present day staggered fermion simulations are run at parameter values where flavor symmetry is nearly restored. In Fig. 9 I show a plot of the ratio of the mass of one of the non-Goldstone pions to the mass of the would-be Goldstone pion, for quenched and dynamic data. The quenched $\beta = 6.2$ data from the Staggered collaboration shows flavor symmetry breaking reduced to a few per cent.

Simulations of full QCD have focussed on $n_f = 2$ with some $n_f = 4$ work. Most efforts have gone into simulations with staggered dynamical fermions. The pioneer in $n_f = 2$ staggered fermions was the work of the (mainly) Santa Barbara group [28] on 6^3, 8^3 and 10^3 by 24 lattices, quark masses in the range $am_q = 0.1$ to 0.025, and β in the range 5.28 to 5.47. Since then the HEMCGC [17] has run at $\beta = 5.6$ on 12^4, 16^4 and $16^3 \times 32$ lattices with $am_q = .01$ and .025, and the Columbia group [26] has run at $\beta = 5.7$ on $16^3 \times 32$ lattices with dynamic mass $am_q = .01$, .015, .02, and .025.

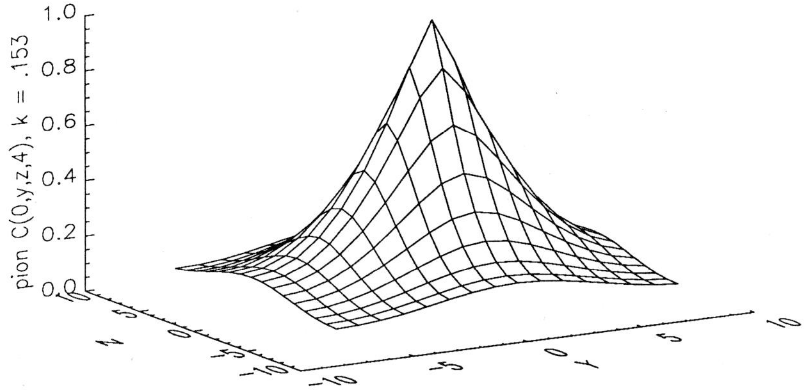

Fig. 6. Coulomb gauge pion's wave function in the plane $z = 0$, from quenched $\beta = 6$ simulations with Wilson fermions at $\kappa = 0.153$.

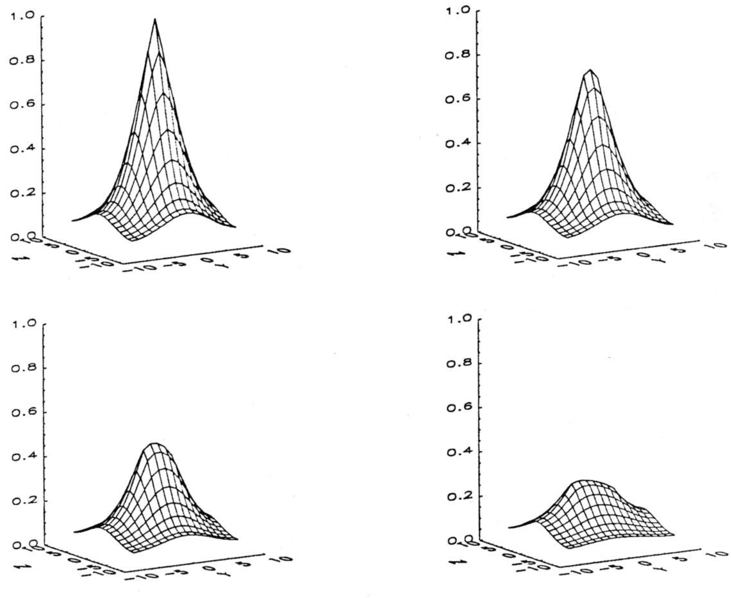

Fig. 7. Coulomb gauge wave function of the d quark in a proton, with the two u quarks offset by 0, 2, 4 and 6 lattice spacings in the y direction, in the plane $z = 0$, from quenched $\beta = 6$ simulations with Wilson fermions at $\kappa = 0.153$.

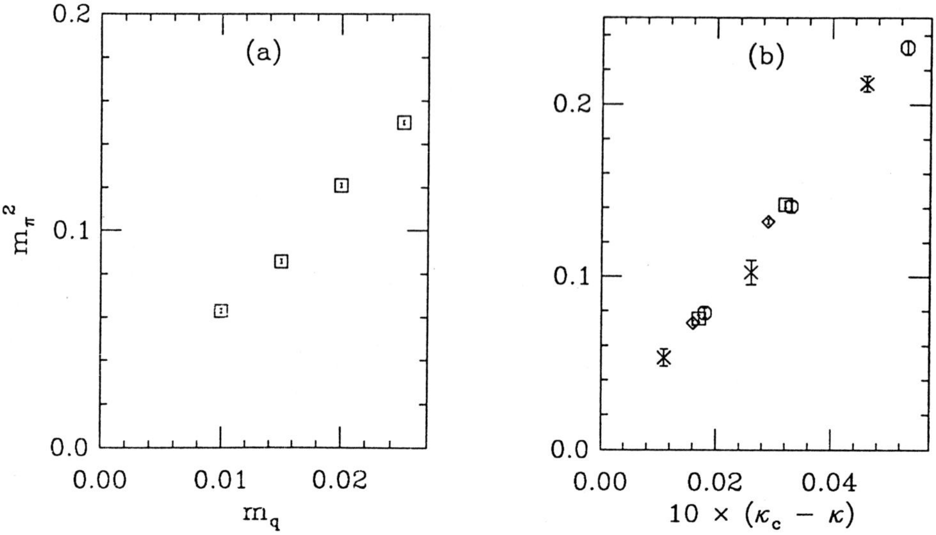

Fig. 8. Variation in pion mass with quark mass for dynamical staggered fermions (a) and with the hopping parameter κ for Wilson fermions (b), where the labels are: crosses and octagons for dynamical staggered fermions at $\beta = 5.6$, $am_q = 0.01$ and 0.025, and squares and diamonds for quenched simulations at $\beta = 5.85$ and 5.95. Staggered data are from the Columbia Collaboration [26] (two flavors, $\beta = 5.7$) and Wilson data from the HEMCGC (Ref. [27]).

The KEK group [29] has also presented spectroscopy at $\beta = 5.7$ on up to 20^4 lattices with $am_q = 0.01$ and 0.02.

All simulations used HMD, and gauge fixed to a smooth gauge and used wall sources and pointlike sinks. All calculations feature high statistics and careful error analysis. They are of unprecedented size and accuracy in lattice QCD. In Fig. 10 I show the Edinburgh plot for these simulations. The errors are tiny even compared to quenched simulations.

The 20^3 masses of the KEK group [29] are consistently smaller than the 16^3 masses of the Columbia collaboration. Presumably the Columbia lattices were too small, and the hadron was being squeezed by the simulation volume. This is a serious concern for all lattice simulations.

Two groups are currently active in $n_f = 2$ dynamic Wilson Fermion simulations, both using HMC and both with heavier quarks than in the staggered simulations. The first is a Los Alamos-based collaboration [16]. They mainly run on 16^4 lattices, $\beta = 5.4 - 5.6$, and low statistics, T about 600 units, about 30 lattices analyzed. They use a gauge invariant extended trial wave function. The second group is the HEMCGC, [11] which has done a simulation at $\beta = 5.3$, $\kappa = .1670$ (so $m_\pi = .45$) on a $16^3 \times 32$ lattice. Their results are still preliminary. The Edinburgh plot is shown in Fig. 11.

We can contrast these results to some quenched spectroscopy. Three groups have published large staggered fermion simulations: the APE collaboration [30] at $\beta = 6.0$

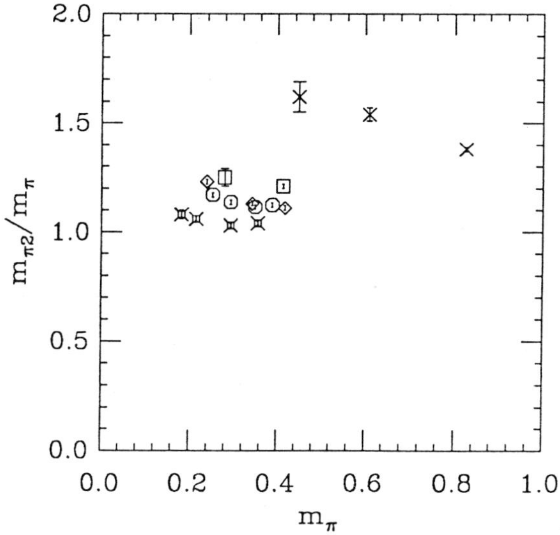

Fig. 9. Ratio of non-Goldstone pion to Goldstone pion mass for $n_f = 2$ dynamic simulations (Columbia (circle), HEMCGC (squares) and the old Santa Barbara results [28] (crosses)) and quenched results from the Staggered collaboration at $\beta = 6.0$ (diamonds) and $\beta = 6.2$ (decorated squares).

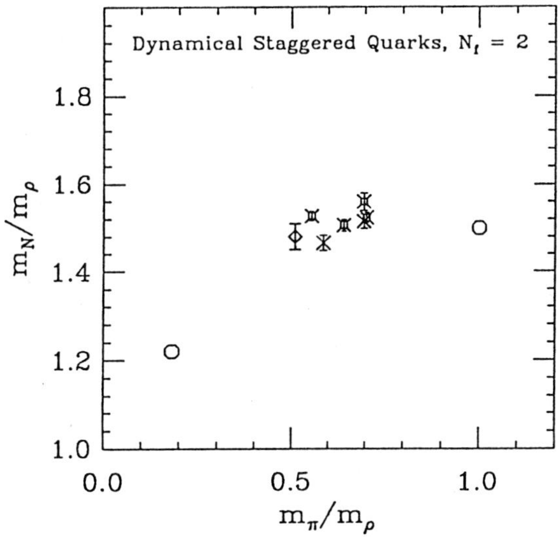

Fig. 10. Edinburgh plot for $n_f = 2$ dynamic staggered fermions. In all Edinburgh plots the circles are the physical ratio and infinite quark mass limit. Here data are from HEMCGC (on a 16^3 lattice) (diamond), Columbia data (squares), Tsukuba data (crosses).

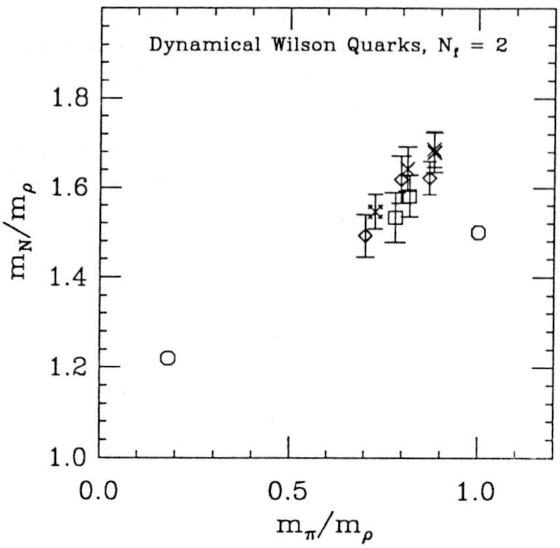

Fig. 11. Edinburgh plot for $n_f = 2$ dynamic Wilson fermions. Data are from the Los Alamos collaboration: $\beta = 5.4$ (crosses), $\beta = 5.5$ (diamonds), $\beta = 5.6$ (squares), and the HEMCGC, $\beta = 5.3$ (fancy cross).

on $24^3 \times 32$ lattices, the Staggered collaboration, [18] mainly at $\beta = 6.0$ and 6.2, on $16^3 \times 40$ and $24^3 \times 40$ lattices, and the HEMCGC [27] at $\beta = 5.85$ and 5.95 on $16^3 \times 32$ lattices (they are trying to match their dynamic simulations to quenched data). The data are all in general agreement with each other. All use extended sources (APE uses cubes, the others, walls). The Edinburgh plot is shown in Fig. 12.

Finally, I show the results of two large scale simulations with dynamical Wilson fermions: APE [31] ($\beta = 5.7$, $12^3 \times 24$ lattices and $\beta = 6.0$, $24^3 \times 32$ lattices) and HEMCGC [27] ($16^3 \times 32$ lattices at $\beta = 5.85$ and 5.95). APE uses cube sources and HEMCGC used wall sources. Fig. 13 is an Edinburgh plot.

3.4 Theory of the Edinburgh Plot

One can apply quark models to lattice spectroscopy as well as to real experimental spectroscopy. For heavy quarks, the nonrelativistic quark model should be a good guide. There, the mass of a hadron is given by a sum of constituent quark masses plus a color hyperfine term,

$$M_H = \sum_i m_i + \xi_H \sum_{ij} \frac{\sigma_i \cdot \sigma_j}{m_i m_j} \tag{69}$$

where ξ is twice as great for mesons as for baryons because of color. At small quark mass chiral models require that the pion mass vanishes as the square root of the quark mass; all other particles extrapolate linearly in the quark mass. One feature which is usually missing is the presence of structure in the rho mass as it crosses $\pi\pi$ thresholds. A "cartoon" which shows all these effects is shown in Fig. 14. It is clear

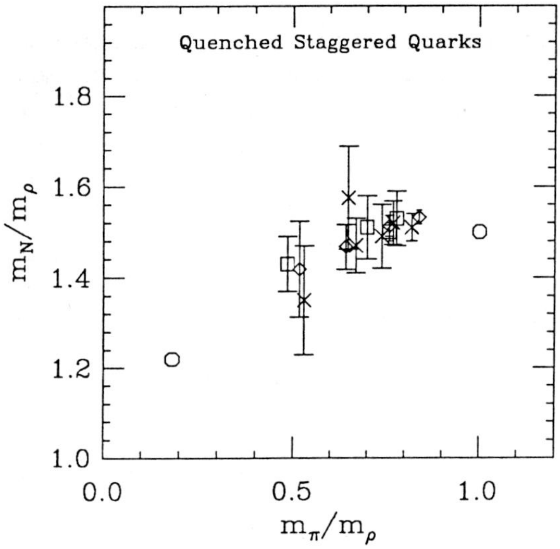

Fig. 12. Edinburgh plot for quenched staggered fermion simulations. Data are APE (diamonds), Staggered collaboration (crosses) and HEMCGC (squares)

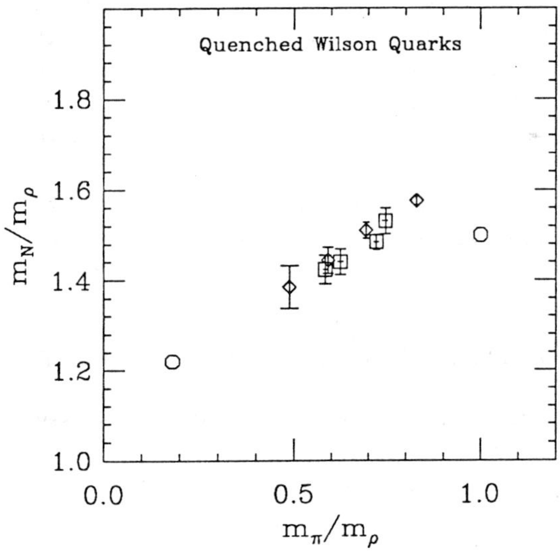

Fig. 13. Edinburgh plot for quenched Wilson fermion simulations. Data are APE (diamonds), HEMCGC (squares).

that one could tune parameters in these models and reproduce the results of lattice QCD simulations. (Try it yourself!)

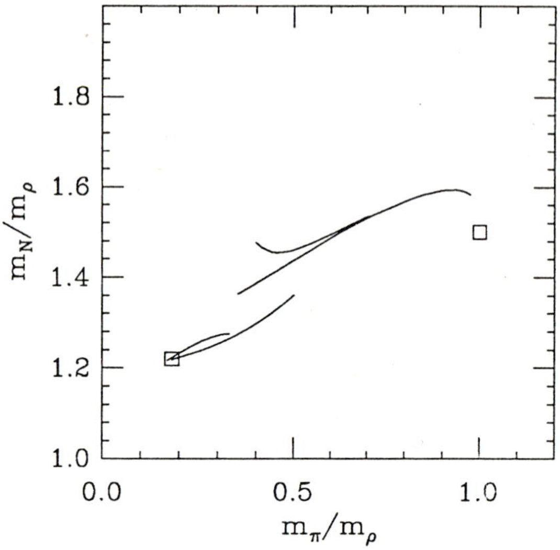

Fig. 14. A "theoretical" Edinburgh plot. The two lines are a simple quark model which includes an avoided $\rho\pi\pi$ level crossing, and a chiral model.

The main question one is trying to address with dynamic fermion spectroscopy simulations is, "Do the sea quarks make a difference in spectroscopy?" Unfortunately, the answer to that question is still No. Other effects (lattice volume or just statistics) are still large enough to mask any obvious differences between quenched and dynamic quark spectroscopy at dynamic fermion mass values studied to date. As a fairly dramatic example, Fig. 15 shows spectroscopy from two Wilson simulations I am involved in, a quenched simulation at $\beta = 6.0$, and Wilson valence spectroscopy with dynamical staggered fermions at $\beta = 5.6$, from the HEMCGC [17] I have plotted the rho, proton, and delta mass in lattice units as a function of the lattice pion mass. Finding observables which change when one passes from the quenched approximation to full QCD remains an open problem.

3.5 Other Particles

There is other spectroscopy information beyond the Edinburgh plot: One can display a plot of baryon hyperfine splitting vs. meson hyperfine splitting (a "Boulder" plot), Fig. 16. Here I compare the two dimensionless quantities

$$R_M = \frac{m_\rho - m_\pi}{3m_\rho + m_\pi} \tag{70}$$

and

$$R_B = \frac{m_\Delta - m_p}{m_\Delta + m_p}. \tag{71}$$

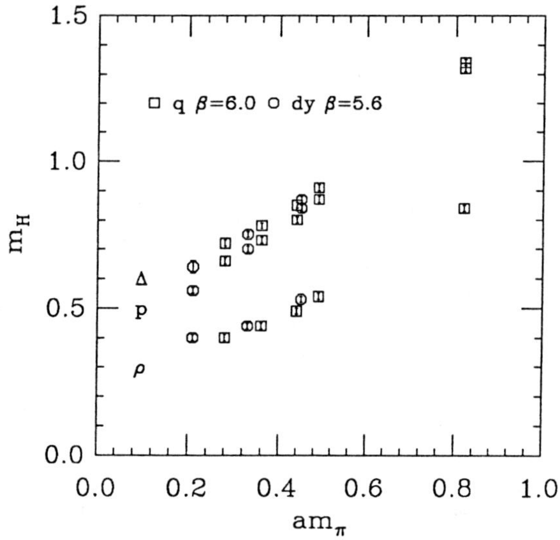

Fig. 15. Comparison of quenched Wilson spectroscopy at $\beta = 6.0$ (squares) with Wilson valence spectroscopy from a dynamical staggered fermion simulation at at $\beta = 5.6$ (octagons).

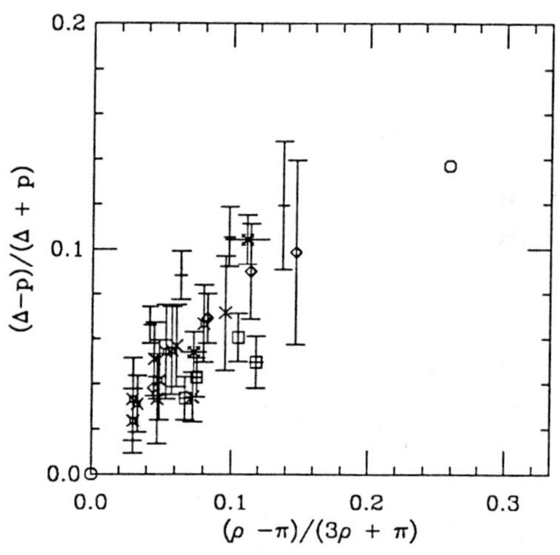

Fig. 16. Baryon hyperfine splitting vs. meson hyperfine splitting, normalized to the center of mass of the multiplet. Data are HEMCGC Wilson valence quarks with staggered sea quarks (fancy cross) Wilson sea quarks (burst) and quenched (squares), APE (diamonds), APE quenched Wilson quarks (diamonds) and staggered quarks (plus signs), Gupta, et. al. (Ref. [16]) Wilson sea quarks (fancy square) and the staggered fermion result of the Staggered collaboration (crosses).

Each of these quantities is the ratio of hyperfine splitting in a multiplet divided by the center of mass of the multiplet. Error bars are still too large for much phenomenology.

Very few calculations have attempted to look at the full flavor SU(3) spectrum of the lattice. In my opinion the rather scanty data appears to reproduce the gross features of the Gell-Mann Okubo formula for octets and the equal-spacing rule for decuplets, but the splittings (often as extrapolated to the zero quark mass limit) are smaller than experiment. I have never seen a lattice simulation with a statistically significant $\Sigma - \Lambda$ mass splitting.

3.6 Excited States

So far, QCD simulations have been restricted almost entirely to studies of the spectroscopy of the lightest hadronic states– the π, ρ, p, and Δ. Some P-wave states' masses have been measured in staggered simulations because they are the odd parity partners of "ordinary" states: the a_1 and ρ are examples of such pairs. In NRQCD, Lepage and Thacker [32] have measured the masses of χ_C and χ_B states (without including spin effects).

Calculations of excited states are difficult for several reasons: First, one needs a source with nonzero overlap onto the desired L sector and zero overlap on $L = 0$, otherwise the signal will be dominated at large t by the lighter $L = 0$ states. Second, the signal is noisy. Recalling the discussion of Sec. 3.1, the lightest state in the "squared source" channel is two (three) pions for mesons (baryons), so the fractional fluctuation in these channels scales like (54) and (55). Finally, there are a lot of states: in the P-wave sector there are ${}^3P_2$, ${}^3P_1$, ${}^3P_0$, and ${}^1P_1$ mesons and a whole SU(6) $L = 1$ [70] of baryons [33]. In the nonstrange $L = 1$ sector, there are five distinct nucleons, one of $j = 5/2$ and two each of $j = 3/2$ and $j = 1/2$, and $j = 3/2$ and $j = 1/2$ Δ's. I don't see how it will be possible to extract the mass of both nucleons of the same j since the lightest one will dominate the correlator in that channel.

M. Hecht and I [34] are making an attempt to see some of the $L = 1$ states. We take as our source a shell model wave function with the quark in an S-wave and the antiquark in a P-wave. This state is a linear superposition of a $\mathbf{p} = 0$ $L = 1$ orbital excitation and a state whose center of mass momentum is nonzero. Convoluting quark propagators as in Sec. 3.2 projects out the $\mathbf{p} \neq 0$ state and gives us the wave function of the $L = 1$ state, whose exponential falloff with t gives a mass. A picture of a ${}^3P_2$ meson is shown in Fig. 17. S and P-wave spectroscopy versus hopping parameter is shown in Fig. 18. We also have data for D-waves, not yet analyzed.

4 QCD at Nonzero Temperature

There has long been a belief that at high temperature ($T > 200$ MeV) or density ($\rho >$ a few quarks/fm^3) strongly interacting matter might undergo a transition from its low temperature hadronic phase into a new phase of matter, the quark gluon plasma. This phase might have existed in the early universe, and might be produced in ultrarelativistic nucleus-nucleus collisions. Since lattice methods allow us to study nonperturbative behavior in QCD they are a natural framework to explore the properties of a new phase and the existence of any transitions.

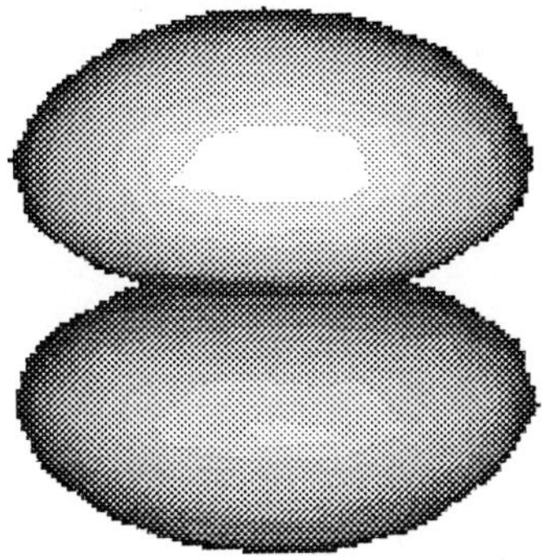

Fig. 17. A contour of constant absolute value for the real part of a Coulomb gauge wave function of a $^3P_2$ meson from quenched $\beta = 6$, $\kappa = 0.145$ simulations.

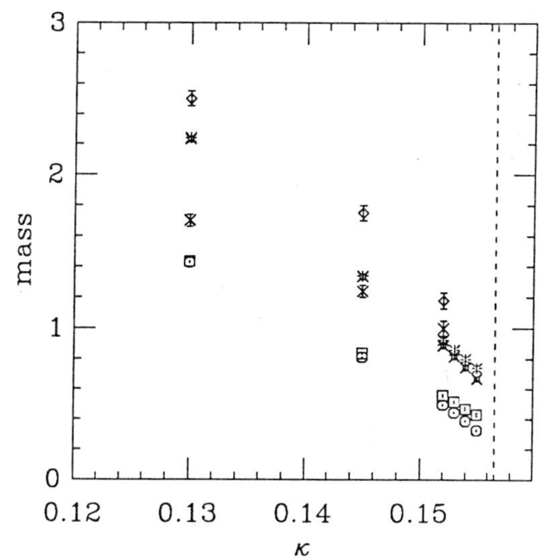

Fig. 18. Masses of S-wave and P-wave hadrons as a function of hopping parameter. Pseudoscalar, vector, nucleon, and delta states are labelled by octagons, squares, decorated squares, and bursts respectively. P-wave mesons and baryons are labelled by crosses and diamonds.

And yet, there is a great mismatch between lattice calculations and experiment. Lattice calculations probe the equilibrium behavior of QCD matter and lattice measurements are (so far) restricted to bulk properties and static correlation functions.

In contrast, high temperature hadronic matter is thought to be produced in a central collision of two nuclei. If the size of the nuclei is many mean free paths the quarks and gluons in the nuclei might approach thermal equilibrium. However, the system will expand and cool rapidly, and experimentalists will see particles produced throughout the history of the event in their detectors. A unique signal for plasma formation is still unknown. It is probably irrelevant to experiment whether there is an actual phase transition or just some sort of rapid crossover to the high temperature state.

Numerical studies of $T \neq 0$ QCD are interesting because they allow us to investigate the qualitative features of a new phase of matter, but the best we can hope to do at present for experiment is to provide input and possibly boundary conditions for phenomenological models of hot QCD matter.

In the absence of dynamical fermions QCD has an order parameter which distinguishes between confinement and deconfinement. (For a review, see Ref. [35].) The partition function for a $T \neq 0$ gauge theory is defined on a space which is infinite in the spatial directions but finite in Euclidean time: $0 < \tau < \beta$ where $\beta = 1/kT$. On the lattice, for fixed lattice spacing a, $N_t a = 1/kT$ and the lattice spacing is given by the coupling $g(a)$. The integration over field variables is restricted to A_μ's which are periodic in τ: $A_\mu(\mathbf{x}, \tau + \beta) = A_\mu(\mathbf{x}, \tau)$.

The order parameter which tests for confinement is the "Wilson line" or "Polyakov loop." It is a degenerate L by T Wilson loop in which the length in Euclidean time is taken to be β; the loop is broken into two separate loops and the free energy of a $q\bar{q}$ pair is given by

$$\exp(-\beta F_{q\bar{q}}) = \langle L(x) L(0) \rangle \tag{72}$$

where

$$L(x) = \mathrm{Tr}\Omega(\mathrm{x}) \tag{73}$$

$$\Omega(x) = T \exp i \int_0^\beta A_0^\alpha(x, t) T^\alpha dt; \tag{74}$$

the T^α's are the generators of the Lie algebra. The free energy of an isolated quark is

$$\exp(-\beta F_q) = \langle L \rangle. \tag{75}$$

On the lattice $\Omega(x)$ is a trace of a product of timelike links around the lattice. The free energy can be interpreted as the potential between two static quarks

$$\exp(-\beta F) = \exp(-\beta V_{singlet}) + (N^2 - 1)\exp(-\beta V_{adjoint}) \tag{76}$$

but one can only make this separation after a complete gauge fixing. Notice that under a gauge transformation parameterized by V

$$L(x) \rightarrow \mathrm{Tr}\mathrm{V}(\mathrm{x}, 0)\Omega(\mathrm{x})\mathrm{V}^\dagger(\mathrm{x}, \beta). \tag{77}$$

The Euclidean action is invariant under a larger group than purely periodic gauge transformations. In the Abelian theory, where $A_0 \rightarrow A_0 + \partial_0 \Lambda$, one can have $\Lambda(x, \tau + \beta) = \Lambda(x, \tau) + \theta$. For example, $\Lambda = \theta t/\beta$. Since A_i is single-valued, $\partial_i \theta = 0$, so θ is a global parameter. In the non-Abelian case one can have $V(x, \tau + \beta) = zV(x, \tau)$, where z is an element of the center of the group (for $SU(N)$, $z = \exp(i2\pi n/N) \in Z(n)$). Local observables are invariant under these aperiodic transformations, but the Wilson line is not: $L \to zL$. Thus the high-temperature system has a global symmetry which may be broken: either $\langle L \rangle = 0$ and the free energy of an isolated quark is infinite (confinement) or $\langle L \rangle \neq 0$ and the free energy of an isolated quark is finite (deconfinement). If the zero-temperature theory is confining one expects a transition to a deconfined phase at some critical temperature. If the transition is continuous, the renormalization group requires that its critical exponents are in the same universality class as other $Z(N)$ globally symmetric models of the same spatial dimensionality. For the physically interesting case $d = 3$, $N = 3$, no spin models with local ferromagnetic interactions supporting continuous transitions are known, and simple mean field arguments ($V(L) = L^2 + L^3 + L^4$) suggest the transition is first order [36]. This prediction has been confirmed by Monte Carlo.

At zero quark mass there should be a transition separating a low temperature chirally broken phase from a high temperature chirally symmetric phase. It is hard to come up with reliable models in this limit. In strong coupling, one can derive an effective Hamiltonian which is an antiferromagnet in the flavor symmetry group of the fermions and has the appropriate order disorder transition. Pisarski and Wilczek [37] performed weak coupling renormalization group studies of $SU(N_f) \otimes SU(N_f)$ chiral Lagrangians and found a first order chirally restoring transition for $N_f \geq 3$.

Away from these limits the situation is murky. A finite quark mass breaks both chiral symmetry and the $Z(N)$ symmetry and would smooth out a continuous transition in either case. If the transitions were first order, they would persist, but could vanish at some critical quark mass. There is no good reason for the two transitions to be connected.

Monte Carlo calculations have been performed for almost every combination of lattice fermion and flavor number you could imagine. The transition in the pure gauge theory is definitely first order and was the first (and so far only) QCD measurement to show asymptotic scaling. I won't discuss dynamic Wilson fermion simulations. They are too confusing. Staggered fermions possess a relic chiral symmetry on the lattice and at the large lattice spacings used today might be more "physical" if chiral dynamics are crucial. So I will focus on them.

Most work concentrates on $n_f = 2$ and 4, with an increasing component of 2 light plus 1 heavier flavor. $n_f = 2$ is interesting if we believe that the transition is dominated by the dynamics of the light quarks. $n_f = 4$ allows an exact lattice algorithm. The biggest lattices are $16^3 \times 8$ and the lightest quarks have $m_q a = 0.01$. People look for transitions by making long simulations at some (β, am_q) and looking at time histories of order parameters like $\langle L \rangle$ or $\bar{\psi}\psi$ for tunneling or crossover. A "typical" set of time histories from a recent simulation [38] ($n_f = 2$, $12^3 \times 6$ lattices) is shown in Fig. 19. From these histories Toussaint invented a variation of the Ferrenberg-Swendsen histogram technique and interpolated $\langle \bar{\psi}\psi \rangle$ in $6/g^2$ and in am_q (Fig. 20). There does not appear to be a sharp transition, but there does appear to be a rapid crossover.

At $n_f = 4$ all recent groups report a first-order transition at small am_q. It seems to persist out to $m/T_c \simeq 0.1$ or so. At larger am_q there is a region of parameter space with no obvious transition.

At $n_f = 3$ there appears to be [39] a first order transition if the quark masses are degenerate. It seems to go away if $m_s > 4m_{ud}$ ($am_{ud} = .025$). These calculations have just begun, and with the extra parameter they are very time consuming.

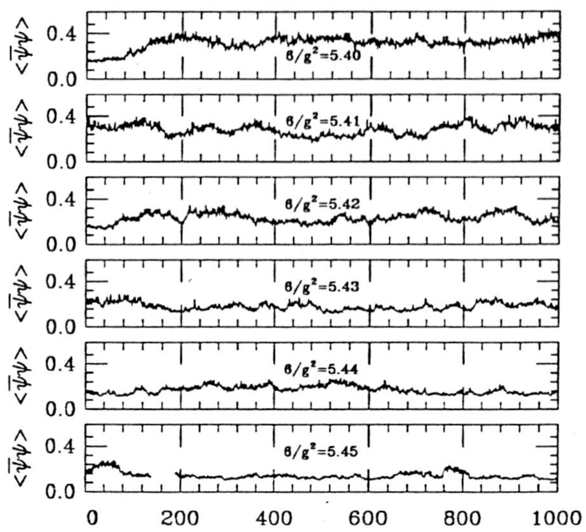

Fig. 19. Time histories of the chiral order parameter, $\bar{\psi}\psi$, for six runs at $m_q = 0.0125a^{-1}$, from Ref. [38].

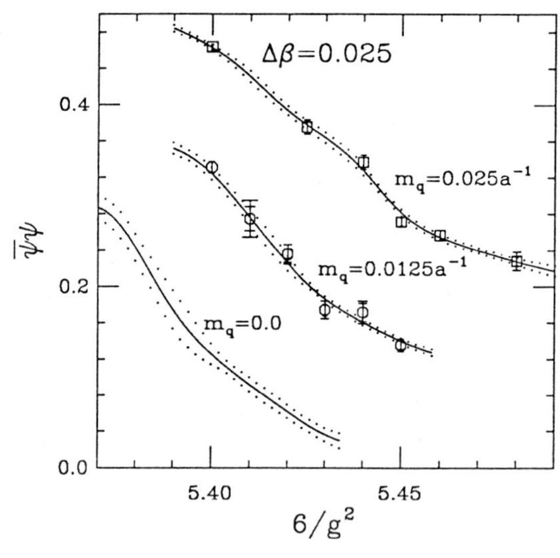

Fig. 20. The chiral order parameter. The symbols are the results of the individual runs, and the continuous curves are the results reconstructed from the estimate of the density of states, from Ref. [38].

In the absence of asymptotic scaling, one determines a transition temperature on the lattice from

$$T_c = \frac{T_c}{M_H} \times M_H = \frac{aT_c}{aM_H} \times M_H = \frac{(1/N_t)}{aM_H} \times M_H. \tag{78}$$

M_H is a physical mass (the rho?) and aM_H the lattice mass from a zero temperature simulation at the critical coupling g_c. Fig. 21 is a summary made by Gottlieb from the world's data. The critical temperature is rather strongly dependent on n_f.

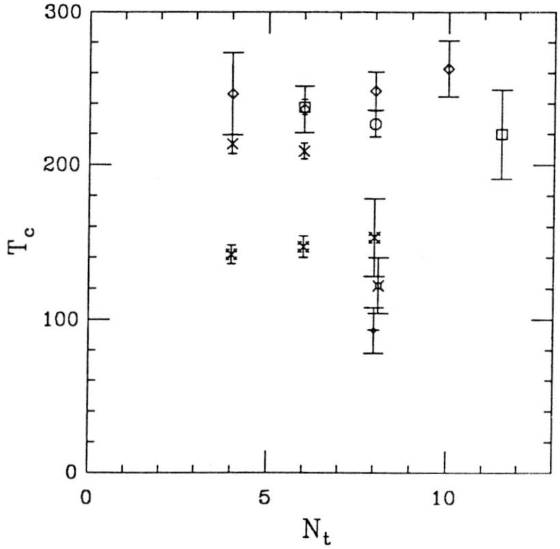

Fig. 21. The crossover temperature T_c in MeV as a function of N_t. The quenched calculations are highest (above 200 MeV) as reported by Ref. [40], using data from Ref. [41] and [42]. The fancy crosses are for two flavors of dynamical quarks. The point at $N_t = 4$ is from Ref. [43] and the one at $N_t = 6$ from Ref. [38]. The point at $N_t = 8$ is a preliminary result of the $16^3 \times 8$ study of Ref. [44] combined with an extrapolation of the spectrum from Ref. [45]. The fancy square is for two flavors and the burst, three flavors, (Columbia), and the fancy diamond, four flavors, from MT_c, Ref. [46].

4.2 Structure of the Plasma Phase

A much more interesting question is "What is the nature of the high temperature phase?" Lattice calculations can hint at answers to this question (at least as far as static observables are concerned) through bulk observables and static correlation functions.

The energy density and pressure are related to differences in the spacelike and timelike plaquette averages in the $T \neq 0$ simulation,

$$\epsilon - 3p = -36\beta(g)/g^3(P_s + P_t) \tag{79}$$

$$\epsilon + p = 4\beta c(\beta)(P_t - P_s) \tag{80}$$

through the beta function $\beta(g)$ and another perturbative coefficient $c(\beta)$ [49]. A picture of ϵ for the pure gauge theory from an $N_t = 4$ lattice is shown in Fig. 22. For a

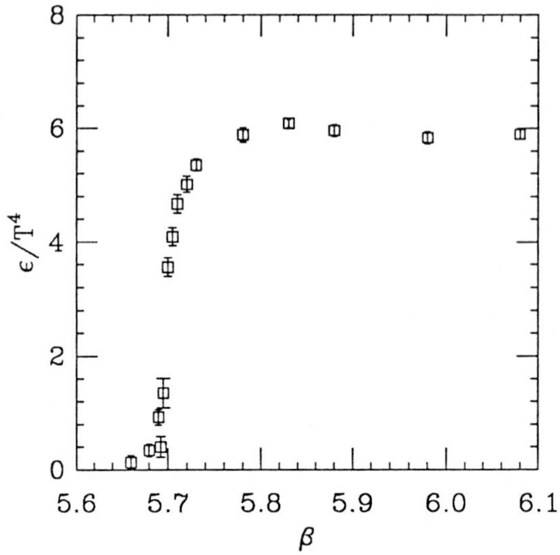

Fig. 22. Energy density vs. temperature for pure gauge $SU(3)$, from Ref. [47].

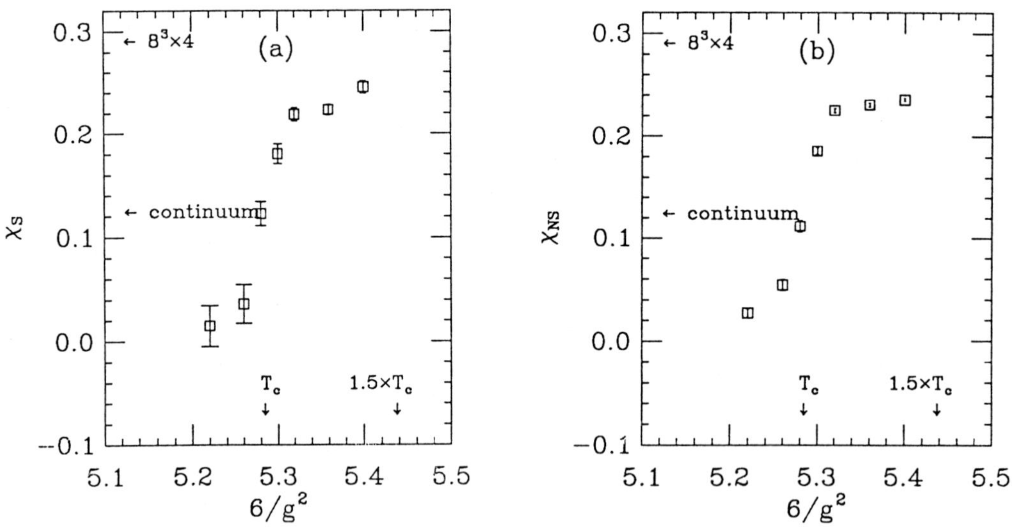

Fig. 23. Quark susceptibilities from $n_f = 2$ simulations, from Ref. [48].

gas of eight massless gluons $\epsilon/T^4 = 8\pi^2/15 = 5.3$, so the curve is suggestive of the naivest interpretation of the plasma. The ratio $(\epsilon - 3p)/\epsilon$ falls slowly to zero at very high T. Second, while one cannot do simulations at nonzero chemical potential μ, one can measure the variation in quark number density with chemical potential, at $\mu = 0$. The singlet and nonsinglet susceptibilities, χ_s and χ_{ns}, measure the response of the total quark number density and the third component of the isospin density. They are defined as

$$\chi_s = (\frac{\partial}{\partial\mu_u} + \frac{\partial}{\partial\mu_d})(n_u + n_d) \tag{81}$$

$$\chi_{ns} = (\frac{\partial}{\partial\mu_u} - \frac{\partial}{\partial\mu_d})(n_u - n_d). \tag{82}$$

Gottlieb, et. al. [48] measured these quantities in $n_f = 2$ simulations on $8^3 \times 4$ lattices. They are shown in Fig. 23. The labels "continuum" and "$8^3 \times 4$" show the expected values for a gas of free massless quarks in the continuum or in the simulation volume.

Next, one can measure static correlation functions, from correlators of Wilson lines to objects which carry interesting quantum numbers. These correlators are the analogs of ones which give hadron masses in zero temperature simulations. They typically show exponential falloff with distance ($C(z) \simeq \exp(-\mu z)$) and the quantity μ is called the "screening mass" in a channel. I do not know of any direct connection between these masses and the masses of excitations which participate in real-time processes in the plasma, but dynamic correlators must reduce to static ones after long time averages. Some of these correlators can be calculated in perturbation theory at high T.

The first correlators studied were glueball-like [50]. They do not look like the exchange of two massless gluons, but that is not expected anyway. In perturbation theory gluons develop a Debye-screening mass at high T of order gT (electric gluons) or g^2T (magnetic gluons).

Next, DeTar and Kogut [51] measured screening masses of correlators with the quantum numbers of the π, σ, ρ, a_1, and nucleon. They saw behavior characteristic of the restoration of chiral symmetry: the π and σ, and ρ and a_1 channels become degenerate at high T, and the nucleon approaches the mass of a positive parity nucleon resonance. A sample of this behavior (from the MT_c collaboration [46] – $n_f = 4$ on a $16^3 \times 8$ lattice) is shown in Fig. 24. Notice again that the masses are not a few times am_q ($\simeq 0$).

A possible connection between these masses and a scenario of deconfinement was recently proposed by the MT_c collaboration. In Euclidean quantum field theory fermion fields are antiperiodic in the τ direction, and so a free fermion at $T \neq 0$ has an Euclidean "energy" of πT, $3\pi T, \ldots$ Thus a $\bar{q}q$ free particle state would show a screening mass of roughly $2\pi T$ and a qqq state would be at $3\pi T$. These predictions, modified slightly by finite volume effects, are shown as dotted lines on the right side of Fig. 24.

More recently Bernard, et. al. [52] looked at the spatial correlation of a quark and an antiquark in a plane transverse to z, the direction of the static correlator. This is a simple variation on the Monte Carlo reconstruction of wave functions described in Chapter 3. The quarks are produced at $z = 0$ with an amplitude which is uniform in the transverse plane, so that if the correlator really involved free quarks, it would be a constant independent of their relative separation. However, Fig. 25 shows a comparison of the $T = 0$ wave function with the $T \neq 0$ correlator, for the pseudoscalar and vector channels. They are not very different!

One can imagine several possibilities for the $q\bar{q}$ structure of the static exchange:

1. Free quarks, perhaps with a temperature-dependent mass. In this case $\psi(\mathbf{x})$ would be constant. This case seems to be ruled out.

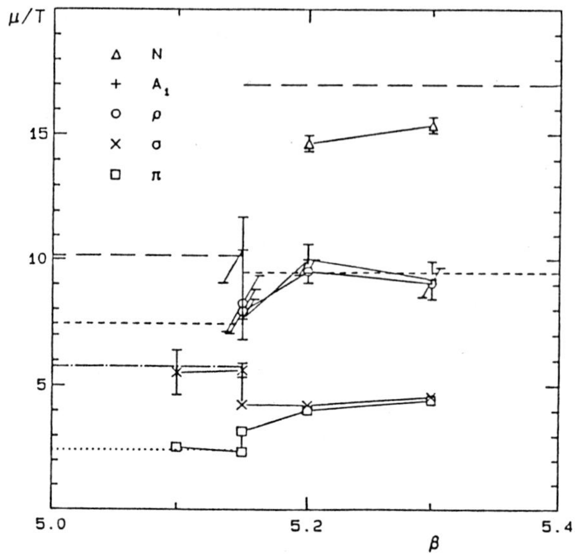

Fig. 24. Screening masses in high temperature QCD, from Ref. [46].

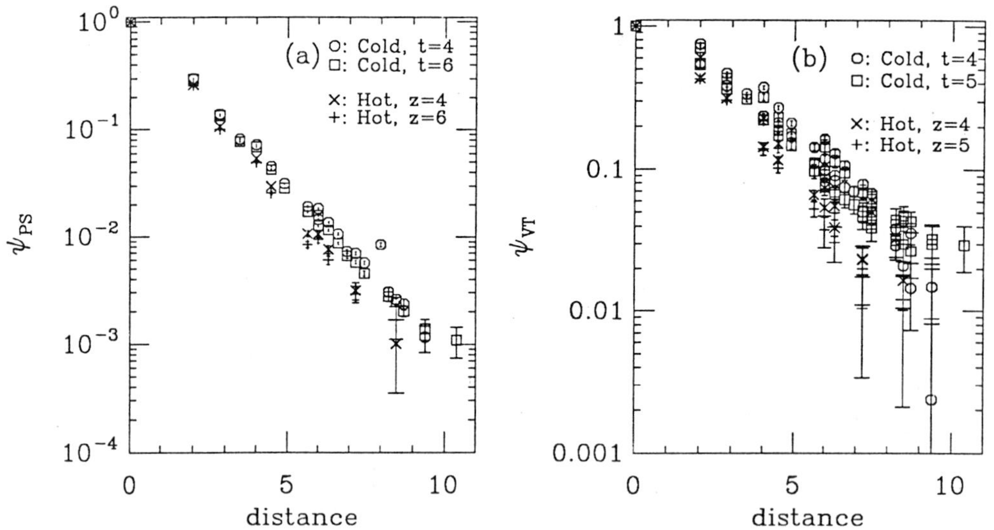

Fig. 25. The pion (a) and rho (b) wave function at $T = 0$ (cold) and the static correlators at $T \simeq 1.5T_c$ (hot). I show the yz slice with $x = 0$. The wave function is defined on the doubled lattice, so the 16^3 spatial lattice yielded an 8×8 wave function.

2. Unbound quarks in an attractive screened potential. In this case we would expect that $\psi(\mathbf{x})$ would go to a constant at large $\mathbf{x}$. The data limit such behavior.
3. Quarks as bound states of a screened potential.

4. Quarks bound in a confining potential.

The last choice is expected from other analytical and numerical studies [53]. Briefly, at high temperatures our lattice is short in one direction, effectively becoming three dimensional for long distance observables. Thus we are measuring wave functions of the zero temperature states of QCD on the compactified spatial manifold $S^1 \times R^2$. In the pure Yang-Mills version of such a theory, large spatial Wilson loops are expected to show an area law, and separated point sources, a linearly rising potential. The very same evidence is offered for confinement at zero temperature, suggesting that the compactified theory is also confining. From this point of view it is not surprising that the screening correlators we have measured should be similar to the zero temperature (four dimensional) wave functions.

So we are in a quandary. If the high temperature phase is made of free quarks, then why are the quarks so strongly correlated? (If there is no line of transitions separating the two phases, why are they different?) If the high temperature phase is made of color singlet objects, why do the energy density and susceptibilities jump to near their free field values?

5 Matrix Elements – A Case Study

The long term goal of lattice QCD calculations is the computation of hadronic matrix elements. At short distances, where the strong interactions are weak, one can calculate hadronic processes using perturbation theory. That is not the case for physics which takes place at long distances. We need to be able to compute there too, either to complete the calculation of short-distance processes (find structure functions at some reference momentum transfer), or to calculate intrinsically long-distance phenomena (like $\rho \rightarrow \pi\pi$). Often we want to compute a matrix element in order to study properties of the standard model (determine the K-M matrix) or go beyond (find the proton lifetime in a grand unified theory).

There is no way to cover this field in one lecture. I will make a few general remarks, and then describe a case study which illustrates most of the complications of a matrix element calculation–the decay constants of D and B mesons. This subject is very technical and I hope I haven't made too many mistakes.

Most of the matrix elements measured on the lattice are expectation values of local operators composed of quark and gluon fields. The mechanical part of the lattice calculation begins by writing down some unamputated N-point function which contains the local operator (call it $J(x)$) and somehow extracting the matrix element. For example, if one wanted $\langle 0|J(x)|h\rangle$ one could look at the two-point function

$$C_{JO}(t) = \sum_x \langle 0|J(x,t)O(0,0)|0\rangle \tag{83}.$$

Inserting a complete set of states

$$1 = \frac{1}{L^3}\sum_{A,\mathbf{p}} \frac{|A,\mathbf{p}\rangle\langle A,\mathbf{p}|}{2E_A(p)} \tag{84}$$

and using translational invariance and going to large t gives

$$C_{JO}(t) = e^{-m_A t} \frac{\langle 0|J|A\rangle\langle A|O|0\rangle}{2m_A}. \tag{85}$$

A second calculation of

$$C_{OO}(t) = \sum_x \langle 0|O(x,t)O(0,0)|0\rangle = e^{-m_A t} \frac{|\langle 0|O|A|\rangle|^2}{2m_A} \tag{86}$$

can be used to extract $\langle 0|J|A\rangle$ through its determination of $\langle 0|O|A\rangle$.

(As you recall from Sec. 3.1 (Equation (60)), everything measured on the same lattice is highly correlated. Good error analysis is even more important for calculations of matrix elements. It is, of course, wrong to fit the ratios of the correlators directly – the ratio of two Gaussian-distributed random distributions is not Gaussian-distributed. One correct thing to do is to make a correlated fit to the two propagators with three parameters – the two matrix elements and the mass. Then one gets a meaningful χ^2 and can measure the correlation between the different fit parameters.)

Similarly, a matrix element $\langle h|J|h'\rangle$ can be gotten from

$$C_{AB}(t,t') = \sum_x \langle 0|O_A(t)J(x,t')O_B(0)|0\rangle. \tag{87}$$

(Can you see how?)

These lattice matrix elements are not yet the continuum matrix elements. The lattice is a UV regulator and changing from the lattice cutoff to a continuum regulator (like $\overline{MS}$) introduces a shift,

$$O^{cont}(\mu) = Z(\mu a, g(a))O^{latt}(a). \tag{88}$$

This happens because the renormalization of an operator is slightly different in the two schemes,

$$\langle f|O^{latt}(a)|i\rangle = O^{(0)}[1 + \frac{\alpha_s}{4\pi}(-\gamma \ln\ \lambda^2 a^2 + C_{latt}) + \ldots] \tag{89}$$

$$\langle f|O^{cont}(\mu)|i\rangle_{\overline{MS}} = O^{(0)}[1 + \frac{\alpha_s}{4\pi}(-\gamma \ln\ \lambda^2/\mu^2 + C_{\overline{MS}}) + \ldots] \tag{90}$$

so

$$\langle f|O^{cont}(\mu = 1/a)|i\rangle_{\overline{MS}} = (1 + \frac{\alpha_s}{4\pi}(C_{\overline{MS}} - C_{latt}) + \ldots)\langle f|O^{latt}(a)|i\rangle. \tag{91}$$

If the lattice spacing is small enough one should be able to compute the renormalization factor Z in perturbation theory. Notice that one could pick a different constant of proportionality in the logarithms: not $\log(1/a)$ but $\log(C/a)$. This shifts the renormalization factor by a constant but might result in a more convergent perturbation theory [54].

Any expert will tell you this review is much too naive. The current we are trying to measure can mix with higher dimensional operators, with a mixing coefficient proportional to some power of the lattice spacing a. One can remove this mixing by dropping the lattice spacing or by inventing "improved" [55] lattice actions or operators which do not have this mixing. Because the lattice breaks rotational invariance, one's operator can mix with a lower dimensional operator with a mixing coefficient $1/a$. These so-called "power divergences" can only be removed by taking linear combinations of the operator's Lorentz indices which do not mix– usually these combinations involve cancellations [56]. You have to keep your wits about you if you want to study matrix elements!

5.1 A Case Study–Heavy Meson Decay Constants

The decay constant f_M of a pseudoscalar meson M is defined as

$$\langle 0|\bar{\psi}\gamma_0\gamma_5\psi|M\rangle = f_M m_M. \tag{92}$$

(Warning: in this convention $f_\pi = 132$ MeV; half the references in the literature include a $\sqrt{2}$ and then $f_\pi = 93$ MeV.) Decay constants are interesting because some of them (π and K) are measured and provide a benchmark for lattice calculations, while some of them are not measured and allow predictions (D, D_s, and B). They probe very simple properties of the wave function: in the nonrelativistic quark model [58]

$$f_M = \frac{\psi(0)}{\sqrt{m_M}} \tag{93}$$

where $\psi(0)$ is the $\bar{q}q$ wave function at the origin. For a heavy quark (Q) light quark (q) system $\psi(0)$ should become independent of the heavy quark's mass as the Q mass goes to infinity, and in that limit one can show in QCD that $\sqrt{m_M}f_M$ approaches a constant. It is believed that CP nonconserving amplitudes are proportional to f_M^2 and so knowledge of f_B provides information about CP nonconservation in the B system.

Almost all calculations in the literature use Wilson fermions. Most lattice simulations quote f_π near 132 MeV (for example $f_\pi = 145(30)$ from Ref. [59]) but in my opinion arbitrariness in the choice of lattice spacing (from the rho mass?) put at least a 20 % systematic uncertainty on this number.

It is not possible to calculate f_B directly on present day lattices. The reason is that the lattice spacing is much greater than the b quark's Compton wavelength (or the UV cutoff is below m_b). In this limit the b quark is strongly affected by lattice artifacts as it propagates. If the scaling formula (91) were true, one could extrapolate from f_D to f_B. But is it true?

There is another approximation which can be used for f_B. That is to make m_b infinite on the lattice and determine the combination $\sqrt{m_B}f_B$ in the limit. On the lattice an infinite mass quark is pinned to a site and its propagator is just a product of links:

$$Q(t,\mathbf{x}) = \prod_{i=0}^{N_t} U_0(t_i,\mathbf{x}). \tag{94}$$

The light quark has its usual propagator. Then one can extrapolate down to the B mass. Do the two extrapolations up and down give the same result?

For a long time the answer was No. The early calculations used correlators of two local currents and never saw a single exponential in $C(t)$. People have since done calculations with extended shell-model sources and as of the November 1991 lattice conference the situation looked more encouraging.

One problem the static calculations face is a noisy correlator. The correlation function falls like $\exp(-\epsilon t)$ where ϵ is the energy of the qQ state. The noise in the channel involves a $q\bar{q}Q\bar{Q}$ correlator. The Q and $\bar{Q}$ propagators are Wilson lines and cancel; the $q\bar{q}$ piece can propagate as a pion. Thus the signal to noise ratio falls like $\exp((m_\pi/2 - \epsilon)t)$ at large t.

The two calculations are by Bernard, Labrenz, and Soni, [57] on a $24^3 \times 55$ lattice at $\beta = 6.3$, and the European Lattice Collaboration, [59] on a $24^3 \times 30$ lattice (doubled to $N_t = 60$ for propagators) at $\beta = 6.4$. The data sets are not large: 20 and 15 lattices, respectively. Both groups calculate heavy-light decay constants on the lattice for several values of the light quark mass and extrapolate to the zero mass limit, and I will show that data. (When the light quark has a small but nonzero mass, the numbers are only a little different.)

Part of the source of problems in the older calculations is shown in Fig. 26: one needs a really big trial wave function to couple to the lightest heavy-light pseudoscalar. Otherwise the effective mass never goes to a constant (the correlator is never dominated by the lightest state) and one's lattice decay constant is contaminated by excited state contributions. I think that one did not expect such a big size for a good trial wave function was needed.

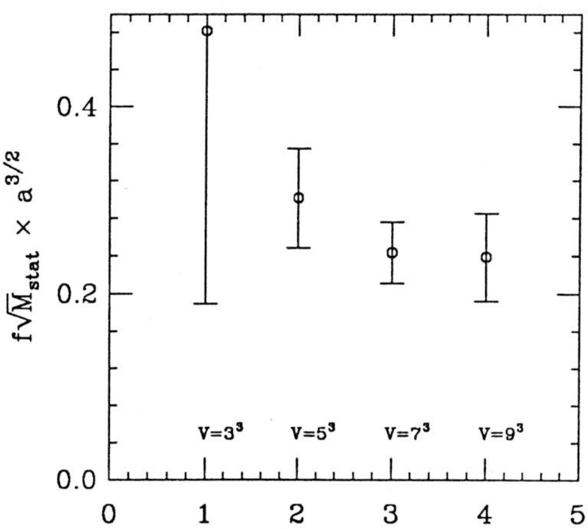

Fig. 26. Dependence of the lattice decay constant on trial wave function size, for cube sources in Landau gauge, from $\beta = 6$ data from Ref. [57].

Once a lattice result is obtained, there are several ways to fix the lattice spacing. The ELC present two choices: first, normalize using f_π itself, and second, use the rho mass. The former choice has the advantage of removing the dependence of the answer on the axial current renormalization factor, Z_A. It does affect the mass scale for the meson, however. The authors then try to extract a value for f_B by extrapolating their lattice data via ("linear" or "quadratic" fits)

$$f_P\sqrt{M_P} = A + \frac{B}{M_P} \tag{95}$$

or

$$f_P\sqrt{M_P} = A + \frac{B}{M_P} + \frac{C}{M_P^2} \quad . \tag{96}$$

After extrapolating, they multiply the dynamic quark result by the factor $(\alpha_s(M_P)/\alpha_s(M_B))^{6/33}$ and the static results by $(\alpha_s(1/a)/\alpha_s(M_B))^{6/33}$ to take the anomalous dimension of the decay constants into account. The two results are shown in Figs. 27 and 28. Normalizing by f_π appears to result in a consistency between the static and dynamic calculations. Normalizing with the rho mass does not.

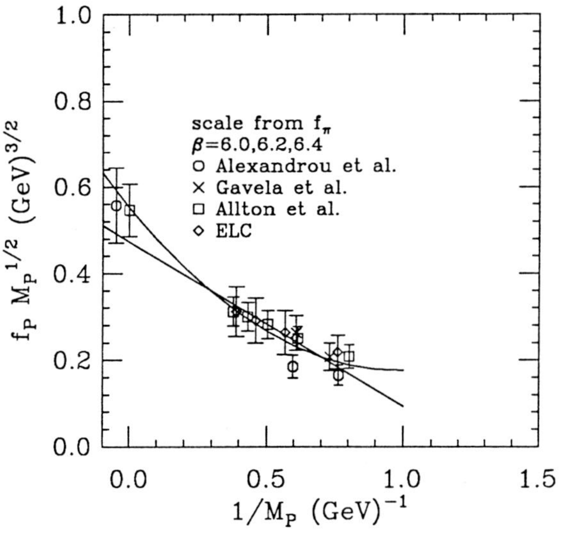

Fig. 27. The quantity $f_P\sqrt{M_P}$ as a function of the inverse pseudoscalar mass, from several calculations with propagating and static quarks: circles are Ref. [60], squares are Ref. [61], crosses are Ref. [62], and diamonds are Ref. [59]. The scale is set by f_π. The curves are the linear and quadratic fits described in the text. The vertical line identifies the point corresponding to f_B.

Now we turn to Ref. [57]. Part of their analysis is quite similar to the ELC's. However, they add a twist. The normalization of fermion fields on the lattice is different from the continuum. The free-field Wilson propagator is

$$G(p) = \frac{1 - 2\kappa\sum_\mu \cos p_\mu - 2i\kappa\sum_\mu \gamma_\mu \sin p_\mu}{(1 - 2\kappa\sum_\mu \cos p_\mu)^2 + 4\kappa^2\sum_\mu \sin^2 p_\mu} \tag{97}$$

while the continuum propagator is

$$G(p) = \frac{\gamma\cdot p - m}{p^2 + m^2}. \tag{98}$$

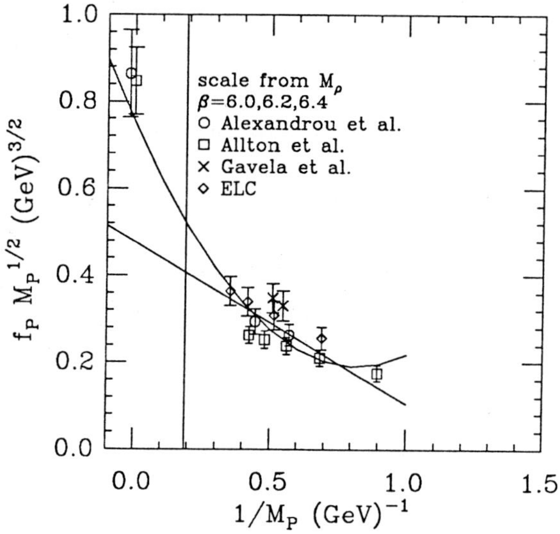

Fig. 28. Same as Fig. 27, but this time the lattice spacing is set by the rho mass.

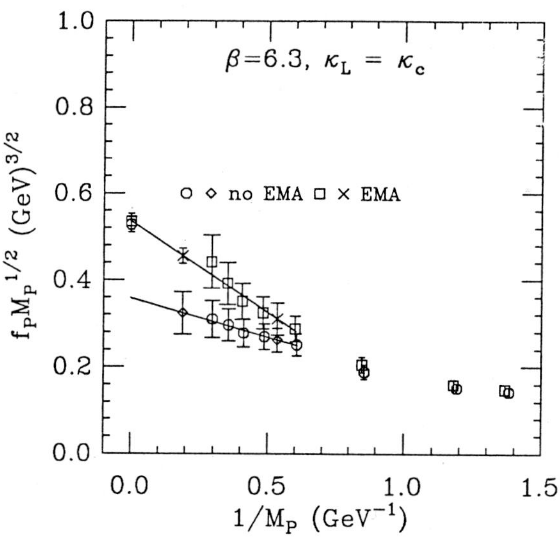

Fig. 29. The quantity $f_P\sqrt{M_P}$ as a function of the inverse pseudoscalar mass, $\beta = 6.3$ data from Ref. [57]. The labels "EMA" and "no EMA" refer to the extra normalization factor described in the text. The diamonds and crosses are the extrapolated values of f_D and f_B.

We can define the field normalization as the square root of the residue of the pole at $\mathbf{p} = 0$. For massless quarks, there is a relative factor of $\sqrt{2\kappa}$ between the lattice and continuum regularizations. For heavier quarks, this relation is not quite right, and if the quark is very heavy, the relative factor is (the square root of) $2\kappa \exp(ma)$ where to

sufficient accuracy (for me, anyway) $ma = 1/2(1/\kappa - 1/\kappa_c)$. This is all true for free field theory and the authors of Ref. [57] apply it to their lattice simulations to normalize f_P, replacing κ_c by its measured value. Doing this is apparently controversial. A good reason for doing it is that there are certainly artifacts of the lattice spacing as the quark mass gets heavy. This renormalization includes some of them. As the quark mass becomes infinite, it "heals" the Wilson propagator smoothly onto the static limit. On the other hand, it can't be the whole story. Interactions renormalize κ_c and also the coefficient 1/2 in the $ma - \kappa$ relation [63]. Away from $\mathbf{p} = 0$ the renormalization is changed, and the quarks in the meson have nonzero momentum by virtue of being bound. How does this affect the renormalization? On the balance, it does include some known effects. At a minimum, it does include some known finite mass effects, and as such shows the level of lattice systematics on the result. I wish I'd thought of it first!

I have only seen this group's data with a normalization to f_π. Their results are shown in Fig. 29. The two lines are linear extrapolations. The EMA data are extrapolated using the static point, the non-EMA data are not. The non-EMA data overlap with the ELC data where they coincide in mass. Otherwise, there is a ten per cent systematic difference between the two calculations at the D, which grows to thirty per cent at the B, where a non-static extrapolation gives about 145 MeV for f_B, according to Labrenz (private communication).

In my opinion the situation is encouraging but not complete. The latest results are shown in Table I. These numbers are already interesting for comparisons with experiment. Presumably data at different lattice spacings will appear so that one can check scaling, and if the lattice spacing can be pushed down, perturbative calculations of corrections, $Z-$factors, and the like will become more reliable. This could all turn into a real can of worms if that doesn't happen.

Table I. Predictions for heavy-light decay constants.

particle	f_P, GeV	Ref.
D	.225(25)	[57]
	.210(40)	[59]
	$< .290$	expt.
D_s	.240(20)	[57]
	.230(50)	[59]
B	.200(10)	[57]
B	.205(40)	[59]
B_s	.205(10)	[57]

6 The Future of Lattice QCD?

There are two sides to the future of lattice calculations. First, large scale simulations will get larger and larger. Larger volume is needed both to decrease the lattice spacing (to approach continuum physics) and to increase the physical volume (to make certain that the states are not being squeezed). Present day simulations have barely enough statistics to allow one to begin to worry about the inherent long autocorrelation times of the algorithms. Dynamical fermion simulations need to be run at lighter, more physical, values of the quark mass. All these changes are computationally very costly.

Progress is going to continue to be incremental. It is important to isolate and quantify the effects of the different sources of systematic error. To this end it is necessary to carry out high precision spectrum calculations in which the quark mass, lattice spacing, and physical size of the lattice are varied independently.

For the newcomer, the increasing size of simulations should not present an impossible barrier. Most of the computer time goes into generating a library of configurations which in many cases are available for use by others. It is fairly easy to obtain a set of basic programs from an established group–the minimum is something to generate gauge configurations, a gauge fixer, and something to construct propagators–and in some cases the code is actually legible! It is quite possible for one or two people with a good idea to do very interesting science with a few hundred hours of time on a Connection Machine (or equivalent) and in America this much time is readily available from the national supercomputer centers. Finally, the best collaborations are actually rather loosely organized, and most projects are done by a few people working together under the umbrella.

And yet–one cannot view the increasing size of QCD collaborations as a good thing. When I think about these large collaborations I am reminded of a quotation by a famous former professor at the University of Graz:

"How striking is the difference between the great works of industry and the modest workshops of science! How imposing are the colossal ocean steamers! But the frequent traveler will have noticed that the officers and crew are always doing the same routine jobs, over and over again. In the passenger salons the same people are always talking about the same things, lounging in the same chairs and playing the same game of shuffleboard. Gargantuan masses, but not one new thought! Admittedly, in science too some things have been accomplished by large-scale effort (as we saw at the Lick Observatory). But the truly great scientific advance (our Ministry of Education mustn't hear this) is always made with the smallest means.

"It must be splendid to command millions of people in great national ventures, to lead a hundred thousand to victory in battle. But it seems to me better still to discover fundamental truths in a very modest room with very modest means–truths that will still be foundations of human knowledge when the memory of these battles is painstakingly preserved only in the archives of the historian. " [64]

The intellectual future of lattice calculations will come from exploiting large simulations in new and clever ways. There is an enormous amount of information in the simulations, waiting for someone to figure out how to extract it. The important lesson of these lectures is that the more fine-grained the observable, the more unambiguous

its interpretation. Studying QCD on a deeper and deeper level is the real future of lattice methods. That is also where the fun is!

Acknowledgements

I would like to thank my friends and colleagues whose discussions and insights I have included in this review: Claude Bernard, Steve Gottlieb, Rajan Gupta, Anna Hasenfratz, Matthew Hecht, Peter Lepage, Rich Loft, Bob Sugar, Steve Sharpe, and Doug Toussaint. I would like to thank Oliver Pene and Jim Labrenz for providing me with pre-publication copies of their work. I would like to thank the conference organizers for creating such a pleasant environment for discussions. It was a real pleasure to participate in the school. Finally, I would also like to thank Reba Katz for a careful reading of the manuscript. This work was supported by the U.S. Department of Energy.

References

1 Wilson, K.: Phys. Rev. D **10** (1974) 2445

2 Elitzur, S.: Phys. Rev. D **12** (1975) 3978

3 This particular formulation of the problem was shown to me by Toussaint, D.

4 See the discussion by Flensberg, M., and Peterson, C.: Nucl. Phys. **B283** (1987) 141

5 A good reference for strong coupling expansions is Drouffe, J.M., Itzykson, C.: Phys. Rept. **38C** (1978) 133

6 Golterman, M., Smit, J.: Nucl. Phys. **B255** (1985) 328

7 Andersen, H.C.: J. Chem. Phys. **72** (1980) 2384; Duane, S.: Nucl. Phys. **B257** (1985) 652; Duane, S., Kogut, K.: Phys. Rev. Lett. **55** (1985) 2774; Gottlieb, S., Liu, W., Toussaint, A.D., Renken, R., Sugar, R.: Phys. Rev. D **35** (1987) 2531

8 Duane, S., Kennedy, A., Pendleton, B., Roweth, D.: Phys. Lett. **194B** (1987) 271

9 Brown, F., Woch, T.: Phys. Rev. Lett. **58** (1987) 2394 ; Creutz, M.: Phys. Rev. D **36** (1987) 55 For a review, see Adler, S.: in the Proceedings of the 1988 Symposium on Lattice Field Theory, Kronfeld, A., Mackenzie, P.: eds., Nucl. Phys. **B (Proc. Suppl.) 9** (1989)

10 This discussion is stolen from Lepage, G.P.: in "From Actions to Answers–Proceedings of the 1989 Theoretical Advanced Summer Institute in Particle Physics," DeGrand, T., Toussaint, D.: eds., (World, 1990)

11 Bitar, K.M., et al.: in preparation.

12 For a good introduction to error analysis see Toussaint, D.: in "From Actions to Answers–Proceedings of the 1989 Theoretical Advanced Summer Institute in Particle Physics," DeGrand, T., Toussaint, D.: eds., (World, 1990)

13 A good gauge fixing algorithm is overrelaxation and is described in Mandula, J.E., Ogilvie, M.C.: Phys. Lett. **248B** (1990) 156

14 Marinari, E.: in the Proceedings of the 1988 Symposium on Lattice Field Theory, Kronfeld, A., Mackenzie, P.: eds., Nucl. Phys. **B (Proc. Suppl) 9** (1989) 209; Bacilieri, P., et al.: Phys. Lett. **214B** (1988) 115

15 Gusken, S., et al.: Phys. Lett. **227B** (1989) 266 ; Gusken, S.: Nucl. Phys. **B (Proc. Suppl) 17** (1989) 361

16 Gupta, R., Baillie, C., Brickner, R., Kilcup, G., Patel, A., Sharpe, S.: Phys. Rev. D **44** (1991) 3272

17 Bitar, K.M., et al.: Phys. Rev. Lett. **65** (1990) 2106, Phys. Rev. D **42** (1990) 3794

18 Gupta, R., Guralnik, G., Kilcup, G., Sharpe, S.: Phys. Rev. D **43** (1991) 2003
19 Kronfeld, A.: in the Proceedings of the 1988 Symposium on Lattice Field Theory, Kronfeld, A.,Mackenzie, P. eds., Nucl. Phys. **B (Proc. Suppl.) 9** (1989)
20 Compare Falcioni, M., Paciello, M., Parisi, G., Taglienti, B.: Nucl. Phys. **B251** (1985) 624; Teper, M.: Phys. Lett. **183B** (1987) 345; Phys. Lett. **185B** (1987) 121; DeGrand, T.: Phys. Rev. D **36** (1987) 176
21 Velikson B.,Weingarten, D.: Nucl. Phys. **B249** (1985) 433
22 Gottlieb, S.: in "Advances in Lattice Gauge Theory," Duke D., Owens, J.: eds. (World Scientific, 1985)
23 Chu, M.-C., Lassia, M., Negele, J.W.: Nucl. Phys. **B360** (1991) 31
24 Brodsky, S., Lepage, G.P.: Phys. Lett. **87B** (1979) 359; Phys. Rev. D **22** (1980) 2157
25 Weingarten, D.,Petcher, D.: Phys. Lett. **99B** (1981) 333; Hamber, H., Parisi, G.: Phys. Rev. Lett. **47** (1981) 1792
26 Brown, F., et al.: Phys. Rev. Lett. **67** (1991) 1062
27 Bitar, K., et al.: in the Proceedings of Lattice '90, U. Heller, A. Kennedy and S. Sanielevici, eds. Nucl. Phys. **B (Proc. Suppl) 20** (1991) 362
28 Gottlieb, S., Liu, W., Toussaint, D., Renken, R., Sugar, R.: Phys. Rev. D **38** (1988) 2245
29 Fukugita, M., Mino, H., Okawa, M., Ukawa, A.: preprint KEK-TH-271 (1990)
30 Bacilieri, P., et al.: Nucl. Phys. **B343** (1990) 228; Cabasino, S., et al.: Phys. Lett. **258B** (1991) 202
31 Bacilieri, P., et al.: Nucl. Phys. **B317** (1989) 509; Cabasino, S., et al.: Phys. Lett. **258B** (1991) 195
32 Thacker, B., Lepage, G.P.: Phys. Rev. D **43** (1991) 196
33 Kokkedee, J.: "The Quark Model," (W. A. Benjamin, New York, 1969)
34 Hecht, M., DeGrand, T.: Phys. Lett. B, in press
35 Svetitsky, B.: Phys. Rep. **132** (1986) 1
36 Svetitsky, B., Yaffe, L.: Phys. Rev. D **26** (1982) 963, Nucl. Phys. **B210** (1982) 423
37 Pisarski, R., Wilczek, F.: Phys. Rev. D **29** (1984) 338
38 Bernard, C., DeGrand, T., DeTar, C., Gottlieb, S., Krasnitz, A., Ogilvie, M.C., Sugar, R.L., Toussaint, D.: University of California, Santa Barbara report UCSB TH–91-52 (1991)
39 Brown, F., et al.: Phys. Rev. Lett. **65** (1990) 2491
40 Gottlieb, S.: Nucl. Physics B (Proc. Suppl.) **20** (1991) 247
41 Kennedy, A.D., Kuti, J., Meyer, S., Pendleton, B.J.: Phys. Rev. Lett. **54** (1985) 87; Gottlieb, S.A., Kuti, J., Toussaint, D., Kennedy, A.D., Meyer, S., Pendleton, B.J., Sugar, R.L.: Phys. Rev. Lett. **55** (1985) 1958
42 Bacilieri, P., et al.: Nucl. Phys. **B343** (1990) 228; Cabasino, S., et al.: Nucl. Physics B (Proc. Suppl.) **20** (1991) 399; Gupta, R., et al.: Phys. Rev. D **43** (1991) 2003; Bitar, K.M., et al.: Nucl. Physics B (Proc. Suppl.) **20** (1991) 362
43 Gottlieb, S., Liu, W., Toussaint, D., Renken, R.L., Sugar, R.L.: Phys. Rev. Lett. **59** (1987) 1513
44 Gottlieb, S., Heller, U.M.: Kennedy, A.D., Kogut, J.B., Krasnitz, A., Liu, W., Renken, R.L., Sinclair, D.K., Sugar, R.L., Toussaint, D., Wang, K.C.: to be published
45 The HEMCGC Collaboration, Bitar, K.M., et al.: Phys. Rev. Lett. **65** (1990) 2106; Phys. Rev. D **42** (1990) 3794; Brown, F.R., Butler, F.P., Chen, H., et al.: Phys. Rev. Lett. **67** (1991) 1062
46 Born, K.D., et al.: Phys. Rev. Lett. **67** (1991) 302
47 Deng, Y.: in the Proceedings of the 1988 Symposium on Lattice Field Theory,Kronfeld, A., Mackenzie, P.: eds., (Nucl. Phys. B (Proc. Suppl.) 9 (1989),334
48 Gottlieb, S., Liu, W., Renken, R., Sugar, R., Toussaint, D.: Phys. Rev. Lett. **59** (1987) 2247; Phys. Rev. D **38** (1988) 2888

49 Karsch, F.: Nucl. Phys. **B205[FS5]** (1982) 285

50 DeGrand, T., DeTar, C.: Phys. Rev. D **34** (1986) 2469

51 DeTar, C., Kogut, J.: Phys. Rev. Lett. **59** (1987) 399

52 Bernard, C., et al.: preprint AZPH–TH/91–60 (1991)

53 Borgs, C.: Nucl. Phys. **B261** (1991) 455; E. Manousakis and J. Polonyi, Phys. Rev. Lett. **58** (1987) 847

54 LePage, G.P., MacKenzie, P.: Nucl. Phys. B (Proc. Suppl) **20** (1991) 173

55 These techniques go back to Symanzik, K.: Nucl. Phys. **B226** (1983) 187 Recently these techniques have been applied to QCD simulations of matrix elements by Martinelli, G., Sachrajda, C., Vladikas, A.: Nucl. Phys. B (Proc. Suppl) **20** (1991) 448 and Heattlie, G., Martinelli, G., Pittori, C., Rossi, G.C., Sachrajda, C.: Nucl. Phys. **B352** (1991) 266

56 For a review, see Sachrajda, C.: Nucl. Phys. B (Proc. Suppl) **9** (1989) 121

57 Bernard, C., Labrenz, J., Soni, A.: contribution to Lattice '91; Labrenz, J.: private communication

58 Van Royen, R., Weisskopf, V.F.: Nuovo Cim. **50** (1967) 17

59 Abada, A., et al.: preprint CERN-TH. 6271/91 and contribution to Lattice '91, Pene, O.: private communication

60 Alexandrou, C., et al.: Phys. Lett. **256B** (1991) 60

61 Allton, C., et al.: Nucl. Phys. **B349** (1991) 598

62 Gavela, M.B., et al.: Phys. Lett. **206B** (1988) 113 Maiani, L., G. Martinelli, G.: Phys. Lett. **178B** (1986) 265

64 Boltzmann, L.: “Reise eines deutschen Professors in Eldorado,” in “Populäre Schriften,” ed. Broda, E.: reprinted 1979, Vieweg, F.: & Sohn, Weisbaden, and reprinted in English translation in Physics Today, January 1992, p. 44

This article was processed by the author using the TEX macro package from Springer-Verlag.

Effective Field Theories

G. Mack, T. Kalkreuter, G. Palma⋆*, M. Speh*

II. Institut für Theoretische Physik, Universität Hamburg,
Luruper Chaussee 149, 2000 Hamburg 50

Abstract. Effective field theories encode the predictions of a quantum field theory at low energy. The effective theory has a fairly low ultraviolet cutoff. As a result, loop corrections are small, at least if the effective action contains a term which is quadratic in the fields, and physical predictions can be read straight from the effective Lagrangian.

Methods will be discussed how to compute an effective low energy action from a given fundamental action, either analytically or numerically, or by a combination of both methods. Basically, the idea is to integrate out the high frequency components of fields. This requires the choice of a "blockspin", i.e. the specification of a low frequency field as a function of the fundamental fields. These blockspins will be the fields of the effective field theory. The blockspin needs not be a field of the same type as one of the fundamental fields, and it may be composite. Special features of blockspins in nonabelian gauge theories will be discussed in some detail.

In analytical work and in multigrid updating schemes one needs interpolation kernels $\mathcal{A}$ from coarse to fine grid in addition to the averaging kernels C which determines the blockspin. A neural net strategy for finding optimal kernels is presented.

Numerical methods are applicable to obtain actions of effective theories on lattices of finite volume. The special case of a "lattice" with a single site (the constraint effective potential) is of particular interest. In a Higgs model, the effective action reduces in this case to the free energy, considered as a function of a gauge covariant magnetization. Its shape determines the phase structure of the theory. Its loop expansion with and without gauge fields can be used to determine finite size corrections to numerical data.

1 What are Effective Field Theories?

We work in the context of Euclidean field theory.

- An effective field theory encodes the predictions of a quantum field theory at *low energies* .
- It is a *field theory* with a *low UV-cutoff* a_0^{-1} - of the order of the relevant particle masses. For instance it may live on a *lattice* of lattice spacing a_0.

⋆ On leave of absence from Universidad de Santiago de Chile, Casilla 307, Correo 2, Santiago-Chile

- Ideally one should be able to read the physics straight from the *effective action* (= action of the effective theory) because loop corrections to tree amplitudes are small at sufficiently low UV-cutoff.

Problem to be discussed: Start from a given *fundamental action* (on the continuum or on a lattice of small lattice spacing $a \ll a_0$)

- How does one *define and compute the effective action*?

Sometimes effective actions are guessed on the basis of known properties.

Example: chiral theories (nonlinear σ-models) as effective theories of mesons. They embody the low energy theorems implied by PCAC.[2] These low energy theorems are sufficiently restrictive to fix the effective action at really low energy (i.e. low cutoff) up to a few coupling constants. By the definition of an effective theory, the question of the removal of the UV-cutoff does not pose itself. Nonrenormalizable σ-models are therefore acceptable.

But ideally, one would like not to guess the effective theory, but to justify it from the given fundamental theory.

Definition of the effective action in terms of the fundamental theory

The effective action will be a function of *low frequency fields* Φ - for instance, fields Φ on a lattice of lattice spacing a_0. They should be defined as *functions of the fundamental fields* φ,

$$\Phi = C(\varphi) \ ,$$

If C is linear, we write instead

$$\Phi = C\varphi \ .$$

C will be called the averaging operator and Φ is called *blockspin.* Later on we will consider fundamental theories which are gauge theories, and we will distinguish in notation between matter fields φ and gauge fields U. In general the low frequency matter fields will be functions not only of the matter fields φ but will also depend on the gauge fields,

$$\Phi = C(U,\varphi) \quad , \text{ or } \quad \Phi = C(U)\varphi \ .$$

The definition of the effective action $\mathcal{H}_{eff}$ in terms of the fundamental Euclidean action $\mathcal{H}(\varphi)$ reads [3]

$$e^{-\mathcal{H}_{eff}(\Phi)} = \int \mathcal{D}\varphi \ \delta(\Phi - C(\varphi)) \ e^{-\mathcal{H}(\varphi)}. \tag{1}$$

If the effective theory lives on a lattice Λ_0, then

$$\delta(\Phi - C(\varphi)) = \prod_{\alpha} \prod_{x \in \Lambda_0} \delta(\Phi_\alpha(x) - C_\alpha(\varphi)(x))$$

---

[2] PCAC = partially conserved axial vector current.

[3] Euclidean field theory may be regarded as a classical statistical mechanics. In this language, the action becomes the Hamiltonian. Therefore we use the letter $\mathcal{H}$.

when Φ has components Φ_α. The 1-dimensional δ-function admits the integral representation

$$\delta(\xi) = \frac{1}{2\pi} \int_{-\infty}^{\infty} dk\ e^{ik\xi} .$$

This formula, or (better) an approximation of the δ-function by a Gaussian

$$\delta_\beta(\xi) = \left(\frac{2\pi}{\beta}\right) e^{-\xi^2/2\beta} = \frac{1}{2\pi} \int_{-\infty}^{\infty} dk\ e^{ik\xi}\ e^{-\beta k^2/2} \quad (\beta \text{ small})$$

can also be used when there are *Fermi fields.*

Problems

- On what *number and kinds of fields* Φ_α should $\mathcal{H}_{eff}$ depend ?
- How to choose the *averaging operator* C.
  Averaging over space time regions will be involved in getting low frequency fields.
- How to *compute the functional integral* (1) by analytical or numerical means.
- How to *detect a bad choice of* C *or* Φ.

The answer to the last question is given by the requirement that *the effective action must have good locality properties.*

Let us emphasize that $\mathcal{H}_{eff}$ is *exact* in the sense that it yields exact expectation values for observables O which depend on the fundamental fields φ only through $\Phi_\alpha = C_\alpha(\varphi)$ so that $O = F(C(\varphi))$.

$$\begin{aligned}\langle O\rangle &= \frac{1}{Z}\int \mathcal{D}\varphi\ F(C(\varphi))\ e^{-\mathcal{H}(\varphi)} \\ &= \frac{1}{Z}\int \mathcal{D}\Phi\ F(\Phi)\ e^{-\mathcal{H}_{eff}(\Phi)} .\end{aligned} \tag{2}$$

We illustrate the issues involved as well as the dangers of a bad choice of Φ or C at an example which is well understood.

Example: Discrete Gaussian model.

We consider first the case where the fundamental theory is a discrete Gaussian model in 3 dimensions. It lives on a lattice Λ of lattice spacing a. This model is the dual transform of a $U(1)$ lattice gauge theory in 3 dimensions. An exact computation of its effective action by convergent series expansions was used in [1] to give a rigorous proof that the 3-dimensional $U(1)$-lattice gauge theory shows linear confinement for all values of the gauge coupling constant g,with $\beta = 4\pi^2/g^2$. [4]

The points of the fundamental lattice will be denoted by letters z. The fundamental field is *integer valued* ,

$$n(z) \in \mathbb{Z} ,$$

[4] The convergent expansions are iterated Mayer expansions. They combine expansions at various length scales. Similar expansions can be written down which live on a multigrid [2] - cp. later.

and the fundamental action is [5]

$$\mathcal{H}(n) = \frac{1}{2\beta} \int_z [\nabla_\mu n(z)]^2 \ .$$

The difficult case is when β/a *is large* (i.e. weak gauge coupling g.) Note that β/a is dimensionless in 3 dimensions.

A choice was made to use a Pauli-Villars cutoff for the effective theory, so that it lives on the original lattice Λ, to begin with. The cutoff a_0^{-1} was chosen to be of the order of the ultimate physical mass m of the theory, in agreement with the philosophy of effective field theory. But it is essential to choose as the low frequency field a *real field*

$$\Phi(z) \text{ real} \ .$$

This illustrates the point that *the fields Φ of the effective theory need not be fields of the same type as those in the fundamental theory* .

One finds that a local sine-Gordon action is an excellent approximation to the effective action when β is large

$$\mathcal{H}_{eff}(\Phi) = \int_z \left(\frac{1}{2\beta} [\nabla_\mu \Phi(z)]^2 + 2\mathbf{z}\,[1 - \cos \Phi(z)] \right) \tag{3}$$

with

$$\begin{aligned} \mathbf{z} &= a^{-3} \exp[-\beta v_{Cb}(0)] \quad \text{for large } \beta \ , \\ v_{Cb}(0) &= 0.2527 \ a^{-1} \ . \end{aligned} \tag{4}$$

$v_{Cb}(0)$ is the Coulomb potential at zero distance on the 3-dimensional lattice. (We chose not to write the terms in $\mathcal{H}_{eff}$ which implement the Pauli-Villars cutoff.)

We proceed to a discussion of the physics of the model on the basis of of the effective action.

1) Mass: The $[1 - \cos]$ potential has its minima at $\Phi = 2\pi\cdot$integer. We shall see in a moment that tunneling between different minima is very much suppressed, so that the global symmetry

$$\Phi(x) \mapsto \Phi(x) + 2\pi N \ , \quad N \text{ integer}$$

is spontaneously broken, and we may restrict attention to one of the minima, e.g. $\Phi = 0$, and expand around it. Set

$$\begin{aligned} \Phi &= \beta^{1/2} \ \Psi \quad , \\ m^2 &= \mathbf{z} \ 2\beta = 2\beta \ a^{-3} \exp[-\beta v_{Cb}(0)] \ . \end{aligned} \tag{5}$$

Then the effective action takes the form

$$\mathcal{H}_{eff} = \frac{1}{2} \int_z \left([\nabla_\mu \Psi(z)]^2 + m^2 \ \frac{1}{2} \ \Psi(z)^2 + [\cdots] \right) \ . \tag{6}$$

---

[5] We use lattice notations as follows on a d-dimensional lattice Λ of lattice spacing a: $\int_z = a^d \sum_{z\in\Lambda}$ and $\nabla_\mu f(z) = a^{-1}[f(z + e_\mu) - f(z)]$, e_μ= lattice vector of length a in μ-direction.

The nonlinear terms $[\cdots]$ are small for large β (provided Ψ is sufficiently near to the selected minimum so that the effective action is sufficiently small to give Ψ a nonnegligible probability). The effective theory is therefore very nearly a free field theory with a nonvanishing mass m which is given by (5). Numerical simulations confirm this [3].

The low lying excitations of the model are therefore spin waves with a finite mass which is exponentially small in units of the original inverse lattice spacing a^{-1}. This is easy to read off the effective action, but is not at all apparent from the fundamental action.

2) Tunneling: Because of the UV-cutoff we can qualitatively think of the theory as living on a lattice Λ_0 of lattice spacing $a_0 = \text{ const. } \cdot m^{-1}$. The effective action becomes

$$\mathcal{H}_{eff} = \sum_{x \in \Lambda_0} \frac{a_0^3}{2\beta} \left[\nabla_\mu \Phi(x)\right]^2 + 2\mathbf{z}_0 \left[1 - \cos\Phi(x)\right] \tag{7}$$

$$\mathbf{z}_0 = a_0^3 \mathbf{z} = \text{ const. }' \, m^{-3}\mathbf{z} \tag{8}$$

$$= \text{ const }'' \beta^{-3/2} \exp\left[\beta v_{Cb}(0)/2\right] \tag{9}$$

We see that $\mathbf{z}_0 \mapsto \infty$ when $\beta \mapsto \infty$. Therefore the hills of the $2\mathbf{z}_0\left[1 - \cos\Phi\right]$ - potential become very high when β becomes large. As a result, tunneling becomes very much suppressed.

The height and widths of the hills of the potential determine the surface tension of the discrete Gaussian model (the free energy per area of an interphase between domains where $\langle\Phi\rangle = 2\pi N$ with different N). This surface tension is equal to the string tension α of the dual $U(1)$-gauge theory. Treating the tunneling in a semiclassical approximation results in

$$\alpha = 8m\beta^{-1} \, . \tag{10}$$

Note that

$$\alpha/m^2 \mapsto \infty \quad \text{as } \beta \mapsto \infty \, .$$

Thus *there are two different physical mass scales in this theory.*

3) 2-dimensional discrete Gaussian model: Next we turn to the *2-dimensional discrete Gaussian model.* This model is the dual transform of the XY-model (plane rotator) in 2 dimensions with Villain action.

A rigorous computation of the effective action of this model has not been reported in the literature, but the method used for the 3-dimensional model can be expected to carry over for large β. In 2 dimensions, β is dimensionless.

The effective action comes out the same as in 3 dimension, except for the appearance of different powers of a (for dimensional reasons) and the following difference in the expression for $\mathbf{z}$. Expression (4) results from an approximation which is legitimate in 3 dimensions, but not in 2. Originally there stands the Yukawa potential $v_\mu(0)$ at zero distance in place of $v_{Cb}(0)$, with mass μ equal to the Pauli-Villars cutoff a_0^{-1}. We are interested in the limit where μa is very small. In 3 dimensions, the zero mass limit exists, and we may approximate $v_\mu(0)$ by $v_{Cb}(0)$. But in 2 dimensions, the zero mass limit does not exist. Instead, $v_\mu(0)$ increases logarithmically with $(\mu a)^{-1}$ as the cutoff $\mu = a_0^{-1}$ is lowered. As a result,

$$\mathbf{z}_0 = a_0^2 \, \mathbf{z} = (a\mu)^{-2} \exp\left[-\beta v_\mu(0)/2\right] \tag{11}$$

decreases when the cutoff is lowered, if β is large enough, and the mass in tree approximation decreases faster than the cutoff μ. Therefore, the cutoff can be lowered indefinitely. Since $\mathbf{z}_0$ decreases indefinitely, the effective action tends to a massless free field theory in the limit of very low cutoff, provided β is large enough. This is the Kosterlitz-Thouless phase of the model.

There is no particular reason to believe that the formula (7) for the effective action remains accurate for intermediate values of β. But the nature of the Kosterlitz-Thouless phase phase transition could be read off, if one had an accurate expression for the effective action at intermediate β.

Lessons from the model

- The physics can readily be read from the effective action, but is not visible in the fundamental action.
- A good choice of the blockspin Φ (real as opposed to integer) is crucial.

If we had chosen an integer blockspin, the spin waves of zero or small mass m as the dominant low energy excitations would not be represented by fields in the action. Moreover, one would meet a disaster while calculating the effective action. If one tried to compute it by lowering the cutoff step by step (which is what an iterated Mayer expansion does), the action would become nonlocal over distances m^{-1} while the UV-cutoff μ is still very much higher than m. A good choice of the blockspin is

$$\Phi(z) \approx \text{average of } n(w) \text{ over a cube of side } a_0$$

using a smoothened step function for the cube.

Stages of complexity of the effective action

To be specific, let us start from the fundamental action of a Higgs model with non-abelian gauge fields living on Λ. Λ may be either the continuum or a lattice of small lattice spacing. We seek an effective theory on a lattice Λ_0 of lattice spacing a_0.

We identify the points $x \in \Lambda_0$ with hypercubes in the fundamental lattice Λ. *These hypercubes are called blocks.* Sometimes we will also regard Λ_0 as a sublattice of Λ, so that $x \in \Lambda_0$ fixes a site in Λ, e.g. the block center.

One may envisage different kinds of and approximations to the effective action which require different kinds of blockspins

1. Ignore both the gauge field U and the fermion fields ψ.
   Approximate the fundamental theory by a $\lambda\varphi^4$-theory. Choose the block average of the scalar field as blockspin [4]

$$\Phi_\alpha(x) = \int_z C(x,z)\, \varphi_\alpha(z) = Z_0^{-1/2}(a_0)\; av_{z\in x}\varphi_\alpha(z)\ ,$$

   with $C(x,z) \propto$ step function. A cutoff dependent factor $Z_0^{1/2}(a_0)$ (wave function renormalization) has been included with the aim to normalize the kinetic term in the effective action in the standard way.

2. Ignore the gauge field, but consider fermions.
   An effective action which has chiral symmetry requires $2^{d/2}$ flavours f in d dimensions (for even d). Otherwise the effective action will inevitably become non-local (Nielsen-Ninomiya theorem [5]). Given fundamental fermion fields with $2^{d/2}$ flavours, we may block to staggered fermions on Λ_0 of lattice spacing $\frac{1}{2}a_0$. Λ_0 decomposes into elementary cells of sidelength a_0 containing 2^d sites each. Only translations by integer multiples of a_0 are true translations. Sites within one elementary cell carry different pseudoflavour (a combination of spin and flavour). A suitable blockspin is of the form
$$\Psi(x) = \int_z \sum_{f,a} C_a^f(x,z)\, \psi_a^f(z) \tag{12}$$
   where a, f are spinor and flavour indices. C_a^f averages over a hypercube of sidelength a_0 centered at x, and is only nonzero when (a, f) matches with the pseudoflavour of x. Details will be given later on.
3. Admit dynamical gauge fields.
   The blockspin for the Higgs field will take the form
$$\Phi_\alpha(x) = \int_z C(U|x,z)\, \varphi_\alpha(z) \tag{13}$$
   similarly as in 1., but with an averaging kernel C which depends on the gauge field U for reasons of covariance. A formula like (12) with U-dependent kernels C_a^f will define the blocked Fermi fields. In addition one will need to define blocked gauge fields. This will be discussed later on.
4. Composite Higgs.
   There may be no fundamental Higgs field in the fundamental action. In this case one may try to define a composite Higgs field Φ in the effective action by a formula of the following kind
$$\Phi_\alpha(x) = \int_z \int_w \sum \overline{\psi}_a^f\, C_{ab}^{fg}(U|x,zw)\, \lambda_\alpha \psi_b^g(w)\,.$$
   with matrices λ_α in colour space (colour indices of the Fermi fields ψ are omitted).
5. Fully effective theory.
   In this case, the fields in the effective action are all interpolating fields for physical particles. They can be defined as composite fields similar to 4., but they will have to be gauge invariant. For instance, the physical electron field could take the form
$$\Psi(x) = \int_z \int_w \psi_a^f(z)\, C_a^f(U|x,zw)\, \varphi(w)$$
   where ψ_a^f are the left and right handed fundamental electron field and φ is the fundamental Higgs field.

One may wish to compute the ultimate effective action in a sequence of steps, lowering the cutoff in stages. In this case, one may have to proceed to different kinds of definitions of blockspins at different values of the UV-cutoff corresponding to different length scales of compositeness.

2 The Constraint Effective Potential

Starting from a fundamental theory on a finite volume, the constraint effective potential is defined as the special case of the effective action in which the lattice Λ_0 consists of a single point. In this case, the block spins Φ may be vectors in colour or flavour space, but they have no space time argument x.

Consider for instance a fundamental theory which is a Higgs model with or without dynamical gauge fields. For simplicity, we do not include fermions. One introduces

$$\Phi = \text{(gauge covariant) magnetization}$$

by a formula similar as before. The constraint effective potential $V_{eff}(\Phi)$ was introduced in [6] and is defined by

$$e^{-V_{eff}(\Phi)} = \int \mathcal{D}\varphi\, \mathcal{D}U\ \delta(\Phi - C\varphi)\ e^{-\mathcal{H}(U,\varphi)} \tag{14}$$

and similarly in the absence of gauge fields. C depends on U in the presence of gauge fields.

$\exp\{-V_{eff}(\Phi)\}$ gives us the probability density of the magnetization Φ. But

$$V_{eff} \sim \text{ volume} \quad .$$

Therefore the minima of V_{eff} become extremely sharp in the limit of large volume, and in the infinite volume limit, the only possible values which Φ may assume are absolute minima of V_{eff}. Therefore, the shape of V_{eff} in the large volume limit gives information about the phase structure of the theory. One has to look for minima. If there are several absolute minima [6] which are transformed into each other by a symmetry, then that symmetry is spontaneously broken. For plots which show the crossover of the effective potential from a single shape (symmetric phase) to double well behaviour (broken phase) see Fig. 1.

If the height of different relative minima changes when one varies a parameter in the theory in such a way that the absolute minimum jumps, then the theory has a first order phase transition.

3 Renormalization Group Picture

In the renormalization group approach [7] one composes the blocking

$$\begin{aligned} \Lambda &\mapsto \Lambda_0 \\ a &\mapsto a_0 \\ \mathcal{H} &\mapsto \mathcal{H}_{eff} \end{aligned}$$

from a sequence of blockings, lowering the cutoff in steps

$$\begin{aligned} \Lambda \equiv \Lambda_N &\mapsto \Lambda_{N-1} \mapsto \ldots \mapsto \Lambda_1 \mapsto \Lambda_0 \\ a \equiv a_N &\mapsto a_{N-1} \mapsto \ldots \mapsto a_1 \mapsto a_0 \\ \mathcal{H} \equiv \mathcal{H}^N &\mapsto \mathcal{H}^{N-1} \mapsto \ldots \mapsto \mathcal{H}^1 \mapsto \mathcal{H}^0 \equiv \mathcal{H}_{eff} \end{aligned}$$

[6] $\mathcal{V}_{eff}$/volume becomes convex in the infinite volume limit because the height of maxima of V_{eff} between minima increases less fast than the volume.

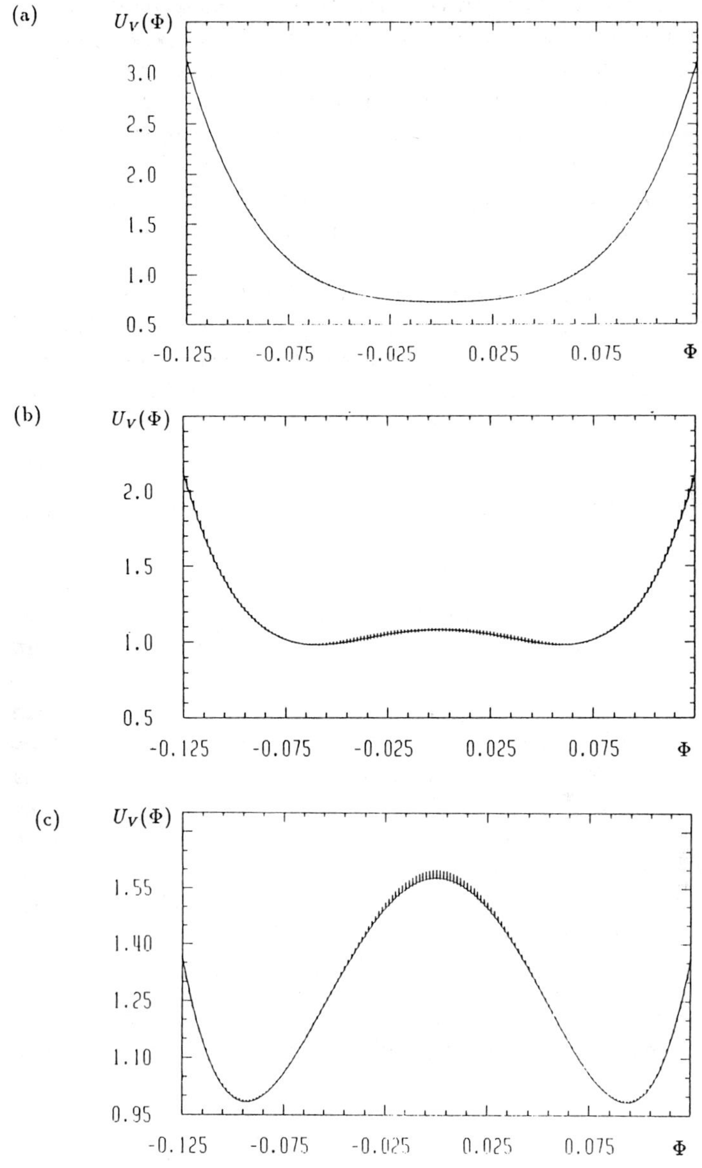

Fig. 1. Multigrid Monte Carlo data for $\lambda\varphi^4$-theory [15] with error bars as a function of the magnetization Φ for bare coupling $\lambda_u = 16.376$ and fit with the one loop formula for the constraint effective potential for three values of the bare mass a) $m_u^2 = -1.14$, b) $m_u^2 = -1.15$, c) $m_u^2 = -1.16$.

where $a_{j-1} = La_j$ with $L = 2, 3$ or so.

The theory with action $\mathcal{H}^j$ lives on a lattice of lattice spacing a_j. Considering one such theory at a time, one goes to the unit lattice by setting $a_j = 1$. In this way one obtains a sequence of actions $\widetilde{\mathcal{H}}^j$,

$$\widetilde{\mathcal{H}}^N \mapsto \widetilde{\mathcal{H}}^{N-1} \mapsto \ldots \mapsto \widetilde{\mathcal{H}}^1 \mapsto \widetilde{\mathcal{H}}^0 \equiv \widetilde{\mathcal{H}}_{eff}. \tag{15}$$

These actions can be compared, since they all live on the same lattice with lattice spacing 1. This yields the

$$\text{renormalization group flow: } \widetilde{\mathcal{H}}^j \mapsto \widetilde{\mathcal{H}}^{j-1} \ .$$

Imagine that the actions are somehow parametrized. One distinguishes between relevant, marginal, and irrelevant parameters. These parameters are also called "running coupling constants".

Relevant: A small change in a relevant coupling constant in $\mathcal{H}^j$ changes $\mathcal{H}^{j-n}$ very much for large n.

Irrelevant: A small change in an irrelevant coupling constant changes $\mathcal{H}^{j-n}$ arbitrarily little for large n.

Marginal: in between.

In the limit of small a, many renormalization group steps will be needed. As a result, knowledge of all relevant and marginal coupling constants in $\mathcal{H}$ suffices to determine $\mathcal{H}_{eff}$. Theories with actions $\mathcal{H}$ whose relevant and marginal parameters all agree are said to belong to the same universality class.

Let us remark that one can also lower an UV-cutoff (Pauli-Villars, for instance) infinitesimally in one step. In this case the renormalization group flow is determined by a functional differential equation known as the renormalization group differential equation [8].

In the conventional approach, all the actions $\mathcal{H}^j$ depend on the same type of fields. But we noted before that more flexibility should be left. It may be necessary to change the kind of fields at some length scales (of compositeness) a_j. In particular, $\mathcal{H}^j$ may depend on different kinds of fields than the fundamental action $\mathcal{H}$ when $j < N$.

Effective observables

In principle, expectation values $\langle O\rangle_{\mathcal{H}}$ of arbitrary observables in the fundamental theory may be translated into expectation values $\langle O_{eff}\rangle_{\mathcal{H}_{eff}}$ in the effective theory, such that

$$\langle O\rangle_{\mathcal{H}} = \langle O_{eff}\rangle_{\mathcal{H}_{eff}} \ .$$

But this requires that one computes effective observables in addition to the effective Hamiltonian [9],

$$O_{eff}(\Phi) = e^{\mathcal{H}_{eff}(\Phi)} \int \mathcal{D}\varphi \; O(\varphi) \; \delta(\Phi - C(\varphi)) \; e^{-\mathcal{H}(\varphi)} \quad . \tag{16}$$

4 Computation of Effective Actions by Perturbation Theory (Loop Expansion)

We discuss linear averaging maps C in theories with or without gauge fields.

Scalar free field theory

$$\mathcal{H}_{\text{free}}(\varphi) = \frac{1}{2}\int_z \varphi(z)[-\Delta + m^2]\varphi(z) = \frac{1}{2}(\varphi, [-\Delta + m^2]\varphi) .$$

Herein, Δ may be either the ordinary Laplacian, or the covariant Laplacian in some given external gauge field, and the original theory may live either on the continuum or on a lattice Λ. Its sites are denoted by z, w. Our aim is to compute an effective action on a lattice Λ_0 of lattice spacing $a_0 \leq O(m^{-1})$. Sites in Λ_0 are denoted by x, y.

We assume a linear averaging map C,

$$\Phi(x) = \int_z C(x,z)\varphi(z) \equiv C\varphi(x) .$$

In the absence of a gauge field, we may choose

$$a_0^d\ C(z,x) = \begin{cases} 1 & \text{if } z \in x \\ 0 & \text{otherwise} . \end{cases} \tag{17}$$

The original lattice Λ of sites z is covered by a block lattice Λ_0, whose sites x are blocks with side length a_0, so we write $z \in x$ if the point z is in the block x. More generally, C may depend on an (external) gauge field.

Following Kupiainen and Gawedzki [4,10], one splits the field φ into a low frequency part ψ, called the "background field", and a high frequency part ζ, called the "fluctuation field". The low frequency field is supposed to be obtainable from the block spin Φ on Λ_0 by interpolation with an interpolation kernel $\mathcal{A}$.

$$\begin{aligned} \varphi(z) &= \psi(z) + \zeta(z) \\ &= \mathcal{A}\Phi(z) + \zeta(z) , \quad \text{with } \mathcal{A}\Phi(z) = \int_{x\in\Lambda_0} \mathcal{A}(z,x)\Phi(x) . \end{aligned} \tag{18}$$

If we demand that

$$C\mathcal{A} = \mathbb{1} , \tag{19}$$

then it follows that the fluctuation field has zero block average,

$$C\zeta = 0 . \tag{20}$$

Now we insert the field split into $\mathcal{H}_{\text{free}}$. Let us consider the case $m = 0$. Generalization is obvious.

$$(\varphi, \Delta\varphi) = (\mathcal{A}\Phi, \Delta\mathcal{A}\Phi) + (\zeta, \Delta\zeta) + 2(\zeta, \Delta\mathcal{A}\Phi) . \tag{21}$$

We demand that $\mathcal{A}$ is chosen in such a way that the mixed term vanishes,

$$(\zeta, \Delta\mathcal{A}\Phi) = 0 . \tag{22}$$

This gives

$$\mathcal{H}_{\text{free}} = \frac{1}{2}(\Phi, -\Delta_{eff}\Phi) + \frac{1}{2}(\zeta, \Delta\zeta) , \tag{23}$$

with

$$\Delta_{eff} = \mathcal{A}^* \Delta \mathcal{A} . \tag{24}$$

Since $C\zeta = 0$ it follows that the mixed term (22) vanishes as desired, provided

$$- \Delta\mathcal{A} = C^* u^{-1} \tag{25}$$

for some u^{-1}.

Applying $\mathcal{A}^*$ to both sides of (25) it follows that

$$u^{-1} = -\Delta_{eff} . \tag{26}$$

The operator Δ_{eff} acts according to

$$\Delta_{eff} f(x) = \int_{y\in\Lambda_0} \Delta_{eff}(x,y) f(y) \tag{27}$$

with kernel

$$\Delta_{eff}(y,x) = \int_z \mathcal{A}(z,y)^\dagger \Delta\mathcal{A}(z,x). \tag{28}$$

Here, Δ acts on the argument z, and † denotes the hermitian conjugate of a matrix, or the transpose, if the matrix is real.

(25) becomes

$$\Delta\mathcal{A}(z,x) = \int_y C(y,z)^\dagger \Delta_{eff}(y,x) . \tag{29}$$

If $C(x,z)$ is the step function (17) then (29) says that $\Delta\mathcal{A}(z,x)$ should be constant on blocks y as a function of z. The kernels $\mathcal{A}$ and Δ_{eff} are determined as solutions of (29) together with the subsidiary condition (19).

Remark: The field $\psi(z) = \mathcal{A}\Phi(z)$ minimizes $\mathcal{H}_{\text{free}}(\psi)$ subject to the subsidiary condition that ψ has the prescribed block average Φ, viz.

$$C\psi(x) = \Phi(x) .$$

This remark can be used to compute the kernel $\mathcal{A}$ by standard optimization algorithms.

Numerical results confirm that the kernel $\mathcal{A}(z,x)$ decays exponentially with the distance of z from the center of the block x, with decay length a_0. This is true with and without an external gauge field in Δ, see for instance Fig. 5 of Sect. 8. As a result, $\Delta_{eff}(x,y)$ decays exponentially as a function of $|x-y|$ with decay length one lattice spacing a_0.

Now we are ready to compute the effective action. By definition

$$e^{-\mathcal{H}_{eff}(\Phi)} = \int \mathcal{D}\varphi\delta(\Phi - C\varphi)e^{-\mathcal{H}_{\text{free}}(\varphi)} . \tag{30}$$

We shift the field φ by an amount depending on Φ, using $\zeta = \varphi - \mathcal{A}\Phi$ as a new variable of integration. Since $C\mathcal{A} = \mathbb{1}$ by assumption, it follows that

$$e^{-\mathcal{H}_{eff}(\Phi)} = e^{\frac{1}{2}(\Phi,\Delta_{eff}\Phi)} \int \mathcal{D}\zeta\ \delta(C\zeta)\ e^{-\mathcal{H}_{\mathrm{free}}(\zeta)} \ . \tag{31}$$

The ζ-integral merely produces a constant independent of Φ. Therefore

$$\mathcal{H}_{eff}(\Phi) = \frac{1}{2}(\Phi, -\Delta_{eff}\Phi) + \text{ const.} \ . \tag{32}$$

$\mathcal{H}_{eff}$ has good locality properties because of the aforementioned decay properties of $\Delta_{eff}(x,y)$. There are exponential tails, but this is tolerable and in general inevitable.

Split of the propagator

The split of the field φ induces a split of the free propagator $v = (-\Delta)^{-1}$ into a low frequency propagator v_{low} and a high frequency propagator (fluctuation field propagator) Γ. For proper choice of the normalization factor $\mathcal{N}$,

$$d\mu_\Gamma(\zeta) = \mathcal{N}^{-1}\mathcal{D}\zeta\ e^{-(\zeta,-\Delta\zeta)/2}\ \delta(C\zeta) \tag{33}$$

is a normalized Gaussian measure with covariance

$$\Gamma(z,w) = \mathcal{N}^{-1} \int \mathcal{D}\zeta\ \delta(C\zeta)\ e^{-(\zeta,-\Delta\zeta)/2}\ \zeta(z)\ \zeta(w) \ .$$

Thinking of the δ-function as a limit of a Gaussian, one sees that

$$\Gamma = \lim_{\kappa\mapsto\infty} (-\Delta + \kappa C^* C)^{-1}. \tag{34}$$

Since ζ is supposed to be the high frequency part of the field φ, it is a field with an infrared cutoff (equal to the UV-cutoff of the effective theory). Therefore one expects that $\Gamma(z,w)$ decays exponentially with the distance of z from w, with decay length a_0. Analytical and numerical results confirm that this is indeed the case. This remains true when there is a a gauge field in Δ, provided the kernel C is chosen in the right way. A proper choice will be introduced later.

Let us now compute the 2-point function,

$$\langle\varphi(z)\varphi(w)\rangle \propto \int \mathcal{D}\Phi\ \mathcal{D}\varphi\ \delta(\Phi - C\varphi)\ e^{-\mathcal{H}_{\mathrm{free}}(\varphi)}\ \varphi(z)\ \varphi(w) \tag{35}$$

$$= \int \mathcal{D}\Phi\ \mathcal{D}\zeta\ e^{(\Phi,-\Delta_{eff}\Phi)+(\zeta,-\Delta\zeta)}$$
$$\delta(C\zeta)\big(\mathcal{A}\Phi(z)+\zeta(z)\big)\big(\mathcal{A}\Phi(w)+\zeta(w)\big). \tag{36}$$

Since $u = (-\Delta_{eff})^{-1}$ it follows that

$$\mathcal{N}_\Phi \cdot \int \mathcal{D}\Phi\ e^{(\Phi,-\Delta_{eff}\Phi)/2}\ \Phi(x)\ \Phi(y) = u(x,y)$$

for a proper choice of the normalization factor $\mathcal{N}_\Phi$.
As a result we get the following final result for $\langle\varphi(z)\varphi(w)\rangle \equiv v(z,w)$

$$v(z,w) = v_{low}(z,w) + \Gamma(z,w) \ , \tag{37}$$

with

$$v_{low} = \mathcal{A}\ u\ \mathcal{A}^* \quad , \quad u = -\Delta_{eff}^{-1} \ . \tag{38}$$

More explicitly, (37) reads

$$\Gamma(z,w) = v(z,w) - \int\int_{x,y\in\Lambda^1} \mathcal{A}(z,x)u(x,y)\mathcal{A}^\dagger(w,y) \tag{39}$$

$$v = -\Delta^{-1} \quad \text{and} \quad u = -\Delta_{eff}^{-1}. \tag{40}$$

Perturbation theory

We will now show how to compute effective actions by perturbation theory. To be specific, let us consider an action for a scalar field φ of the form

$$\mathcal{H}(\varphi) = \int_z \{-\frac{1}{2}\varphi(z)\Delta\varphi(z) + \mathcal{V}(\varphi(z))\} = (\mathcal{H}_{\text{free}} + V)(\varphi) \ . \tag{41}$$

with

$$\mathcal{V}(\xi) = \frac{1}{2}m_u\ \xi^2 + \frac{1}{4!}\ \lambda_u\ \xi^4 \ , \tag{42}$$

for instance. Introduce

$$V_\Phi(\zeta) = \int_z \mathcal{V}\big(\mathcal{A}\Phi(z) + \zeta(z)\big) \ . \tag{43}$$

This determines a self interaction of the high frequency field ζ which depends parametrically on the block spin Φ.

Proceeding as before one finds the following expression for the effective action

$$\mathcal{H}_{eff}(\Phi) = -\frac{1}{2}(\Phi, \Delta_{eff}\Phi) + \widehat{V}_{eff}(\Phi) \tag{44}$$

with

$$e^{-\widehat{V}_{eff}(\Phi)} = \int d\mu_\Gamma\ e^{-V_\Phi(\zeta)} \ . \tag{45}$$

We remember that $d\mu_\Gamma$ is the normalized free field measure with propagator Γ. The logarithm of the ζ-integral can be computed by standard perturbation theory. It is given by a sum of Feynman diagrams, whose lines represent propagators Γ, and whose vertices depend parametrically on Φ.

The resulting effective action $\mathcal{H}_{eff}$ will have good locality properties, provided the fluctuation field propagator $\Gamma(z,w)$ and the interpolation kernel $\mathcal{A}(x,z)$ decay exponentially with decay length a_0.

Perturbation expansions of the kind described here were first done by Kupiainen and Gawedzki [4,10], as part of a rigorous renormalization group analysis. Constraint effective potentials were also computed by another method in [11].

Constraint effective potential

We specialize the results of the previous subsection to the case of the constraint effective potential. In this case, Λ_0 has only one point x. Therefore the kernel of Δ_{eff} can only have (x, x) as its only pair of variables. We admit the possibility that Δ is the covariant Laplacian in an external gauge field. Of course, Δ_{eff} will also depend on U in this case. We abbreviate

$$\Delta_{eff}(x, x) = \epsilon_0 \ .$$

Eq. (44) simplifies to

$$\mathcal{H}_{eff}(\Phi) \equiv V_{eff}(\Phi) = \Phi \ \epsilon_0 \ \Phi + \widehat{V}_{eff}(\Phi) \ . \tag{46}$$

with $\widehat{V}_{eff}$ from (45).

In the presence of an external gauge field U, ϵ_0 depends on U. $\widehat{V}_{eff}$ will also depend on U, because the high frequency propagator Γ does, and the kernel $\mathcal{A}$ as well. In addition there is a gauge field dependence in the normalization factors of the Gaussian measures. This leads to a Φ-independent, but U-dependent term in $\widehat{V}_{eff}$ which will be ignored for now, but which has to be taken into account later on when we consider dynamical gauge fields.

Without gauge field, $\epsilon_0 = 0$. But in the presence of a gauge field which is not a pure gauge, $-\Delta$ is strictly positive, and therefore ϵ_0 is also strictly positive.

Let us emphasize that $\mathcal{H}_{eff}$ has no infrared problems because Γ has a built-in infrared cutoff.

1-loop approximation

In 1-loop approximation, the contribution $\widehat{V}_{eff}$ to the constraint effective potential (46) becomes [12,13,14]

$$\widehat{V}_{eff}(\Phi) = \int\int_{z,w} \mathcal{V}'(\mathcal{A}\Phi(z))\Gamma(z, w)\mathcal{V}'(\mathcal{A}\Phi(w)) + \frac{1}{2}\mathrm{Tr}\ \ln(-\Delta + \mathcal{V}''(\mathcal{A}\Phi)) \tag{47}$$

where a prime ($'$) indicates a derivative with respect to the argument, and $\mathcal{V}''$ is the multiplication operator defined by

$$(\mathcal{V}''(\mathcal{A}\Phi)f)(z) = \mathcal{V}''(\mathcal{A}\Phi(z))f(z) \ . \tag{48}$$

In graphical notation

$$\mathrm{Tr}\ \ln(-\Delta + \mathcal{V}''(\mathcal{A}\Phi)) = \sum_{n\geq 1} \frac{(-)^n}{n} \int_{z_1} \cdots \int_{z_n}$$

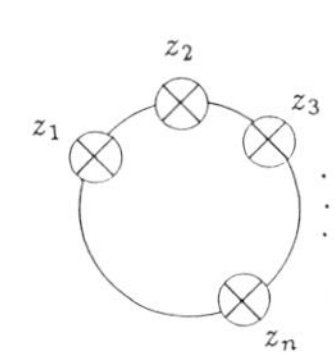

with Feynman rules

$$z \ \text{———} \ w = \Gamma(z, w) \quad , \qquad \otimes_z = \mathcal{V}''(\mathcal{A}\Phi(z)) \ .$$

The first term in expression (47) is not 1-particle irreducible. This term vanishes in the absence of gauge fields, and more generally it vanishes up to second order in the gauge coupling constant.

The contribution $\widehat{V}_{eff}$ to the full effective action (44) on a lattice Λ_0 of more than one point has the same form. The 1-particle reducible contribution is important in this case [10].

Use of the 1-loop formula to compute finite size corrections to the constraint effective potential

For definiteness, let us consider $\lambda\varphi^4$-theory.

From the constraint effective potential/volume in the infinite volume limit one can extract renormalized coupling constants, provided one knows the wave function renormalization constant from an independent calculation.

In fact, according to the standard renormalization of the $\lambda\varphi^4$-theory, one needs to define the three quantities m_r, λ_r and Z_φ (renormalized mass, self-coupling constant and wave function renormalization constant respectively). This can be achieved in the symmetric phase through the second and the fourth derivatives of the constraint effective potential at its minimum, viz.

$$m_r^2 = \frac{U''_{eff}(\Phi_{\min})}{Z_\varphi^{-1}} \tag{49}$$

and for the self-coupling constant

$$\lambda_r = \frac{U''''_{eff}(\Phi_{\min})}{Z_\varphi^{-2}}. \tag{50}$$

In the broken phase it is useful to define the renormalized square mass also through (49) but the self-coupling constant λ_r is now defined through the tree level relation:

$$\lambda_r = 3\,m_r^2\,/v_r^2 \tag{51}$$

where v_r is the vacuum expectation value of the field $\varphi_r = Z_\varphi^{-1/2}\varphi$ and $U_{eff} = V_{eff}/\text{volume}$.

One can obtain a renormalized loop expansion for the constraint effective potential by manipulating the bare expansion. Eqs. (49) - (51) determine the renormalized coupling constants as functions of the bare parameters. Inverting this relations, one obtains the bare quantities as a function of the renormalized ones. One may use them to express $\mathcal{V}$ in expression (47) (or its corresponding n-loop generalization) for the constraint effective potential as a function of the renormalized coupling constants. Reordering the resulting expansion in number of loops gives the desired expansion in terms of λ_r, m_r.

In practice, numerical computation of constraint effective potentials is only feasible on finite and not too large lattices.

The following strategy, which was first proposed by one of us (G.P.) in [12], eliminates the need to use very large lattices. It makes it possible to extract renormalized coupling constants from numerical data for values of the bare coupling constants which are very close to the critical line, and for lattice sizes which need not be large compared to the correlation length.

Triviality of $\lambda\varphi^4$-theory implies that the renormalized coupling constant λ_r is necessarily small when one is close to a critical point, provided the UV-cutoff is high enough, i.e. provided $a \ll$ correlation length.

Therefore, a perturbative calculation of the constraint effective potential on a *finite lattice* as a function of the renormalized coupling constants will yield accurate results.

In the practical application [12,13] it was verified that the two loop corrections are indeed very small.

The result of this computation is fitted to numerical data taken at various values of the bare coupling constants to extract the renormalized coupling constants as functions of the bare ones.

Some results of this kind are shown below. Numerical results of Meyer and Mack [15] are shown in Fig. 1 together with a fit by a renormalized 1-loop formula for the constraint effective potential. Numerical results for the constraint effective potential for 4-component ϕ^4-theory at stronger coupling but further away from the critical line were presented in [16].

The results are compared with analytical results of Lüscher and Weisz [17] in Table 1 .

Table 1. Renormalized parameters from the fits of Fig. 1 and the corresponding values obtained from Lüscher and Weisz (^{LW}).

	$m_r(\infty)$	$\lambda_r(\infty)$	m_r^{LW}	λ_r^{LW}
$m_0^2 = -1.14$	0.079 ± 0.003	7.5 ± 0.6	0.08	7.4 ± 0.2
$m_0^2 = -1.15$	0.06 ± 0.005	7.1 ± 0.1	0.06	7.2 ± 0.2
$m_0^2 = -1.16$	0.13 ± 0.008	7.8 ± 0.1	0.13	7.6 ± 0.2

Perfect agreement is found. In this way, the nonperturbative part of the analytical computation of Lüscher and Weisz is successfully tested.

5 Averaging and Interpolation Kernels C and $\mathcal{A}$ in the Presence of Gauge Fields, for Bosons

In the presence of a gauge field U, the definition of a blockspin Φ for a (n-component) scalar field φ and the split of this field into low frequency and high frequency part read

$$\begin{aligned} \Phi &= C(U)\varphi \, , \\ \varphi &= \mathcal{A}(U)\Phi + \zeta \, , \\ C(U)\mathcal{A}(U) &= \mathbb{1} \, . \end{aligned} \tag{52}$$

The operators C and $\mathcal{A}$ have kernels

$$\mathcal{A}(U)\Phi(z) = \int_{x\in\Lambda_0} \mathcal{A}(U|z,x)\, \Phi(x) \, , \tag{53}$$

$$C(U)\varphi(z) = \int_{z\in\Lambda} C(U|x,z)\, \varphi(z) \, . \tag{54}$$

As usual we denote points of the block lattice Λ^0 by x, y and points of the fundamental lattice Λ by z, w.

In the absence of gauge fields, a good choice of C is given by

$$C^*(z,x) \equiv C(x,z)^\dagger = \begin{cases} 1a_0^{-d} & \text{if } z \in x \\ 0 & \text{otherwise} \end{cases} .$$

(Here, † denotes the hermitian conjugate of a matrix.) This satisfies

$$-\Delta_{N,x} C^*(z,x) = \begin{cases} \epsilon_0(x)\ C^*(z,x) & \text{if } z \in x \\ 0 & \text{otherwise} \end{cases} . \tag{55}$$

$\Delta_{N,x}$ is the lattice version of the Laplacian with Neumann boundary conditions on the boundary of the block x, and $\epsilon_0(x) = 0$ is its lowest eigenvalue. In the presence of a gauge field, the Laplacian with Neumann boundary conditions takes the form

$$\Delta_{N,x} f(z) = a^{-2} \sum_{w\ nn.z, w\in x} [U(z,w)f(w) - f(z)] . \tag{56}$$

$U(z,w)$ is the lattice gauge field attached to the link from w to z, the summation runs over those nearest neighbours w of z which are inside the block x.

Proposal: [18] Retain the eigenvalue equation (55) for C in the presence of gauge fields, with covariant Laplacian $\Delta_{N,x}$ as defined in (56), and $\epsilon_0(x)$ equal to its lowest eigenvalue. Of course, both C and $\epsilon_0(x)$ will depend on the gauge field.

The interpolation kernel $\mathcal{A}$ may also be defined by the same equation as before. There should be $u(y,x)$ such that

$$-\Delta\mathcal{A}(z,x) = \int C^*(z,y)\ u(y,x) . \tag{57}$$

Here Δ is the gauge covariant Laplacian on the whole lattice. u is a Lagrange multiplier which is determined by the subsidiary condition (52).

In the special case when the block lattice Λ_0 consists of a single point x, the Neumann boundary conditions are inoperative, because the block x has no boundary when periodic boundary conditions for Λ are in force. Therefore $\Delta_{N,x} = \Delta$. The equation for $\mathcal{A}$ is therefore satisfied by

$$\mathcal{A} = C^* \quad \text{if } \Lambda_0 = \text{ single site} . \tag{58}$$

Let us next discuss the arbitrariness in the solution of (55). Consider for example the gauge group $G = SU(2)$. For fixed x, the 2×2 matrix $C(z,x)$ may be thought to be composed of two 2-vectors, each of which is a solution of the eigenvalue equation (55). (It can be proven that there exist two linearly independent solutions with the same eigenvalue.) Arbitrary linear combinations of these two solutions will also be solutions. This reflects itself in the arbitrariness

$$C^*(z,x) \mapsto C^*(z,x)\ \Omega(x) \tag{59}$$

of the 2×2-matrix solution C of (55), where $\Omega(x)$ are arbitrary 2×2-matrices. To eliminate this arbitrariness, we may impose subsidiary conditions

1. Normalization: $CC^* = \mathbb{1}$.
   This restricts the arbitrariness to unitary matrices $\Omega(x)$.

2. Gauge transformation property:
Select a distinguished site $\widehat{x}$ in the block x – for instance the center of the block – and demand that $C^*(\widehat{x}, x) > 0$ as a matrix. This makes the kernel C unique. It will transform under a gauge transformation [7] according to

$$C^*(z,x) \mapsto g(z)C^*(z,x)g(\widehat{x})^{-1} . \tag{60}$$

It follows that the blockspin $\Phi(x)$ transforms like a field sitting at the block center $\widehat{x}$,

$$\Phi(x) \mapsto g(\widehat{x})\Phi(x) . \tag{61}$$

Given C, the kernel $\mathcal{A}$ is also unique and it transforms under gauge transformations exactly like C^*.

There exists an efficient algorithm for computing the gauge field dependent kernels C and $\mathcal{A}$ [18].

Let us consider the implications for the effective potential in the presence of dynamical gauge fields, (14). Since the U-dependent kernel C will depend on the choice of $\widehat{x}$, the gauge covariant magnetization will also depend on $\widehat{x}$. But it is readily verified that the effective potential (14) is independent of the choice of $\widehat{x}$ [14]

6 Effective Potential for a Higgs Model with Dynamical Gauge Fields

Let us consider a Higgs model with dynamical gauge fields but without fermions, with a gauge group $SU(2)$ and a 2-component scalar field φ. Let

$$\Phi = \text{gauge covariant magnetization} = \int_z C(z)\varphi(z) , \tag{62}$$

defined with the help of the gauge covariant kernel $C(x,z)$ which was described in the last section. The argument x is redundant because there will be only one point x in Λ_0. Of course, C and Φ will depend on the gauge field U, and so does $\mathcal{A} = C^*$. By definition,

$$\begin{aligned} e^{-V_{eff}(\Phi)} &= \int \mathcal{D}U \; e^{-\widetilde{V}(U,\Phi)} \\ e^{-\widetilde{V}(U,\Phi)} &= \int \mathcal{D}\varphi \; \delta(\Phi - C\varphi) \; e^{-\mathcal{H}(U,\varphi)} . \end{aligned} \tag{63}$$

$\mathcal{H}$ is the standard action for a Higgs model. Using an appropriate lattice definition of the field strength tensor $F_{\mu\nu}$ [19], the Wilson action for the lattice gauge field can be written in standard $F_{\mu\nu}F_{\mu\nu}$-form, and

$$\begin{aligned} \mathcal{H} &= \int_z \left[\frac{1}{4} F_{\mu\nu}F^*_{\mu\nu} + \frac{1}{2}\langle\varphi, -\Delta\varphi\rangle + \mathcal{V}(\varphi(z)) \right] \\ \mathcal{V}(\varphi(z)) &= \frac{1}{2} m_u^2 \varphi^2(z) + \frac{1}{4!}\lambda_u \varphi^4(z) . \end{aligned} \tag{64}$$

[7] Remember that under a gauge transformation $\varphi(z) \mapsto g(z)\varphi(z)$ and $U(z,w) \mapsto g(z)U(z,w)g(w)^{-1}$. The covariant Laplacian $\Delta_{N,x}$ is gauge covariant. Therefore the eigenvalue equation (55) is also gauge covariant.

Δ is the gauge covariant Laplacian. It depends on U.

The functional integral will be performed in two steps. (For details of the loop expansion, see [14].) Calculations by a different method were reported in [20] for a $U(1)$-gauge field.)

φ integration

$\mathcal{D}\varphi\ \delta(\Phi - \int_z C(z)\varphi(z))$ is an unnormalized Gaussian measure with nonzero mean. First we shift the field to bring the mean to zero. We introduce a new integration variable ζ in place of φ,

$$\varphi(z) = \eta(z) + \zeta(z) \quad \text{with} \quad \eta(z) = \mathcal{A}(z)\ \Phi. \tag{65}$$

Since Λ_0 is a single point, the choice $\mathcal{A} = C^*$ is in agreement with previous definitions. The interpolation kernel satisfies therefore the eigenvalue equation

$$-\Delta\mathcal{A}(z) = \epsilon_0\ \mathcal{A}(z)\ , \tag{66}$$

$\epsilon_0 =$ lowest eigenvalue, $\Delta =$ covariant Laplacian on the lattice with periodic boundary conditions.

Using $\mathcal{A}^*\mathcal{A} = C\mathcal{A} = \mathbb{1}$, the kinetic term becomes

$$\langle\varphi,\ -\Delta\varphi\rangle = \Phi\epsilon_0\Phi + \langle\zeta, -\Delta\zeta\rangle \tag{67}$$

so that

$$e^{-\widetilde{V}(U,\Phi)} = e^{-\frac{1}{2}\Phi\epsilon_0(U)\Phi} \int \mathcal{D}\zeta\ \delta(\int_z C(z)\zeta(z))$$
$$\exp\left(-\frac{1}{2}\langle\zeta, -\Delta\zeta\rangle + \int_z \mathcal{V}(\eta(z) + \zeta(z)\right)\ . \tag{68}$$

The eigenvalue problem (66) can be solved by time independent perturbation theory which is familiar from elementary quantum mechanics. Set

$$U(z + e_\mu, z) = 1 - gA_\mu(z) + \dots\ . \tag{69}$$

To order g^2 one finds

$$\epsilon_0 = \frac{g^2}{8}\int_z |A_\mu(z)|^2 + \frac{g^2}{4}\sum_{\mu,\nu}\int_z\int_w A_\mu(z)\Gamma(z,w)A_\nu(w)\ . \tag{70}$$

The high frequency propagator Γ is only needed to zeroth order in g and the expression valid in a scalar theory without gauge fields can therefore be used for it at this point.

In conclusion

$$\int\left\{\frac{1}{4}F_{\mu\nu}F^*_{\mu\nu} + \frac{1}{2}\Phi\epsilon_0\Phi\right\} = \int\frac{1}{4}F_{\mu\nu}F^*_{\mu\nu} + \frac{1}{2}m^2(\Phi)\int|A_\mu|^2 + \dots. \tag{71}$$

with

$$m^2(\Phi) = \frac{1}{4}g^2|\Phi|^2\ . \tag{72}$$

We see that a Φ-dependent mass term for the gauge field has appeared.

In the full theory, the fluctuations of the random variable Φ tend to zero in the infinite volume limit (cp. J. Klauders lectures at this school [21]), and its value becomes fixed to the position $\widehat{\Phi}$ of an absolute minimum of $V_{eff}(\Phi)$. As a result

$$m^2(\widehat{\Phi}) = \text{W-meson mass} . \tag{73}$$

The integral

$$\int \mathcal{D}\zeta \exp\left(-\frac{1}{2}\langle\zeta, -\Delta\zeta\rangle + \int \mathcal{V}(\eta(z) + \zeta(z))\right) \tag{74}$$

is computed by expanding the exponent in the gauge coupling constant g and using the standard formula for a Gaussian integral

$$\int \mathcal{D}\zeta \exp\left\{-\sum_z \left\{\frac{1}{2}\langle\zeta|M_z\zeta\rangle + \langle\zeta|J_z\rangle\right\}\right\} =$$
$$(\det M)^{1/2} \exp\left(\frac{1}{2}\sum_z \langle J_z|M_z^{-1}J_z\rangle\right) . \tag{75}$$

Expressing it in terms of the *normalized* Gaussian measure $d\mu_\Gamma$ with covariance (= free propagator) Γ, the integral (74) equals

$$e^{-\frac{1}{2}\mathrm{Tr}_C \ \ln \Gamma} \int d\mu_\Gamma(\zeta)\ e^{-\int_z \mathcal{V}(\eta(z)+\zeta(z))} . \tag{76}$$

The high frequency propagator Γ depends on the gauge field, and so does $\mathcal{A} = C^*$ and therefore η, defined in (65). Γ was defined in (34) and

$$\mathrm{Tr}_C = \text{trace on the space of functions } f \text{ which obey } Cf = 0 .$$

With some effort, $\mathrm{Tr}_C \ \ln \Gamma$ can be evaluated to second order in g. It is a quadratic expression in the gauge fields which will add to the kinetic term of the gauge field.

Gauge field integration

Expanding everything to second order in the gauge coupling constant g, the integration over the gauge fields becomes a Gaussian integration which may be performed. Thanks to term $\Phi\epsilon_0\Phi$ in in (67) and also additional quadratic terms in the vector potential A_μ which resulted from the φ integration, the kinetic term for the gauge field is nondegenerate.

In fact, it turns out that the whole quadratic contribution in the gauge field A_μ in the large volume limit is equal to

$$S_g^q[A] = \frac{1}{2v}\sum_k \sum_{\mu\nu}$$
$$\delta_{rs}\left\{\delta_{\mu\nu}[\widehat{k}^2 + m^2(\Phi)] - \left(1 + \frac{m^2(\Phi)}{\widehat{k}^2}(\Phi)\Theta(\widehat{k}^2)\right)\widehat{k}_\mu\widehat{k}_\nu\right\}|\widetilde{A}_\mu(k)|^2 \tag{77}$$

where $\widetilde{A}_\mu(k)$ is the Fourier transform of the gauge field $A_\mu(z)$ and $\Theta(\widehat{k}^2)$ is equal to one for $\widehat{k}^2 \neq 0$ and the interpolation to continuous $\widehat{k}^2$ vanishes faster than $\widehat{k}^2$ for $\widehat{k}^2 \to 0$.

The quadratic operator of (77) *does* have an inverse (the propagator) and therefore no gauge fixing is necessary! The final result in finite volume is given in (120) in the Appendix.

Comparison with known results [22] for the conventional effective potential reveals that (120) simplifies in the infinite volume limit to

$$V_{eff}(\Phi)/\text{ volume } = \text{conventional effective potential in the Landau gauge.} \quad (78)$$

The conventional effective potential is the Legendre transform of the generating function for Greens functions, with a constant field as its argument. It has been computed before [22] on a finite lattice L^4 in the Landau gauge.

If one performs a formal continuum limit in addition to the infinite volume limit in order to replace sums over k by integrals, the expression (120) simplifies (after inserting counterterms into the action) to the following expression for the one-loop constraint effective potential $U_{eff} = V_{eff}/$ volume [14].

$$\begin{aligned} U_{eff}(\Phi) = U_{cl}(\Phi) + \tfrac{1}{64\pi^2} \Big\{ & (U''_{cl}(\Phi))^2 \left[\ln\left(U''_{cl}(\Phi)\right) - \tfrac{1}{2}\right] \\ & + 3\left(\tfrac{U'_{cl}}{\Phi}(\Phi)\right)^2 \left[\ln\left(\tfrac{U'_{cl}(\Phi)}{\Phi}\right) - \tfrac{1}{2}\right] + 9\mu_W^4(\Phi)\left[\ln\left(\mu_W^2(\Phi)\right) - \tfrac{1}{2}\right]\Big\} \end{aligned} \quad (79)$$

with

$$U_{cl}(\Phi) = \frac{1}{2}\, m_r^2\, \Phi^2 + \frac{1}{4!}\, \lambda_r\, \Phi^4 \quad \text{and} \quad m^2(\Phi) = \frac{1}{4}\, g_r^2\, \Phi^2. \quad (80)$$

The expression (79) is manifest gauge invariant and it agrees with the known result of Anna Hasenfratz and Peter Hasenfratz [23] for the conventional effective potential computed in the Landau gauge. Φ is the unrenormalized covariant magnetization, i.e. no wave function renormalization factor is included in its definition.

Small renormalized couplings $\lambda_r = O(g_r^4)$ are within the domain of validity of the perturbative result of (79). Its shape is shown in Fig. 2. Inspection of Fig. 2 shows that the absolute minimum jumps when the parameters are varied. Therefore there is a first order phase transition at weak couplings. This was noted before by Hasenfratz and Hasenfratz [23], following the pioneering work of Coleman and Weinberg [24] .

7 Blockspin for Fermions

As already discussed in T. DeGrand's lecture [25], the naive discretization of the continuum fermion action results in a lattice model with unphysical low energy modes, called "doublers". To reduce the number of flavours, i.e. the degeneracy of the lattice Dirac operator, one can use one-component "staggered" fields which live on different lattice sites. In d dimensions, a staggered fermion field describes physical fermions with $2^{[d/2]}$ flavours.

We shall first be concerned with blockspins of the form (12), i.e. we ignore gauge fields for now and consider only fermions.

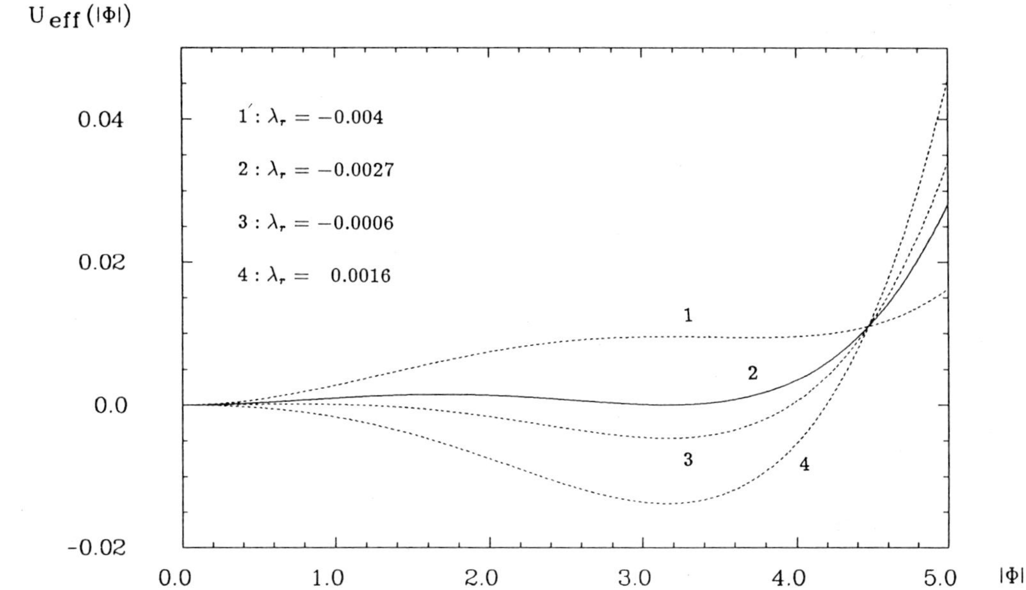

Fig. 2. The Constraint Effective Potential in the presence of gauge fields in the infinite and large cutoff limit. The gauge coupling was fixed to be $g_r^2 = 0.4$, the renormalized mass m_r^2 is given by $m_r^2 = -\frac{2}{3}\frac{\lambda_r}{g_r^2}$ and the scalar self coupling constant assumes values on both sides of the critical line (solid line). We observe that the minimum jumps from zero to Φ_{min} when the parameters are varied from the symmetric to the Higgs phase. This means that the system has a weak first order transition.

Free Staggered Fermions on the Lattice

For simplicity let us first look at free staggered fermions on a d-dimensional lattice $\Lambda_{a/2} = (\frac{1}{2}a\mathbb{Z})^d$ of lattice spacing $a/2$. Only translations by integer multiples $\mathbf{n}a$ of a are regarded as true lattice translations, i.e. survivors of translations in the continuum. These true translations map the sublattice $\Lambda_a = (a\mathbb{Z})^d$ of lattice spacing a into itself.

We imagine that the lattice $\Lambda_{a/2}$ is divided into hypercubes of sidelength a containing 2^d sites each. Sites within one hypercube are distinguished by their "pseudoflavour" H. H is specified by a set of distinct indices μ_i

$$H = \{\mu_1\mu_2\ldots\mu_h, 0 \le h < d, \mu_1 < \mu_2 < \ldots < \mu_h\} \quad . \tag{81}$$

Sites x_H in $\Lambda_{a/2}$ with the same pseudoflavour H are of the form

$$x_H \equiv \hat{x} + \frac{1}{2}e_H, \quad e_H = \sum_{\mu\in H} e_\mu, \quad \hat{x} \in \Lambda_a$$

where e_μ is the lattice vector of length a in μ-direction in Λ_a. The sites x_H form a sublattice Λ_a^H of $\Lambda_{a/2}$ (see Fig. 3), viz. $\Lambda_a^H = \Lambda_a + \frac{1}{2}e_H$.

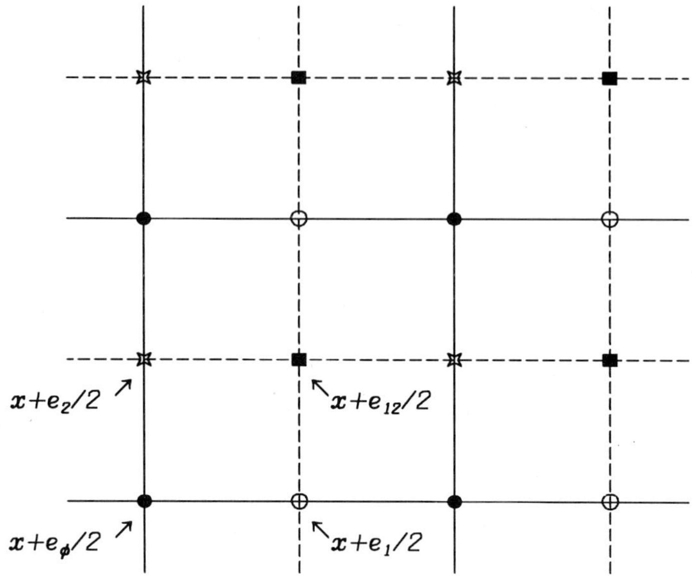

Fig. 3. Part of a pseudoflavour lattice Λ^H for $d = 2$, $H \in \{\emptyset, \{1\}, \{2\}, \{1,2\}\}$, embedded into $\Lambda = \mathbb{Z} \times \mathbb{Z}$.

Pseudoflavour is a combination of spin and flavour in the sense that there is an expansion

$$\sum_H \chi(x_H)\, dx^H = \sum_{\alpha,f} \Phi_\alpha{}^f(x)\; Z_\alpha{}^f \;. \tag{82}$$

which exhibits the field $\chi(x_H)$ at site x_H in a hypercube as a linear combination of field components $\Phi_\alpha{}^f$ with spinor index α and flavour index f ($Z = (Z_\alpha{}^f)$ is a matrix of differential forms). Each index takes values $1, \ldots, 2^{[d/2]}$. For details, see [26].

Blocking Consistent with the Symmetry

We distinguish between two kinds of blockings: from the continuum to some lattice, and from one lattice to a coarser lattice. We wish to block in such a way that as much is preserved of internal and space time symmetries as is possible.

Blocking from the Continuum to the Lattice. Given Fermi fields $\Psi_\alpha^f(z)$ on the continuum with $2^{[d/2]}$ flavours f, let $\chi^H(z)$ be the linear combination of field components Ψ_α^f of given pseudoflavour. The staggered blockspin on the lattice $\Lambda_{a/2}$ has the form

$$\Xi(x) = \int_z C^H(x,z)\; \chi^H(z) \quad \text{if } x \in \Lambda_a^H \subset \Lambda_{a/2} \quad . \tag{83}$$

Here $\int_z$ is an integral over continuous space.

In the absence of gauge fields the kernel of the averaging operator is given by

$$(\frac{a}{2})^d \; C^H(x,z) = \begin{cases} 1 & \text{if } x \in \Lambda_a^H, \max\ |x^\mu - z^\mu| \leq \frac{a}{2} \\ 0 & \text{otherwise} \end{cases} . \tag{84}$$

i.e. one averages over a cube of sidelength a with center x. Note that the cubes around block lattice sites x, y may overlap if x, y have different pseudoflavour (see below). It can be verified that this choice of blockspin leads to good locality properties of the effective free action on the lattice [27].

We proceed with considerations on the staggered fermion symmetry group.

Residual Symmetry on the Lattice. For bosons, any flavour symmetry G in the continuum is preserved on the lattice; continuous space time translations get broken to lattice translations and similarly for Euclidean rotations.

For staggered fermions, the situation is different [28]: blocking as described above from the continuum, the total symmetry $\mathcal{G} = G \times \mathcal{T}$ (e.g. flavor group $G = SU(2)$ for $d = 4$, translation group $\mathcal{T}$) gets broken to a subgroup $\mathcal{G}_a$ on a lattice of lattice spacing $a/2$. $\mathcal{G}_a$ is generated by translations e_μ by integer multiples of a and by certain combinations εd^H of flavour transformations ($\varepsilon = \pm 1$), and translations by $e_H/2$. The action T_a of the lattice symmetry group on a staggered field on the lattice $\Lambda_{a/2}$ reads as follows

$$(T_a(\varepsilon d^K)\chi)(y) = \varepsilon \; \check{\rho}_{H,K} \; \chi(y + \frac{1}{2}e_K) \quad \text{if } y \in \Lambda_a^H \; . \tag{85}$$

The symbols $\check{\rho}_{H,K}$ denote sign factors as in the Clifford product of forms $\vee$, defined by

$$dx^H \vee dx^K = dx^\mu \wedge dx^\nu + \delta^{\mu\nu} \qquad \text{and} \qquad dx^H \vee dx^K = \check{\rho}_{H,K} dx^{H \Delta K} \; ,$$

with the symmetric difference $H \Delta K = (H \cup K) \backslash (H \cap K)$ of sets H, K of pseudoflavour indices.

This agrees with the known symmetry group of standard staggered fermion actions [29]. Note that translations by e_H are true translations.

Fine-to-Coarse Blocking. The fine-to-coarse blocking step maps fields which live on the fine lattice $\Lambda_{a/2}$ onto fields living on the coarser lattice $\Lambda_{L_b a/2}$ of lattice spacing $L_b a/2$, with an integer scale factor $L_b > 1$. It is natural to demand that the symmetry group $\mathcal{G}_{L_b a}$ on the coarse lattice should be a subgroup of the symmetry group $\mathcal{G}_a$ on the finer lattice

$$\mathcal{G}_{L_b a} \subset \mathcal{G}_a \; . \tag{86}$$

In addition to the true translations by $e'_H = L_b e_H$ there will be combinations [8]

$$\varepsilon d'_H \in \mathcal{G}_{L_b a}$$

of flavor transformations and translations by $\frac{1}{2}e'_H$.

[8] A prime ($'$) refers to the coarser lattice.

The symmetry operations d'_H must be a combination of some d_K and a true translation e. From (85), it can be seen that d'_K contains an admixture of translation by $-e'_K/2$. Since d_K contains an admixture of a translation by $\frac{1}{2}e_K$, we must have

$$-\frac{1}{2}e'_H = -\frac{1}{2}e_K + e \quad \text{with} \quad e = \sum_\mu n_\mu e_\mu \ . \tag{87}$$

Since $e'_H = L_b e_H$, the only way how this can happen is that $H = K$ and $e = \frac{1}{2}(L_b - 1)e_H$ = *true translation*. This requires that L_b is odd, hence $L_b = 3, 5, \ldots$ but *not* $L_b = 2$!

This result might have been expected, since only for odd L_b the pseudoflavour (and parity) on the fine lattice $\Lambda_{a/2}$ and on the block lattice $\Lambda_{L_b a/2}$ match nicely when we regard $\Lambda_{L_b a/2}$ as sublattice of $\Lambda_{a/2}$ (see Fig. 4 below).

For a true symmetry it should not matter in which order we apply the blocking step and the symmetry operations: the blocking map C should commute with the action T of the symmetry under $\mathcal{G}_{L_b a} \subset \mathcal{G}_a$, viz.

$$T_{La}(d'_H)\, C(x,z) = C(x,z)\, T_a(d_H)\, T_a(\frac{1}{2}\,(L_b - 1)\, e_H) \ . \tag{88}$$

This is fulfilled if C is translation invariant so that

$$C(x + e'_H, z + L_b e_H) = C(x,z) \tag{89}$$

and if in addition C has the support property

$$C(x,z) = 0 \qquad \text{unless } x, z \text{ carry the same pseudoflavour.} \tag{90}$$

In words, the support property states the following. If $y \in \Lambda^H_{L_b a/2}$ then the (staggered) blockspin $\Xi(y)$ is a weighted sum of fields $\chi(z)$ at sites with the same pseudoflavour H. Blockings with such a property were considered for 2-dimensional $U(1)$ gauge fields by Ben Av et al. [30]. However, they chose a scale factor $L_b = 2$.

Generalization to Nontrivial Gauge Fields

We now admit dynamical gauge fields in our staggered fermion theory; the averaging kernel C will then have to depend on the gauge field for reasons of covariance as in (13). Our interest will gradually shift towards a practical implementation of the renormalization group picture outlined in Sect. 3, known as the *multigrid approach* to numerical computations in lattice field theories [9].

To specify a blocking prescription, we retain blocks as introduced in Sect. 5: the solution of the gauge covariant eigenvalue equation

$$-\Delta_{b.c.} C^*(z,x) = \lambda_0(x) C^*(z,x) \tag{91}$$

yields a gauge covariant kernel C. Results reported below show that (91) produces a good blockspin when $\Delta_{b.c.}$ is the Laplace operator with appropriate boundary conditions (basically Neumann b.c.). The covariant Laplace operator connects only lattice

[9] Those readers who are not familiar with multigrid methods may first want to read Sect. 8; a more detailed mathematical account can be found in [31].

sites of the same pseudoflavour; therefore (91) is consistent with the condition (90) above.

In a multigrid approach, one needs interpolating kernels $\mathcal{A}$ from coarse to fine lattices in addition to the averaging kernels C. The kernels $\mathcal{A}$ should be smooth. In the present context, there are a priori two possible definitions of smoothness.

1. "Laplacian" smoothness. χ possesses Laplacian smoothness if

$$(\nabla_\mu \chi, \nabla_\mu \chi) \equiv (\chi, -\Delta\chi)$$

is small, with the ***covariant Laplacian*** Δ: for our model of staggered fermions this is expected to work if the field strength tensor $F_{\mu\nu}$, conveniently defined on the lattice [19] by

$$F_{\mu\nu}(z) = U(z + \frac{1}{2}e_\mu, \nu)\, U(z,\mu) - U(z + \frac{1}{2}e_\nu, \mu)\, U(z,\nu)\ . \tag{92}$$

is small. This is the standard definition of smoothness.

The problem is that there are no smooth functions in this sense if the gauge fields are disordered, because Δ is strictly positive in this case.

2. "Diracian" smoothness. χ possesses Diracian smoothness if

$$(\not\nabla\chi, \not\nabla\chi) \equiv (\chi, -\not\nabla^2\chi)$$

is small. Since

$$-\not\nabla^2 = -\Delta + \sigma_{\mu\nu}F_{\mu\nu}\ , \tag{93}$$

sites of different pseudoflavour are connected.

When one wants to compute propagators of fermions by use of the multigrid approach, then one should in principle use interpolating kernels $\mathcal{A}$ which possess Diracian smoothness. Unfortunately this is impractical for reasons of storage space. Laplacian smooth kernels $\mathcal{A}$ are much more practical. Possible compromises are discussed in [26].

In Sect. 9 the problem of smoothness of the interpolation kernel $\mathcal{A}$ (see Sect. 5) will be readdressed using optimization wisdom borrowed from neural computation.

Special Choice which yields a Good Blockspin

In the language of Sect. 5, we know that we have made a good choice of blockspin if the effective action remains *local* after arbitrarily many or arbitrarily big blocking steps. The averaging kernel $C(x,z)$ which is nonvanishing only if z and x have the same pseudoflavour, and which satisfies the Laplace equation (90) with Neumann boundary conditions on block boundaries, has proven to be good in that sense. This result extends to blocking from continuum action with $2^{[d/2]}$ flavours to staggered fermion field on the lattice. Details will be given in the next section. Fig. 4 visualizes the overlapping of blocks which do not share sites of the same pseudoflavour as the block centers. In numerical investigations [32] non-overlapping blocks were used as well. But then one does not retain the successful pure gauge limit.

A successful block spin requires $2^{[d/2]}$ flavors. There is no consistent way of putting a single chiral particle on the lattice (this is the statement of the theorem by Nielsen and Ninomiya [5]): Therefore there is no good blockspin either for a single chiral field, as we had already remarked in Sect. 1.

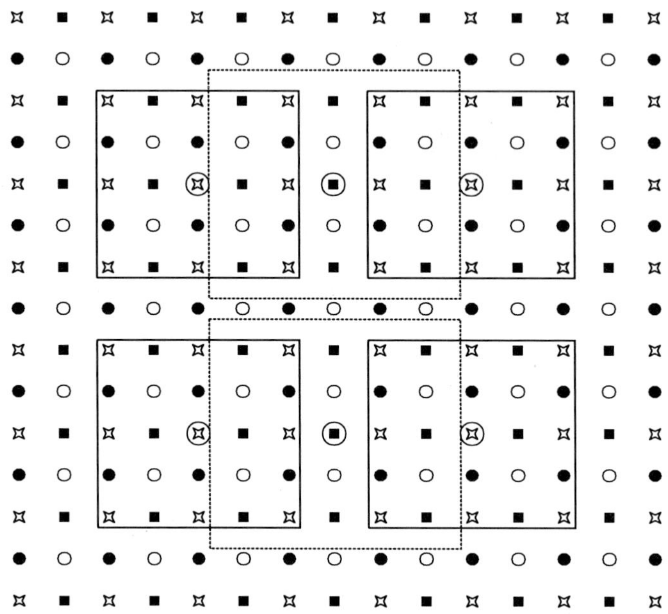

Fig. 4. The good choice of blocks to average over. $C(z,x)$ is only nonvanishing $\hat{x}$ (encircled symbol). Therefore the seemingly overlapping blocks (square) have actually no sites in common.

8 Efficient Computation of Gauge Covariant Propagators

A quenched simulation of lattice QCD requires the evaluation of products of Dirac propagators $S(U)$. S depends on the actual gauge field configuration U. In a Monte Carlo simulation of full QCD, one needs also the fermion determinant

$$\det{}^{-1/2} S(U) = \mathcal{N}^{-1} \int \mathcal{D}\phi \, \exp\left[-(\phi, S(U)\phi)\right] \ .$$

Since for every new gauge field configuration U, $S(U)$ needs to be recomputed or updated, it is important to have an efficient method to do so. The conjugate gradient (CG) or the minimal residual algorithm is state of the art. Great hopes to do better are attached to multigrid (MG) methods [30,32,33,34,35,36,37,38].

Deterministic Multigrid Methods

For convenience, we will use the opposite convention in enumerating the lattices of a multigrid in this section. The fundamental (i.e. finest) lattice is denoted by $\Lambda = \Lambda^0$. There will be a sequence of lattices Λ^0, Λ^1, Λ^2, ... of increasing lattice spacing a_i, viz. $a_{i+1} = L_b a_i$ with $a_0 = a$. We wish to solve an inhomogeneous linear equation

$$D_0 \chi_0 = f_0 \tag{94}$$

on the fundamental lattice $\Lambda = \Lambda^0$, for given f_0. In our case, $D_0 = -\nabla^2 + m^2$. After some relaxation sweeps on Λ^0 one gets an approximation $\widetilde{\chi}_0$ to χ_0 which differs from the exact solution by an error $e_0 = \chi_0 - \widetilde{\chi}_0$. The fundamental idea of the MG to the solution of partial differential (or difference) equations [31,39] is that the error e_0 should become smooth very fast, although it may become small very slowly because of critical slowing down (CSD). The error satisfies the equation

$$D_0 e_0 = r_0 \tag{95}$$

which involves the residual $r_0 = f_0 - D_0\widetilde{\chi}_0$. If e_0 is smooth, it is determined to a very good accuracy by a function e_1 on the next coarser lattice Λ^1, and can be represented in the form

$$e_0 = \mathcal{A} e_1 \tag{96}$$

with an interpolation map $\mathcal{A}$ which should be so chosen that it maps functions on Λ^1 into smooth functions on Λ^0. Conversely, e_1 can be obtained from e_0 with the help of an averaging map C which satisfies

$$C\mathcal{A} = \mathbb{1} \ . \tag{97}$$

It follows that $e_1 = Ce_0$. Inserting (96) into (95) and acting on the result with C, we see that e_1 will satisfy the equation

$$D_1 e_1 = r_1 \tag{98}$$

with

$$D_1 = CD_0\mathcal{A} \quad , \quad r_1 = Cr_0 \ . \tag{99}$$

The problem has been reduced to an equation on the coarser lattice. If there is still too much CSD at this level, one may repeat the procedure, going to coarser and coarser lattices.

Given $\mathcal{A}$, a possible choice of C which satisfies (97) is $C = (\mathcal{A}^*\mathcal{A})^{-1}\mathcal{A}^*$. The kernel $\mathcal{A}(z,x)$ should be a smooth function of z. The appropriate notion of smoothness depends on the dynamics, i. e. on D_0, in general. *Smooth means little contributions from eigenfunctions to high eigenvalues of D_0.* This insight is confirmed by results reported below. The point is important in systems in gauge fields and for other disordered systems.

Gauge Covariance

In a gauge theory, kernels $\mathcal{A}$ and C should be chosen in a gauge covariant fashion; see the discussion in Sect. 5. Under gauge transformations on Λ^0 they transform according to

$$\begin{aligned} \mathcal{A}(z,x) &\mapsto g(z)\mathcal{A}(z,x)g(\widehat{x})^{-1} \ , \\ C(x,z) &\mapsto g(\widehat{x})C(x,z)g(z)^{-1} \ . \end{aligned} \tag{100}$$

This is consistent with $C\mathcal{A} = \mathbb{1}$.

The most general expression of a kernel $\mathcal{A}$ with this covariance property is a weighted sum of parallel transporters $U(\mathcal{C})$ along paths $\mathcal{C}$ from $\widehat{x}$ to z, e. g.

$$\mathcal{A}(z,x) = \sum_{\mathcal{C}:\widehat{x}\mapsto z} \varrho(\mathcal{C})\, U(\mathcal{C}) \ , \tag{101}$$

where $\varrho(\mathcal{C})$ are numbers. And analogously for C. We will prefer not to specify the weights ϱ explicitly, but to determine $\mathcal{A}$, C as solutions of covariant equations.

Ground-State Projection Multigrid

In the ground-state projection MG method the averaging operator C from a grid to the next coarser grid is a projector on the ground-state of a local Hamiltonian. The adjoint of C satisfies a gauge covariant eigenvalue equation (cf. (55), (91)),

$$(-\Delta_{\mathrm{b.c.}} C^*)(z,x) = \epsilon_0(x) C^*(z,x) \ . \tag{102}$$

Remember that the kernel $C^*(z,x)$ which solves (102) is an $N_c \times N_c$ matrix which is in general not an element of the gauge group. (N_c is the number of colours.)

For bosons one may choose Neumann boundary conditions on block boundaries, $-\Delta_{\mathrm{b.c.}} = -\Delta_{N,x}$, as discussed in Sect. 5, (55). The lowest eigenvalue $\epsilon_0(x)$ of $-\Delta_{\mathrm{b.c.}}$ is a gauge invariant quantity which is a measure of disorder. A normalization condition $CC^* = \mathbb{1}$ may be imposed, and the covariance condition $C^*(\widehat{x},x) > 0$ (as a matrix) makes the definition of C unique. Kernels defined by (102) and $C^*(\widehat{x},x) > 0$ enjoy gauge covariance (100).

For staggered fermions, the Laplacian should be regarded as living on the sublattices which consist of sites with the same pseudoflavour. It connects only sites with the same pseudoflavour. Neumann boundary conditions on the Laplacian on the sublattice of sites z with the same pseudoflavour as x must be supplemented by the requirement that $C(x,z) = 0$ unless z and x have the same pseudoflavour. This extra requirement could be regarded as a consequence of supplementary boundary conditions.

There exists an efficient algorithm for computing C [18]. The programs are fully vectorized. In 4-dimensional $SU(2)$ gauge fields, computation of C on the whole lattice costs CPU time of the order of one heatbath sweep for the gauge field. Thus, it is not too expensive to solve (102), and it is not necessary to sacrifice gauge covariance by working with gauge field independent kernels in gauge-fixed U-configurations. (Gauge fixing costs also CPU time.)

Interpolation Kernel $\mathcal{A}$

Given the averaging kernel C, there exists an ideal choice of the interpolation kernel $\mathcal{A}$. It is determined as follows. For every function ("blockspin") Φ on Λ^1, $\phi = \mathcal{A}\Phi$ minimizes the action $\mathcal{H} = (\phi, D_0\phi)$ subject to the constraint $C\phi = \Phi$. Recalling the definition of smoothness, this can be rephrased: $\phi = \mathcal{A}\Phi$ is the smoothest possible function with prescribed block average $\Phi = C\phi$. It follows that $\mathcal{A}$ satisfies (106) below. With this choice of $\mathcal{A}$, the action $\mathcal{H}$ completely decouples into a sum of actions for the different MG layers. Moreover, $D_1 = (CC^*)\mathcal{A}^* D_0 \mathcal{A}$; this is selfadjoint for our C-kernels which satisfy $CC^* = \mathbb{1}$.

A good "choice of blockspin", i.e. of C, is characterized by the fact that the ideal kernel $\mathcal{A}(z,x)$ ($z \in \Lambda^0$, $x \in \Lambda^1$) associated with it has good locality properties. This means that $\mathcal{A}(z,x)$ is big for $z \in x$, and decays exponentially in $|z - \hat{x}|$ with decay length 1 block lattice spacing a_1 (cf. Sect. 4). Numerical computations of the ideal kernel $\mathcal{A}$ were done in quenched 4-dimensional $SU(2)$ gauge fields at various values of $\beta = 4/g^2$, including the case of a completely random gauge field ($\beta = 0$). (For bosons $D_0 = -\Delta + m^2$, while for fermions $D_0 = -\not{\nabla}^2 + m^2$ with $m^2 = 0$ or small). It was verified that for *any* β-value the definition of the averaging kernel C as a solution of a gauge covariant Laplace eigenvalue equation (55), (91) (with Neumann boundary conditions on block boundaries) yields a good choice of blockspin in the sense described above. This is true both for bosons and for staggered fermions; see Figs. 5, 8. $\mathcal{A}$ decays exponentially over distance a_1 (nearly) as fast in the presence of gauge fields as in their absence.

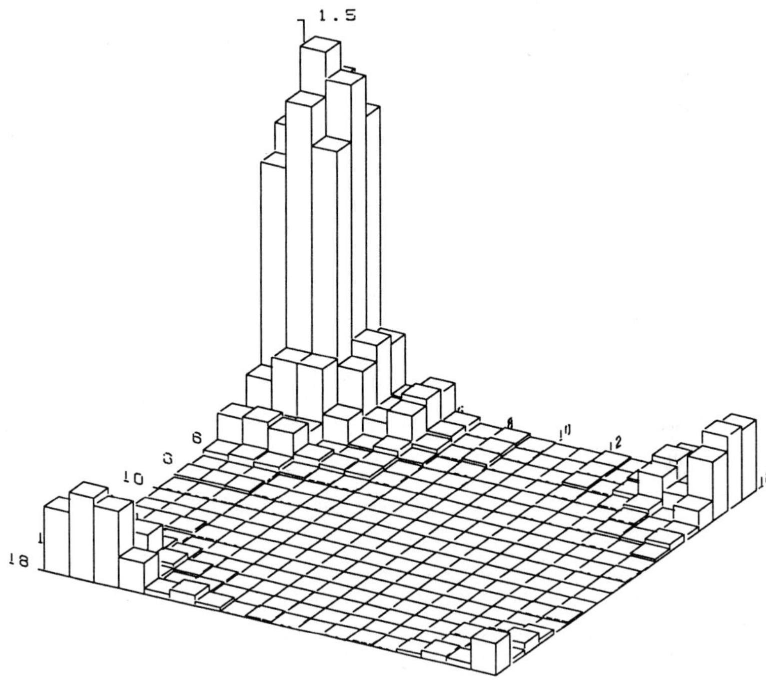

Fig. 5. Optimal interpolation kernel $\mathcal{A}(z,y)$ for bosons on an 18^4 lattice in a quenched $SU(2)$ gauge field equilibrated at $\beta = 2.7$. ($L_b = 3$, periodic b. c.) Shown is a two-dimensional cut through the block center $y = 0$, z_3 and z_4 are fixed. The vertical axis gives the trace norm of $\mathcal{A}(z,y)$. (Remember that $\mathcal{A}(z,y) \in \mathbb{R} \cdot SU(2)$.)

Because of the exponential decay of the optimal kernel $\mathcal{A}$, one may try to approximate it by a simpler choice which fulfills $\mathcal{A}(z,x) = 0$ for $z \notin$ a certain neighborhood of block x. This neighborhood might be taken to be only the block x itself. The coarse grid operator would then be defined through $D_1 = \mathcal{A}^* D_0 \mathcal{A}$.

It follows from the optimization criteria discussed below that $\mathcal{A}$ should have the property that the lowest eigenmode ϕ_0 of D_0 should admit a representation of the form $\phi_0 = \mathcal{A}\Phi$ with a suitable Φ to a very good accuracy. The proposal of [18] to define $\mathcal{A}$ through the eigenvalue equation (102) with $-\Delta_{\text{b.c.}}$ taken to be a local Hamiltonian with Dirichlet instead of Neumann boundary conditions does not meet this requirement.

For periodic boundary conditions the Galerkin choice (also called variational coarsening) $\mathcal{A} = C^*$ is better. In this case D_1 can be defined as $D_1 = \delta^{-1}CD_0C^*$ [38]. The fluctuating length of its matrix elements can be adjusted to fluctuate around the "right" value (of the exact $D_1(x,y)$) by tuning the real parameter δ.

Numerical Results for Propagators Computed by MG

Ground-state projection MG computations for propagators were done in 2-*d* $U(1)$ [34,35] and (approximate ground-state projections) in 2-*d* and 4-*d* $U(1)$ and 2-*d* $SU(2)$ gauge fields [34,36,32]. The first ground-state projection MG computation of gauge covariant propagators in 4-*d* non-Abelian gauge fields (for gauge group $SU(2)$) was presented in Ref. [38]. Besides ground-state projection there is another approach for choosing the averaging kernel C. This approach is called "parallel transported MG". It is based on the representation (101) and is studied in Refs. [30].

Bosonic Propagators in 4-*d* $SU(2)$ Gauge Fields. For bosons, $D_0 = -\Delta + m^2$. The MG method is of interest near criticality, i. e. for slowly decaying propagators. For nontrivial gauge fields we enforce slow decay by choosing m^2 negative and very close to the negative of the lowest eigenvalue $-m^2_{\text{cr}} > 0$ of $-\Delta$.[10] The relaxation time τ – defined by the *asymptotic* exponential decay of the norm of the residual or error – behaves like $\tau \propto (\Delta m^2)^{-z/2}$ for small $\Delta m^2 = m^2 - m^2_{\text{cr}}$.

Results in 4-dimensional $SU(2)$ gauge fields equilibrated with Wilson's action at various values of β [38]:

- The MG algorithm with the ideal interpolation kernel $\mathcal{A}$ eliminates CSD for *any* value of the gauge coupling; see Table 2. These computations were the first without CSD in non-trivial gauge fields. They prove that *ground-state projection is a good choice of* C, and that *the MG method can cope with the "frustration" (disorder) which is inherent in non-Abelian gauge fields.* However, the optimal $\mathcal{A}(z,y)$ is not translational invariant (except for $U \equiv \mathbb{1}$) and has support on all sites z of the fine lattice, for all sites y of the block lattice. Therefore the use of this kernel for production runs is impractical, but it was important to answer questions of principle. In particular this result confirms that there exists an appropriate notion of smoothness in disordered cases, as described above. The ideal $\mathcal{A}$ is the smoothest kernel which obeys $C\mathcal{A} = \mathbb{1}$.

[10] Relaxation times of conventional iterative algorithms depend only on m^2 and *not* on the lattice size $|\Lambda|$; this is contrarily to the Dirichlet problem, for instance. In case of periodic b. c. there is only an implicit (slight) dependence on $|\Lambda|$ through the value of m^2_{cr}.

- The most practical MG algorithm with $\mathcal{A} = C^*$ does not eliminate CSD in non-trivial gauge fields, $z = 2$ remains (as in 1-grid relaxation), although $z = 0$ in pure gauges; see Fig. 6. However, MG is able to outperform the conjugate gradient algorithm; see Fig. 7.
- After the publication of Ref. [38], a practical nonlinear MG algorithm has been developed which eliminates the appearance of CSD when the mass is lowered, some volume dependence of the correlation times still remains, though. Details will be reported elsewhere [40].

Table 2. Results of the idealized MG algorithm with lexicographic SOR (relaxation parameter ω) on 9^4 lattices Λ^0. Kernel $\mathcal{A}$ is the solution of the equation $\left([-\Delta + m_{\text{cr}}^2 + \kappa C^* C]\mathcal{A}\right)(z,y) = \kappa C^*(z,y)$ for large κ. This expression tends to the ideal $\mathcal{A}$-kernel described in the text for large κ.

β	m_{cr}^2	optimal ω	τ for $\Delta m^2 \leq 10^{-3}$
∞	0	1.27	1.6
2.7	−0.8210607	1.38	1.9
2.5	−0.9477085	1.40	1.9
2.2	−1.2218471	1.45	1.9
1.8	−1.7567164	1.57	2.5
0	−2.7480401	1.69	5.2

Propagators for Staggered Fermions in 4-*d* $SU(2)$ Gauge Fields. Of the two different proposals for the averaging kernel C made in [26], only the Laplacian choice has been implemented numerically yet. There, $-\Delta_{\text{b.c.}} = -\Delta_{N,x}$ in (102), where Δ is the fermionic 2-link Laplacian defined in (93). The optimal interpolation kernel $\mathcal{A}$ associated with this C has the desired falloff properties in non-trivial gauge fields; see Fig. 8. Thus, the Laplacian choice for C defines a good blockspin for staggered fermions, and therefore it seems not to be necessary to implement the more intricate proposals mentioned in [26]. However, numerical simulations with the optimal $\mathcal{A}$ have not been performed yet.

MG with variational coarsening $\mathcal{A} = C^*$ works, but up to now we had to restrict ourselves to relatively small lattices (up to 18^4) where this MG algorithm is not competitive with conjugate gradient. The authors of Refs. [30] report that their algorithm performs considerably better when they re-unitarize the blocked gauge field (matrix elements of the effective Dirac operator) such that it is again an element of the gauge group. However, our results indicate that the difference between using a unitary or a dielectric gauge field [19] on coarser layers can be compensated by adjusting the real parameter δ in D_1. By tuning δ, one can re-unitarize the blocked gauge field "on the average".

We are optimistic that the implementation of the nonlinear MG algorithm found for bosons (see above) also gives a practical and competitive MG method for fermions. This implementation is currently investigated. For details refer to [40].

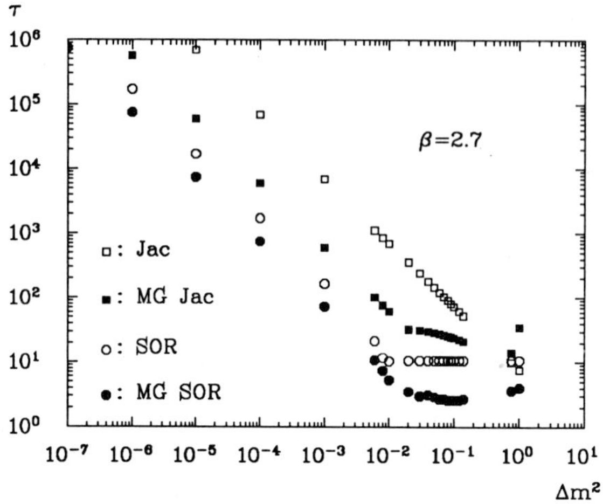

Fig. 6. Computation of bosonic propagators $(-\Delta + m_{cr}^2 + \Delta m^2)^{-1}$. Relaxation times τ (in comparable work units) of iterative algorithms on an 18^4 lattice in a quenched $SU(2)$ gauge field at $\beta = 2.7$ with $m_{\rm cr}^2 = -0.7554339$. (MG with $\mathcal{A} = C^*$; Jac means Jacobi relaxation, while SOR stands for successive over-relaxation.)

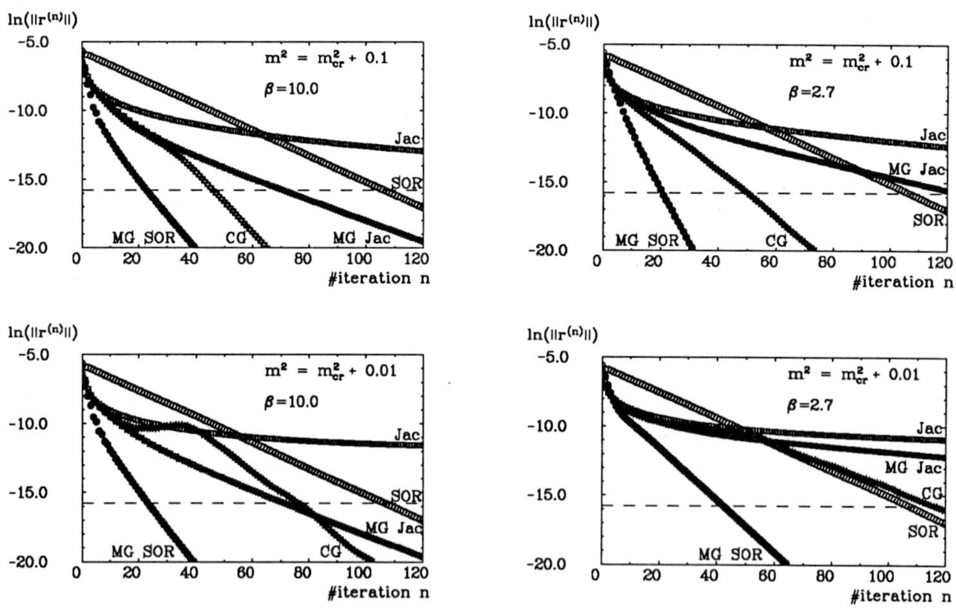

Fig. 7. Computation of bosonic propagators $(-\Delta+m_{cr}^2+\Delta m^2)^{-1}$. Convergence on an 18^4 lattice in quenched $SU(2)$ gauge fields at (a) $\beta = 10.0$, and (b) $\beta = 2.7$, with $m_{\rm cr}^2 = -0.1533739$, resp. -0.7554339. (MG with $\mathcal{A} = C^*$.)

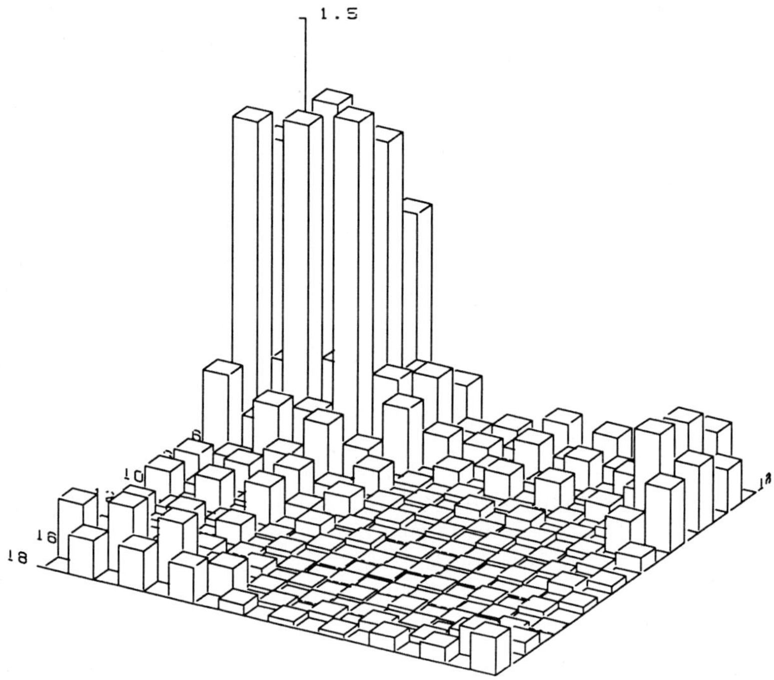

Fig. 8. Optimal interpolation kernel $\mathcal{A}(z,y)$ – associated with the "Laplacian" choice for C – for staggered fermions on an 18^4 lattice in a quenched $SU(2)$ gauge field equilibrated at $\beta = 2.7$. ($L_b = 3$, periodic b. c.) $\mathcal{A}$ is the solution of the equation $\left([-\nabla\!\!\!/^2 + \kappa\, C^* C]\mathcal{A}\right)(z,y) = \kappa\, C^*(z,y)$ for large κ. Shown is a two-dimensional cut through the block center $y = 0$, z_3 and z_4 are fixed. The vertical axis gives the trace norm of $\mathcal{A}(z,y)$. (Remember that $\mathcal{A}(z,y) \in \mathbb{R} \cdot SU(2)$.) Note the support properties of $\mathcal{A}$, it has support on all even lattice sites and is vanishing on all odd sites. This is due to the non-vanishing field strength at finite β. In the limiting case of pure gauges the field strength term $\sigma_{\mu\nu}F_{\mu\nu}$ vanishes and the support of $\mathcal{A}$ is reduced to 1/16-th of $|\Lambda|$. Then we have 16 bosonic $\mathcal{A}$-kernels on 9^4 sublattices.

9 Neural Optimization of Deterministic Multigrids

The results of Sect. 8 confirm that the multigrid approach is able to handle inhomogeneous wave equations in disordered media, e.g. in the presence of gauge fields, but the appropriate notion of smoothness depends on the dynamics, i.e. on the operator D_0 in the equation which one wishes to solve. The interpolation kernels $\mathcal{A}$ should be smooth. Smooth means little contributions from eigenfunctions to high eigenvalues of D_0.

This means that the appropriate kernels $\mathcal{A}$ are dynamically determined and need to be found before or together with the solution of the equation. (The averaging kernel C is often related to $\mathcal{A}$.) The question is how. There are interesting potential applications other than gauge theories [11]. Therefore a general strategy is desired.

[11] For instance, one may admit link variables $U(z,w)$ in the covariant Laplacian Δ which

Criteria for optimality

One may write down a cost functional whose minimum yields the optimal kernels, i.e. the fastest possible convergence of the multigrid iteration.

For definiteness and simplicity, think of a twogrid. Assume that updating consists of standard damped relaxation sweeps on the fundamental lattice to smoothen the errors, followed by an updating step $\chi_0 \mapsto \chi_0 - \mathcal{A}e_1$ which involves the exact solution e_1 of (98) on the coarse grid Λ_1, with some D_1 which is to be determined together with $\mathcal{A}$ and C.

Convergence speed is estimated by the norm of the iteration matrix ρ. The iteration matrix is the product $\rho = \rho_0\rho_1$ of the iteration matrix ρ_0 for the smoothing, and the iteration matrix ρ_1 for the coarse grid updating. The norm

$$\|\rho\| \leq \|D_0\rho_0\| \cdot \|D_0^{-1}\rho_1\| \ .$$

Only the second factor depends on $\mathcal{A}$, C and D_1. Inclusion of the operators $D_0^{\pm 1}$ in the factors reflects the philosophy that smoothing suppresses the high frequency modes of the error [31]. Choosing the trace norm, on is lead to determine $\mathcal{A}$, C and D_1 from the minimality condition for $E_1 = \|D_0^{-1}\rho_1\|^2$,

$$\min = E_1 \ = \int\!\!\int_{z,w\in\Lambda^0} |\Gamma(z,w)|^2 \tag{103}$$

$$\Gamma(z,w) = v(z,w) - \int\!\!\int_{x,y\in\Lambda^1} \mathcal{A}(z,x)\ \widetilde{u}(x,y)\ C(y,w) \tag{104}$$

$$v = D_0^{-1} \ \text{and} \ \ \widetilde{u} = D_1^{-1}. \tag{105}$$

The minimality condition leads to conditions on kernels $\mathcal{A}$, C and on D_1. Suppose first that no constraints of practicality are imposed on the kernels. We assume that D_0 is selfadjoint. Optimizing $\mathcal{A}$, for given C and $\widetilde{u} = D_1^{-1}$, one finds the condition $\Gamma C^* = 0$. That is,

$$D_0\mathcal{A} = C^* u^{-1} \tag{106}$$

with $u = \widetilde{u}CC^*$. This reproduces the familiar condition for an ideal $\mathcal{A}$-kernel. Variation of the cost functional with respect to C yields $\mathcal{A}^*\Gamma = 0$. This is equivalent to a second condition which is obtained from (106) by interchange of the role of $\mathcal{A}$ and C^* and substituting D_1^* for D_1. Variation with respect to D_1 yields no new condition. These conditions leave much freedom. [12]

But practicality requires that one imposes restrictions of the form $\mathcal{A}(z,x) = 0$ except when z is sufficiently close to x, and similarly for C and D_1. (We will regard Λ^1 as a sublattice of Λ^0.) The proposal is to find the optimal kernels which satisfy these constraints by minimizing E_1.

---

take values 0 or 1. Links (z,w) with $U(z,w) = 0$ are considered deleted from the lattice. If deleted links are distributed at random with suitable probability, on obtains a fractal lattice. Solutions of the equation $(-\Delta + M^2)\chi_0 = f_0$ admit a random walk representation on this lattice.

[12] For instance, one may demand that $C = (\mathcal{A}\mathcal{A}^*)^{-1}\mathcal{A}^*$ or, equivalently, $\mathcal{A} = C^*(CC^*)^{-1}$. Then $C\mathcal{A} = \mathbb{1}$, the second condition follows from the first, and Γ agrees with the fluctuation field propagator (40) with $D_1 = -CC^*\Delta_{eff}$ if $D_0 = -\Delta$.

The idea was tested on a baby model, the 1-dimensional Poisson equation

$$(-\Delta + M^2)\,\chi_0 = f_0 \tag{107}$$

on a chain Λ^0 of lattice spacing 1, with very small M. Λ^1 consisted of the even sites in Λ^0. As a condition of practicality it was demanded that

$$C(x,z) = 0 = \mathcal{A}(z,x) \quad \text{unless} \quad |z-x| \le 1 \ .$$

Optimization via gradient descent yielded the expected result, in the limit of small M,

$$C(x,x) = 1 \quad , \quad C(x, x\pm 1) = \frac{1}{2} \quad , \quad D_1 = 2(-\Delta_1 + 4M^2) \tag{108}$$

$$\mathcal{A}(z,x) = C(x,z) \qquad (x \in \Lambda^1, z \in \Lambda^0) \ , \tag{109}$$

modulo normalization conventions. Δ_1 stands for the Laplacian on Λ^1.

Neural net approach

It is interesting to rephrase the problem in the language of neural nets [41].

We regard the layers of the multigrid as layers of a perceptron. The sites of the lattices are called nodes in this context. In the twogrid case there will be an input layer with nodes $z \in \Lambda^0$, a hidden layer with nodes $x \in \Lambda^1$ which are sites of the coarse grid, and an output layer with nodes $w \in \Lambda^0$. Nodes z of the input layer are connected to nodes x of the hidden layer by "neurons" with connection strength $C(x,z)$, so that node x receives as input a linear superposition of the output of the input nodes z. Nodes x of the hidden layer are connected to nodes w of the output layer by "neurons" of connection strength $\mathcal{A}(w,x)$ (see Fig. 9).

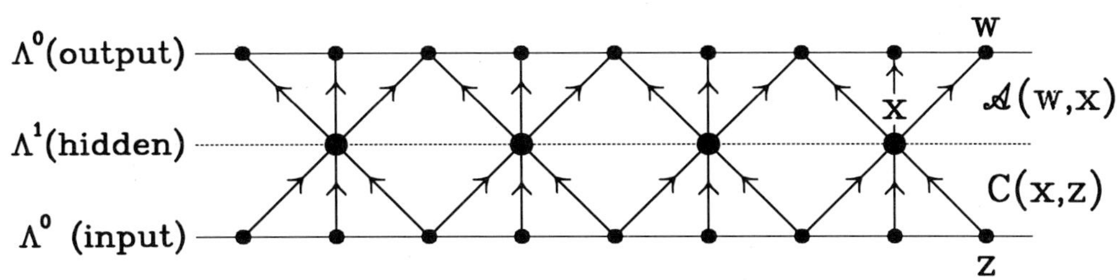

Fig. 9. Neural net from a twogrid (Λ^0, Λ^1) in one dimension. It has one input layer Λ^0, one output layer Λ^0 and one layer Λ^1 of hidden nodes.

The network receives as input an approximation $\xi(z)$ to the solution of the equation (94) which we wish to solve, and performs an iteration step in order to obtain as output an improved solution $O(z)$. The desired output is the exact solution

$\zeta(z) = D_0^{-1} f_0$. If the iteration step involves the exact solution of the coarse grid equation (98) then the network is not strictly feed-forward. [13] This is because the hidden layer needs to do computations more general than the usual computation of some nonlinear function at individual nodes in order to solve the coarse grid equation. This could be remedied if one envisages iterative solution of the coarse grid equation (98). We return to this below.

The network is supposed to learn to compute an output O which is as close as possible to the target ζ. The learning process consists in adjusting the connection strengths $C(x,z)$ and $\mathcal{A}(z,x)$. One envisages that a training set (ξ^μ, ζ^μ) is presented to the network to teach it. It involves a sequence of inputs ξ^μ. The network has its output O^μ compared to the desired answer ζ^μ and receives feedback about the error which is then used to adjust the connection strengths according to some learning rule. Application of typical learning rules amounts to iterative minimization of some cost functional E. The standard cost functional is

$$E = \sum_\mu \sum_z |O^\mu(z) - \zeta^\mu(z)|^2 \ .$$

In our case, the output is a linear function of the input and of f_0 which has the exact solution $D_0^{-1} f_0$ as a fixed point. Therefore

$$O = \rho\, \xi + (1-\rho)\mathcal{D}_0^{-1} f_0 \ ,$$

where ρ is the iteration matrix. The target $\zeta^\mu = D_0^{-1} f_0$ is independent of the input. The cost functional becomes

$$E = \sum_\mu \left\| \rho\, (\xi - D_0^{-1} f_0) \right\|^2$$

where $\|f\|^2 = \langle f, f\rangle = \sum_z |f(z)|^2$. Since the equation is linear, we may imagine without loss of generality that f_0 is arbitrarily small. Assuming the training input is a complete orthonormal set ξ^μ of functions on Λ^0 in the limit of small f_0, the cost functional becomes

$$E = \|\rho\|^2 \equiv \mathrm{Tr}(\rho\rho^*) \ .$$

This is the same cost functional as considered before.

We could restrict our attention to the coarse grid updating whose updating matrix is ρ_1. In this case one wants mainly to suppress the low frequency modes of the error. This motivates us to propose a training set with ξ^μ such that $D_0\xi^\mu$ are orthonormal. As a result one finds that E_1 should be minimized, where $E_1 = \|D_0^{-1}\rho_1\|^2$ as before. Minimization of E_1 leads to less practical learning rules than E because of the presence of the operator D_0^{-1}, though.

Let us return to the question of the implementation within the framework of feed-forward networks. We assume that the input layer knows to compute $r_0(z) = D_0\xi(z) - f_0(z)$ from the input ξ. This is easily implemented within the standard feed-forward framework by adding an extra layer. Initial fine grid relaxation sweeps to

[13] A neural network is called "feed-forward" if the nodes in a layer receive their only input from nodes on preceding layers [41].

smoothen r_0 are also easily implemented. The nodes x of the hidden layer will receive input $r_1(z) = \sum_z C(x,z) r_0(z)$ that is computed from the smoothened version of r_0.

Let us imagine that we do not use the exact solution e_1 to the coarse grid equation (98), but content ourselves with an approximation $\Xi(x)$ to $-e_1(x)$ which comes from a single Jacobi relaxation sweep. Then the node x is asked to compute the output

$$\Xi(x) = -d_1(x)^{-1}\, r_1(x) \ , \tag{110}$$

from its input $r_1(x)$, which it can legally do. d_1 is the diagonal part of D_1, viz.

$$D_1 f(x) = d_1(x) f(x) + \text{ contributions } \propto f(y) \text{ with } y \neq x \ .$$

We imagine that there are additional connections (not shown in figure (9)) of strength 1 from input nodes to output nodes which represent the same lattice point z. The total input to the output node z is then

$$O(z) = \xi(z) + \sum_x \mathcal{A}(z,x)\ \Xi(x) \approx \xi(z) - \sum_x \mathcal{A}(z,x)\ e_1(x) \ , \tag{111}$$

and this is also the output from the output node z.

Updating of connection strengths $C(x,z)$ and $\mathcal{A}(x,z)$ proportional to the gradient of the cost functional E leads to a standard *back-propagation learning rule* as described in [41]. As is discussed there (Sect. 6.1) the same network - or rather a bidirectional version of it - can be used to compute the necessary adjustments.

There is still one nonstandard feature. The network should also learn $d_1(x)^{-1}$. This becomes a standard learning problem (to be solved by the back-propagation learning rule) if we think of duplicating the hidden layer, connecting the duplicates of site x by "neurons" of connection strength $d_1(x)^{-1}$.

We discussed a twogrid for simplicity. Generalization to a multigrid is obvious and will be presented elsewhere [42].

10 Blocked Gauge Fields

Remember that the gauge covariant Laplace operator Δ for gauge field U has kernel $(a = 1)$

$$\Delta(z,w) = \begin{cases} -2d & \text{if } z = w, \\ U(z,w) & \text{if } z \text{ n.n. } w \ . \end{cases} \tag{112}$$

in d dimensions. And similarly for the staggered Dirac operator[14]

$$\not{\nabla}(z,w) = \begin{cases} \eta_\mu(z)\, U(z,w) & \text{if } w = z + \frac{1}{2} e_\mu,\ \mu = -d,\dots,d, \mu \neq 0, \\ 0 & \text{otherwise} \ . \end{cases} \tag{113}$$

A blockspin definition for matter fields gives us the effective operators on coarser layers, $\Delta_{\text{eff}} = \mathcal{A}^* \Delta \mathcal{A}$ and $\not{\nabla}_{\text{eff}} = \mathcal{A}^* \not{\nabla} \mathcal{A}$. Given Δ_{eff} or $\not{\nabla}_{\text{eff}}$, we have candidates for gauge fields $\mathcal{U}$ on the block lattice,

$$\begin{aligned} \mathcal{U}(x,y) &= \Delta_{\text{eff}}(x,y) \quad \text{for } x,y \text{ nearest neighbours} \\ \text{resp. } \mathcal{U}(x,y) &= \not{\nabla}_{\text{eff}}(x,y) \,/\, \eta_\mu(x) \quad \text{if } y = x + L_b \tfrac{1}{2} e_\mu \ . \end{aligned} \tag{114}$$

[14] We use the conventions $\eta_{-\mu} = -\eta_\mu$ and $e_{-\mu} = -e_\mu$. z is called odd if the sum of its integer coordinates is odd.

These expressions for $\mathcal{U}(x,y)$ have the correct gauge covariance property.

We add some comments on the Dirac case. Since we use odd L_b, the staggered block lattice can be regarded as a sublattice of Λ in which the pseudoflavour of a site does not depend on whether we regard it as a site of the fundamental or the block lattice, and the same is true of its property of being even or odd. The kernel $\mathcal{A}$ will always have the property that $\mathcal{A}(z,x)=0$ unless z and x are both even or both odd. Therefore $\not\nabla_{eff}$ connects even to odd sites, The factor $\eta_\mu(x)^{-1}$ can be regarded as defined by regarding x as a point in a sublattice of the fundamental lattice. This agrees with the natural definition on the block lattice. Note also that $\eta_\mu(z)$ is independent of the μ-component z_μ) of z.

The blocked gauge fields $\mathcal{U}(x,y)$ are actually "dielectric" gauge fields [19]. They may be brought back into the gauge group by multiplying with a positive real number (for $U(1)$ and $SU(2)$) or with a positive matrix (for $U(N), N>2$).

A different choice of blockspin for gauge fields was introduced by Balaban and used in rigorous renormalization group work [43].

11 Simulation Methods for the Computation of Effective Actions

The standard Monte Carlo simulation techniques for the computation of functional integrals can be used to compute effective actions, and in particular effective potentials.

There are three methods to take into account the δ-function constraints.

1.) Gaussian blockspins. Use a Gaussian in place of the δ-function in the constraint. This definition of a block spin transformation is very convenient for numerical work. But it has some inherent dangers which have been discussed in the literature (cp. [44]): Fictitious marginal operators may appear and cloud the renormalization group flow. Some numerical results are found in [44].

2.) Simulations at fixed blockspin Φ [44]. Do the updating so that Φ is fixed. Compute

$$\langle \partial\mathcal{H}/\partial\Phi(x)\rangle_\Phi .$$

This determines $\mathcal{H}_{eff}(\Phi)$ up to a constant.

For linear blockspin transformations, the quantity $\partial\mathcal{H}/\partial\Phi(x)$ is defined as follows:

$$\frac{\partial\mathcal{H}}{\partial\Phi(x)}=\int_z\frac{\partial\mathcal{H}(\phi)}{\partial\varphi(z)}\mathcal{A}(z,x) \tag{115}$$

so that the action $\mathcal{H}$ changes by $\delta\mathcal{H}=\frac{\partial\mathcal{H}}{\partial\Phi(x)}\delta\Phi(x)$ when φ is changed by an amount $\delta\varphi(z)=\mathcal{A}(z,x)\delta\Phi(x)$. Simulations with fixed block spin are simulations for an auxiliary theory with an infrared cutoff a_0^{-1}. Therefore the critical slowing down will be governed by the ratio a_0/a which is usually appreciably less than the correlation length in units of a.

This method is general, but it is costly when one wants to compute more than effective potentials.

3.) Method of "fluctuating coupling constants" [9,15] This method is applicable in cases such as Higgs models with and without gauge fields, where the effective action depends only on matter fields Φ which are related by a linear blockspin definition to matter fields which appear *polynomially* in $\mathcal{H}$.

Given φ, U then the high frequency field ζ is defined. A sequence of configurations (φ_a, U_a) yields therefore a sequence (ζ_a, U_a), $a = 1, 2, \ldots, N_{conf}$. Expanding

$$\mathcal{H}(U_a, \zeta_a + \mathcal{A}\Phi) = \sum_m g_a^m \Phi^m \tag{116}$$

defines a sequence g_a^m of "fluctuating coupling constants". They are stored. At the end of the run, the effective potential can be recovered from them. For a 1-component field

$$e^{-\mathcal{H}_{eff}(\Phi)} = N_{conf}^{-1} \sum_a Z_a^{-1} \exp\left[-\sum_m g_a^m \Phi^m\right] \tag{117}$$

where Z_a is the 1-dimensional integral

$$Z_a = \int d\Phi \exp\left[-\sum_m g_a^m \Phi^m\right] . \tag{118}$$

One may choose to expand around another point than $\Phi = 0$ instead.

In principle, fluctuation coupling constants could also be used to encode the information about full effective actions. But in this case g_a^m would be functions of m arguments $x_1, \ldots, x_m$ on the block lattice. Only few of them could possibly be stored.

Multigrid Updating of Matter Fields The multigrid method provides a tool to fight critical slowing down in Monte Carlo simulations. It uses or is motivated by constructs that we discussed. Therefore we would like to mention it at least briefly.

Imagine that the blockspin Φ^j attached to some block lattice Λ^j of lattice spacing a_j is changed at site $x \in \Lambda^j$ by $\delta\Phi^j(x)$. This will induce a change of the matter field

$$\delta\varphi(z) = \mathcal{A}^j(z, x)\delta\Phi^j(x) .$$

$\mathcal{A}^j$ should be thought of as minimizing the kinetic energy subject to some constraint (such as $C\varphi = \Phi$), therefore it should be *smooth in* z (no jumps). If we do not specify the averaging kernel C, there are no further restrictions on $\mathcal{A}$.

Similar updatings are possible for fields subject to constraints such as unit vectors in σ-models, spins in CP^n-models etc. For instance

$$\varphi(z) \mapsto \exp\left[\mathcal{A}(z, x)\ \epsilon\right]\ \varphi(z) , \tag{119}$$

where the matrix ϵ is an element of the Lie algebra of the symmetry of the model.

In the "unigrid" approach of Hasenbusch, Mack and Meyer [45], hypercubes x of different side lengths are considered and updatings of the form (119) are performed with smooth kernels $\mathcal{A}(z, x)$ which vanish for z outside x. The positions of the hypercubes x are chosen at random in the lattice. It turns out to be important for the effectiveness of the method that the selected hypercubes x of one length have some overlap.

The unigrid method proved very successful for 2-dimensional asymptotically free models: The critical exponent z which governs critical slowing down is nearly zero in this case. References are [46] for the CP^3-model and [45] for the $O(3)$−model. It is essential for the success that these models have no mass parameter [47]. The effective potential for 4-dimensional ϕ^4−theory was computed using a similar updating scheme [15], but with disjoint hypercubes x which form a lattice, and with kernels $\mathcal{A}$ which had support on hypercubes x and on their nearest neighbours.

Acknowledgements

Support by Deutsche Forschungsgemeinschaft is gratefully acknowledged. For providing resources and advice, we are indebted to HLRZ/Jülich and its staff. G.M. and M.S. thank the organizers of the 31.IUKT in Schladming for their efforts in creating the pleasant and stimulating atmosphere during the school. One of us (M.S.) thanks Thorsten Ohl for GNUidance.

Appendix: Constraint Effective Potential with Gauge Fields in Finite Volume

The result of the 1-loop calculation of Sect. 6 reads as follows (for details see [48])

$$U_V(\Phi) = U_V^{(0)}(\Phi) + \frac{1}{2V}\,\mathrm{Tr}\left[\ln\left\{\frac{1}{2}\,\frac{\delta}{\delta A_{rz\mu}}\,\frac{\delta}{\delta A_{r\bar{z}\nu}}\left(S_g^{(q)}[\Phi, A]\right)\right\}\right]_{A=0} \tag{120}$$

with

$$U_V^{(0)}(\Phi) = U_{cl}(\Phi) + \frac{1}{2V}\sum_n{}' \ln\left(\widehat{k}_n^2 + U_{cl}''(\Phi)\right) + \frac{3}{2V}\sum_n{}' \ln(\widehat{k}_n^2 + U_{cl}'(\Phi)/\Phi) \tag{121}$$

$$S_g^{(q)}[\Phi, A] = S_g^q[A] + U_V^{(2)}[\Phi, A] \tag{122}$$

where

$$\begin{aligned} U_V^{(2)}(\Phi, A) = {} & \tfrac{1}{4V}(U_{cl}'(\Phi)/\Phi)\sum_n{}' \tfrac{1}{(\widehat{k}_n^2)^2}\sum_r |A_{rno}|^2 + \tfrac{1}{4V}\sum_n{}' (\Gamma_{n\phi_0}(\Phi) + 3\Gamma_{n\phi_r}(\Phi)) \\ & \sum_r (A_{rnn}^2 + \tfrac{1}{\widehat{k}_n^2}|A_{rno}|^2) \\ & -\tfrac{1}{4V^2}\sum_{n,m}{}' (2\Gamma_{n\phi_r}(\Phi)\Gamma_{m\phi_r}(\Phi) + \Gamma_{n\phi_0}(\Phi)\Gamma_{m\phi_r}(\Phi) \\ & +\Gamma_{n\phi_r}(\Phi)\Gamma_{m\phi_0}(\Phi))\sum_r |A_{rnm}|^2 + \tfrac{\lambda_0}{12}\Phi^2\sum_n{}' (4\Gamma_{n\phi_r}(\Phi) + \Gamma_{n\phi_0}(\Phi)) \\ & \tfrac{1}{(\widehat{k}_n^2)^2}\sum_r |A_{rno}|^2 \end{aligned} \tag{123}$$

with

$$\begin{cases} U_{cl}(\Phi) = \frac{1}{2} m_0^2 \Phi^2 + \frac{\lambda_0}{4!}\Phi^4, & \widehat{k}_n^2 = 4\sum_\mu \sin^2(\frac{\pi}{L}n_\mu), \quad \frac{L}{2} < n_\mu \leq \frac{L}{2} \\ \Gamma_{n\phi_0}(\Phi) = (\widehat{k}_n^2 + U_{cl}''(\Phi))^{-1}, & \Gamma_{n\phi_r}(\Phi) = (\widehat{k}_n^2 + U_{cl}'(\Phi)/\Phi)^{-1} \end{cases} \tag{124}$$

Φ the absolute value of the covariant magnetization of (62) and $A_{rno}, A^2_{rnn}, A_{rnm}$ defined by

$$\begin{cases} (A_r^2)_{oo} &= \frac{1}{V}\sum\limits_{z\mu}\psi_0^*(z)A_{rz-\mu\mu}A_{rz-\mu\mu}\psi_0(z-\mu) \\ (A_r^2)_{no} &= \frac{1}{V}\sum\limits_{z\mu}\psi_n^*(z)A_{rz-\mu\mu}A_{rz-\mu\mu}\psi_0(z-\mu) \\ A_{rnm} &= \frac{1}{V}\sum\limits_{z\mu}\psi_n^*(z)A_{rz-\mu\mu}\psi_m(z-\mu) \\ |A_{rno}|^2 &= A_{rno}A_{rno}^* \end{cases} \tag{125}$$

Here, $\psi_n(z) = \exp(\frac{2\phi i}{L}nz)$ and the eigenvalue ϵ_0 is given by (70). U'_{cl} and U''_{cl} are the first and second derivative of U_{cl}, and prime $'$ on the sum means that the $n = 0$ term is omitted.

The trace in (120) involves summation over the discrete $O(4)$-matrix indices and a sum over space-time points on a lattice Λ.

References

1. Göpfert, M., Mack, G.: Commun. Math. Phys. 82 (1982) 545
2. Mack, G., Pordt, M.: Commun. Math. Phys. 97 (1985) 267
3. Karliner, M., Mack, G.: Nucl. Phys. B225 [FS9] (1983) 371
4. Gawędzki, K., Kupiainen, A.: Commun. Math. Phys. 77 (1980) 31
5. Nielsen, H.B., Ninomiya, M.: Phys. Lett. B105 (1981) 219
6. O'Raifeartaigh, L., Wipf, A., Yoneyama, H.: Nucl. Phys. B271 (1986) 653
7. Wilson, K.: Phys. Rev. D3 (1971) 1818; Rev. Mod. Phys. 55 (1983) 583; Kogut, J., Wilson, K.: Phys. Rep. 12C (1974) 75
8. Polchinski, J.: Nucl. Phys. B231 (1984) 269
9. Mack, G.: Multigrid Methods in Quantum Field Theory , Cargèse Lectures July 1987, *in:* 't Hooft, G.: et al., Nonperturbative Quantum Field Theory (Plenum Press, New York, 1989)
10. Gawędzki, K., Kupiainen, A.: Commun. Math. Phys. 99 (1985) 197
11. Ringwald, A., Wetterich, C.: Nucl. Phys. **B 334** (1990) 506;
    Wetterich, C.: Nucl. Phys. **352** (1991) 529;
    Wetterich, C.: The average action for scalar fields near phase transitions, DESY 91-088 (August 91)
12. Palma, G.: Finite Size Scaling Analysis of the Constraint Effective Potential computed from Multigrid Monte Carlo, DESY 90-083 (July 1990)
13. Palma, G.: Renormalized Loop Expansion to Compute Finite Size Effects of the Constraint Effective Potential, to appear in Z. Phys.

14. Palma, G.: Renormalized Loop Expansion for the Gauge Covariant Constraint Effective Higgs Potential on the Lattice, PhD-thesis Hamburg, DESY 92–031 (February 1992)
15. Mack, G., Meyer, S.: The effective action from multigrid Monte Carlo, DESY 89-009 (January 1989) and Nucl. Phys. B (Proc. Suppl.) 17 (1990) 293
16. Lin, L., Kuti, J., Shen, Y.: Nucl. Phys. B (Proc. Suppl.) 9 (1989) 26
17. Lüscher, M., Weisz, P.: Nucl. Phys. B290 [FS20] (1987) 25;
Lüscher, M., Weisz, P.: Nucl. Phys. B295 [FS21] (1988) 65
18. Kalkreuter, T.: Projective block spin transformations in lattice gauge theory, preprint, DESY 90–158 (December 1990) to appear in Nucl. Phys. B
19. Mack, G.: Nucl. Phys. B235 [FS11] (1984) 197
20. Wetterich, C., Reuter, M.: Average Action for the Higgs Model with Abelian Gauge Symmetry, preprint DESY 92–037 (March 1992)
21. Klauder, J.R.: lectures at this school
22. Montvay, I.: Nucl. Phys. B293 (1987) 479
23. Hasenfratz, A., Hasenfratz, P.: Phys. Rev. D34 (1986) 3160
24. Coleman, S., Weinberg, E.: Phys. Rev. D7 (1973) 1888
25. DeGrand, T. : lectures at this school
26. Kalkreuter, T., Mack, G., Speh, M.: Int. J. Mod. Phys. C3 (1992) 121
27. Mai, G.: Ein Blockspin für $2^{d/2}$ Fermionen, diploma thesis, Hamburg (October 1989)
28. Joos, H., Schaefer, M.: Z. Phys. C34 (1987) 465;
Golterman, M.F.L: Nucl. Phys. B278 (1986) 417
29. Smit, J.: Acta Phys. Polon. B 17 (1986) 531
30. Ben-Av, R., Brandt, A., Solomon, S.: Nucl. Phys. B329 (1990) 193;
Ben-Av, R., Brandt, A., Harmatz, M., Katznelson, E., Lauwers, P.G., Solomon, S., Wolowesky, K.: Phys. Lett. B253 (1991) 185; Nucl. Phys. B (Proc. Suppl.) 20 (1991) 102;
Ben-Av, R., Lauwers, P.G., Solomon, S.: Nucl. Phys. B 374 (1992) 249;
Lauwers, P.G., Solomon, S.: Int. J. Mod. Phys. C3 (1992) 149
31. Hackbusch, W.: Multi-grid Methods and Applications (Springer-Verlag, Berlin, 1985)
32. Hulsebos, A., Smit, J., Vink, J.C.: Int. J. Mod. Phys. C3 (1992) 161
Vink, J.C.: Phys. Lett. B272 (1991) 81; Nucl. Phys. B (Proc. Suppl.), to appear (LATTICE '91)
33. Brower, R.C., Moriarty, K.J.M., Myers, E., Rebbi, C.: in: Multigrid Methods, ed. S.F. McCormick (Marcel Dekker, New York, 1988)
34. Brower, R.C., Rebbi, C., Vicari, E.: Phys. Rev. D43 (1991) 1965; Phys. Rev. Lett. 66 (1991) 1263;
Brower, R.C., Moriarty, K.J.M., Rebbi, C., Vicari, E.: Nucl. Phys. B (Proc. Suppl.) 20 (1991) 89; Phys. Rev. D43 (1991) 1974
35. Hulsebos, A., Smit, J., Vink, J.C.: Nucl. Phys. B (Proc. Suppl.) 20 (1991) 94; Nucl. Phys. B368 (1992) 379
36. Brower, R.C., Edwards, R.G., Rebbi, C., Vicari, E.: Nucl. Phys. B366 (1991) 689
37. Vyas, V.: Int. J. Mod. Phys. C3 (1992) 169
38. Kalkreuter, T.: Phys. Lett. B276 (1992) 485
39. Brandt, A.: Multigrid techniques: 1984 guide, GMD-Studie No. 85
40. Kalkreuter, T.: *in preparation*
41. Hertz, J., Krogh, A., Palmer, R.G.: Introduction to the Theory of Neural Computation (Addison-Wesley, Redwood City CA, 1991)
42. Mack, G., Speh, M.: *in preparation*
43. Balaban, T.: Commun. Math. Phys. 89 (1983) 571

44. Pinn, K.: Z. Phys. C45 (1990) 453;
Pinn, K.: Z. Phys. C47 (1990) 325;
Computation of Effective Hamiltonians by Monte Carlo Simulations with Fixed Block-spins, preprint DESY 88–123 (August 1988)
45. Hasenbusch, M., Mack, G., Meyer, S.: Nucl. Phys. B (Proc. Suppl.) 20 (1991) 110
46. Meyer, S., Hasenbusch, M.: Testing accelerated algorithms in the lattice CP^3 model, preprint Kaiserslautern-TH-91-23 (January 1992)
47. Grabenstein, M., Pinn, K.: Acceptance Rates in Multigrid Monte Carlo, preprint DESY 91–157 (February 1992), to appear in Phys. Rev. D
48. Palma, G.: *in preparation.*

This article was processed using the LaTeX macro package with LMAMULT style

Computers in the Design and Analysis of HEP Experiments

F. James

CERN, Geneva, Switzerland

1 Summary

Experimentalists in elementary particle physics have always been among the world's biggest consumers of computing resources, and techniques have now developed to the point where nearly every aspect of our science, from the design of the accelerators to the publishing of the physics results, makes such extensive use of computers that they have become truly indispensable.

We describe the more important computing techniques used for the biggest experiments of both the current generation (CERN's SPS and LEP, and the Fermilab collider) and the next generation (LHC and SSC) of particle accelerators and storage rings. First we cover some of the Software Engineering methods needed to allow hundreds of physicists to work together to produce and maintain the hundreds of thousands of lines of computer code required for the different components of software for the big experiments: (1) the data acquisition and detector control, (2) the event and detector simulation, and (3) the reconstruction and analysis of the data. Then we will see how the experimentalist visualizes either real or simulated events, and how he can analyze Gigabytes of data interactively. To acquire the required reliability and accuracy in the simulation requires new techniques for pseudorandom number generation and we are studying the possibility of using quasi-Monte Carlo techniques.

2 Accelerator Studies

The oldest of the CERN accelerators still in service, the 30 GeV Proton Synchrotron, was designed and built about thirty years ago at a time when commercially available computers were very primitive and of little help, so nearly all the necessary calculations were done by hand. Since that time, accelerators have grown in diameter by nearly two orders of magnitude, and a variety of complicated effects such as nonlinearities in the magnetic elements and beam-beam interactions in colliders must now be taken into account.

We consider here just one of many calculations which are now being performed in design studies for the Large Hadron Collider (LHC) which is proposed for installation in the 27-kilometer long LEP tunnel at CERN. To describe the motion in a proton accelerator with the unavoidable nonlinearities due to magnet imperfections one has to consider at least the coupled horizontal and vertical transverse motion. It is known that under very general conditions, the volume occupied by the beam in the four-dimensional phase space (two transverse space coordinates and two transverse momentum components) is a constant of the motion. The problem is how to depict four-dimensional phase space. A solution is to use a two dimensional projection on a computer screen with an option of rotation, so that a three dimensional picture can be visualized. The fourth dimension can be represented by color, which ranges from blue to red as the most negative value and the most positive value of the fourth coordinate respectively.

Figure 1 is a plot representing in this way the coordinates in four-dimensional phase space of 20,000 points on the trajectory of one particle from a one million-turn simulation of LHC, carried out on the Cray X-MP at CERN. With no nonlinearities we would find a double layered symmetric torus. If however nonlinearities are added, as in the picture shown, these layers separate and rather complicated structures including submanifolds become visible.

Fig. 1. Representation of a proton trajectory in the four-dimensional phase space available in the proposed LHC.

The program which produces this simulation has to access a data base which contains the data necessary to reconstruct the magnetic field maps for all of the 350 magnets in the beam, and yet these programs are typically rather small compared with those necessary for the acquisition and reconstruction and simulation of the data produced by the LHC experiments.

3 LEP Experiments at CERN

Each of the four LEP experimental detectors (Aleph, Delphi, L3 and Opal) cost of the order of one hundred million dollars to build, not counting the salaries of the three to four hundred physicists and many more engineers and technicians who participated in their design, building and operation. With four major experiments all getting beam at exactly the same time, all looking for the same phenomena, the competition has been exceptionally strong. It is clearly of little interest to do an experiment if it is months or even weeks behind the others in producing results, so everything had to be ready from Day One. This of course was a great challenge to those responsible for the detector components (the hardware), but was an even greater challenge to the physicists responsible for the computing aspects, since it was not unheard-of for a detector to be built on time, but no one could recall a major experiment for which the software had functioned from the beginning.

The problem was not lack of time (software planning started five years before Day One) or of manpower (hundreds of people were willing to participate in the effort). The really frightening aspects of this project were:

- The software would have to work correctly on data which no one had ever seen.
- Because of the long lead time, there would be no excuse for failure, even though we could expect that both the computing environment and many features of the experiment would have changed considerably before Day One.

4 Software Engineering

In spite of the traditional suspicion with which Computer Science (now Informatics) is viewed by the community of high energy physicists, several of us decided to use part of the lead time to investigate whether the theoreticians of the computing world had anything to offer which could help us to write real programs.

The basic idea of Software Engineering was very appealing: To try to make the writing of computer programs like an engineering discipline. The analogy with Structural Engineering suggested already that there could be a better way to write programs than what we had always done. Our traditional "methodology" was simply to sit down at a terminal and start typing in lines of Fortran until we had enough to compile into subroutines, and if they didn't do what we wanted, we would "debug" them until they did. Occasionally we would reach a point where the programs got so big we couldn't see just what we were doing, and that would require stopping to think and even do some planning, particularly to get the data structures right, but all this was done in completely random order. Nevertheless it produced working programs in the days of moderate-sized experiments. Would it still be possible with the upcoming giant projects?

It was clear that an engineer faced with the problem of building a bridge, for example, would not proceed in our way. Even if he had already built some bridges before, he could not start by ordering steel and pouring concrete before he had at least calculated how much he needed and of what size. In fact, the necessary steps to be taken, all of which have clear analogies in programming, would be something like the following:

1. *Specification.* Decide what kind of bridge is needed, how many lanes of traffic, what capacity of weight, what access roads, etc.
2. *Design.* Draw up a description of the bridge as you think it should look, with detailed pictures of it, specification of the materials to be used in all parts.
3. *Discussion and Approval.* The bridge is to be paid for and used by people who are not engineers. Your design has to be explained in terms they can understand so they can approve it. You may have to propose several different designs among which they can choose one.
4. *Construction.* Only after the design is fixed and approved can actual construction begin. This step requires the additional discipline of Project Management: intelligent acquisition of materials, hiring and motivation of personnel, continuous quality control, allocation of resources and verification of attainment of planning goals, etc.
5. *Testing and Maintenance.* This step can be the longest and most expensive. Like the other steps, it is greatly facilitated when the preceding steps are carried out with this one in mind.
6. *Modification, Upgrading, and Replacement.* Even the best designed and built bridges may have to be modified in the light of experience or changing environments. This step can also be facilitated by recognizing such a possibility early on.

It was easy to see that the above steps were necessary for building a bridge, and the analogy with building programs was also clear, so we were ready to be convinced that we should use a similar methodology, but the next steps were far from clear. After all, we thought we knew what we wanted the programs to do, but no one had ever considered writing it down in great detail and discussing it. Even less obvious was the important design step. Making a picture of a proposed bridge is one thing, but how could we do that for a program? And even if we could do it, would the non-programmers ever be able to understand and approve it? In fact one of the common techniques of physicist-programmers had always been to write their programs in such an abstruse and complicated way that no one else could understand how they worked — this guaranteed that no one would mess up their programs by trying to modify them. Were we ready to give up this time-honoured practice?

There followed a difficult period of groping for appropriate methods. Some said we were wasting our time when we should be writing code. The coding was in fact starting and we hadn't yet decided how we were going to do the planning. We discovered that the Software Engineering discipline was still young and had not developed time-tested tools, especially the computer-aided design tools which we needed. We ourselves had no experience in this area and were not sufficiently expert to evaluate the the existing tools, let alone explain them to the other physicists. Then, just when it seemed that we would abandon it all, a consensus developed in some collaborations that we must give it a try. My own LEP collaboration Aleph voted to impose two Software Engineering disciplines: a formal design methodology[1] and a formal data model[2]. It turned out to revolutionize the way we thought about and realized our programs. I will give a brief indication of how these two techniques were used.

4.1 The SASD Programming Methodology

We chose to adopt most of the techniques which make up the methodology known as Structured Analysis and Structured Design, or SASD[1]. To implement this decision we took the unprecedented move of hiring a consultant who visited CERN and some of the other institutes in our collaboration to give one-week courses to all the people involved heavily in writing the software. We were pleasantly surprised at how well the courses were received by people who we thought would resist being told to change their ways of programming. Our physicists learned a lot about modern methodology, and the consultants even learned something about high energy physics, which turned out to be very useful when other physics groups also decided to learn SASD.

It is out of the question to go deeply into the details of SASD in the short time we have here, but it should be possible to give some idea of the elements that made the greatest impact on our work: the Data Flow Diagram (DFD) and associated Data Definition Language (DDL). The DFD is in some sense the equivalent of the engineer's blueprints; it is both the global and detailed "picture" of the program, which must be prepared by a specialist, but can then be understood and discussed by people with no experience in programming. Like the mechanical drawing, it consists of a series of diagrams, with the lower level diagrams showing details of the higher level ones. The highest level diagram is like the picture of the whole bridge, which indicates only the overall design concepts by specifying the main data flow.

All the DFD are composed of "bubbles" which represent processes, connected by arrows which represent data. The processes therefore receive data and transform it into other forms to be sent to other processes. This turns out to be an excellent model for the kind of programs we write, and we found no difficulties (once we had finished the week's training) in describing our programs in this way. In the higher level diagrams, the bubbles would correspond to very complex processes like "do pattern recognition" or "reconstruct tracks" which may each represent tens of thousands of lines of code. Then each of these bubbles is decomposed in a lower level diagram into its subprocesses until, at the lowest level, one bubble represents a very simple elementary process like "select track of highest momentum" which corresponds to one DO-loop or a few lines of code.

Note that the DFD is very different from the old "flow chart" which most of us have seen or used. The flow chart is just a representation of the computer code itself, with all its details, IF-statements and DO-loops indicated explicitly. The arrows represent program control, and it is not possible to view the program at different levels; all detail must be present everywhere, which makes the flow chart no easier to read than the program itself.

The arrows between the DFD bubbles represent data which has been processed or created by the previous bubble and is being used as input by the bubble(s) at the end(s) of the arrow(s). The exact content and format of these data must be well understood and documented, and this discipline is imposed on us by the requirement that precise definitions and unique labels for all these data be entered into the Data Dictionary using the Data Definition Language. This Data Dictionary can then be used not only to label the arrows in the DFD, but also by other tools including those which process the actual Fortran code itself, and it thus becomes a crucial part of the experiment's software documentation. This is the key element which enables many people to work nearly independently on different parts of the program, since it defines the interface between them.

Finally, computer tools were developed, both commercially and in-house, to facilitate the creation and maintenance of the DFD and DDL files. The basic idea is that the planning and design are done with these files, not with the Fortran code. That means that during the design stages the DFD are being modified constantly as we seek the best design. We therefore need good graphical tools which enable us to add and modify bubbles and arrows in the DFD, constantly verifying that all the diagrams at different levels remain consistent with each other. The trial DFDs are then discussed in collaboration meetings, and all the physicists including those not involved in the programming make suggestions and eventually agree on what is wanted. When the detailed design is approved, it is an easy matter to write the code, and it necessarily corresponds to the aims of the physicists. In addition, it is already documented before it is written. Later, when the programmer takes on other duties or leaves the collaboration, his successor can easily understand what the program is intended to do, and further modifications are made first in the DFD to assure consistency with others, then the code is modified accordingly.

4.2 The Entity-Relationship Data Model

The potential benefits to be gained from the use of a formal data model in high energy physics were first recognized by Palazzi and Fisher, who proposed the use of the Entity-Relationship Model and organized the ADAMO[2] project to implement the necessary tools to support the use of this model.

The data model describes the formal structure of the data and restricts the data to those kinds which can be formulated according to the model. The Entity-Relationship (ER) Model allows only entities (objects which have a fixed number of attributes) and relationships which are associations between the entities. The attributes are values which are in general restricted to lie within given ranges or domains.

The best way to understand the ER model is perhaps to consider data which do *not* conform to it, and see how to reformulate that data to make it conform. Let us take an entity called a "track", as we may have structured it before we used the ER model. It would have a certain fixed number of attributes such as length, position of first measured point, momentum, etc., and then we might have stuck into the same array the list of all the measured points along the track, of which there would of course be a variable number. This would not conform to ER, since the point measurements are considered attributes and they are not of fixed number. In ER language, the attribute table would have an undetermined number of columns, which is illegal. The way round this problem is to consider as one single track attribute an association to the list of points which makes up that track. Then in the list of point measurements, each point is an entity which itself has a fixed number of attributes like (x, y, z) position and measurement uncertainty.

The benefits of using the ER model were extraordinary. It enabled us to simplify the documentation and even the Fortran naming conventions, since each entity had a fixed number of attributes. We began to think in terms of rectangular tables, which of course are well accommodated in the Fortran language, but also produced pleasant side effects. To take the example just given, there are advantages in keeping a separate table of point measurements rather than considering them as attributes of tracks, since it is then easier to reassign a point to another track during the pattern recognition. Indeed the list of points has to exist anyway before the tracks are found, so it makes no sense to repeat this information in the track descriptions. Such elimination of redundant data is an important feature of all formal data models.

The apparently trivial restriction of tabular data imposed by the ER model had caused us to view the pattern recognition process as an *association* between track entities and point entities, whereas we had previously considered points to be attributes of tracks, which was certainly a confusion since the points themselves have attributes. This kind of subtle change in the way we viewed our data allowed us to simplify and clarify our programs considerably.

5 Event Visualization

For many of the older techniques of particle detection, such as cloud chambers, bubble chambers, spark chambers and emulsions, the raw data was a photograph of the charged-particle tracks. Data reduction then consisted of making measurements on the photographs to obtain the values of angles and curvatures of the tracks in two or three dimensions. Nowadays the process is inverted; the raw data are electronic signals which correspond to coordinates of points on tracks or energy deposits in calorimeters, and if one wants to see a visualization of the event, this must be obtained by a computer reconstruction of these data in graphical form. Figures 2, 3, and 4 show some visualizations of Z^0 decays from e^+e^- collisions in the Aleph detector at LEP.

In principle, it should be possible to do physics with these data as soon as the signals are associated to particle trajectories, without having to see the event like a photograph. Indeed, when physics studies are made, it is seldom with one event, but nearly always by making distributions of kinematical quantities like momenta, rapidities, effective masses, etc., all of which are determined without reference to the visualization of the event. However, there are reasons, and not just nostalgia for the old photographic days, why one often wants to visualize an event rather than know numerical values for kinematical quantities in the data.

One such reason is that, especially during the first few days of data collection, one simply wants to see the events because the eye is still the best tool for recognizing completely unexpected phenomena, and some interesting events can be identified at sight. However, this reason is quickly losing importance as in modern experiments it almost never happens that any interesting effect can be detected by simply viewing events.

The overridingly important reason for wanting to visualize events in modern experiments is to verify the correctness of the programs for data acquisition, event reconstruction, and detector simulation. An important class of program or data errors can be spotted immediately by looking at the visualization of an event. And of course, for our purposes in this lecture, the visualization offers us a convenient way to show what a typical event looks like. We should not however lose sight of the fact that the actual data used for the physics analysis is often more complete and more precise than what we can represent graphically. The challenge to those responsible for

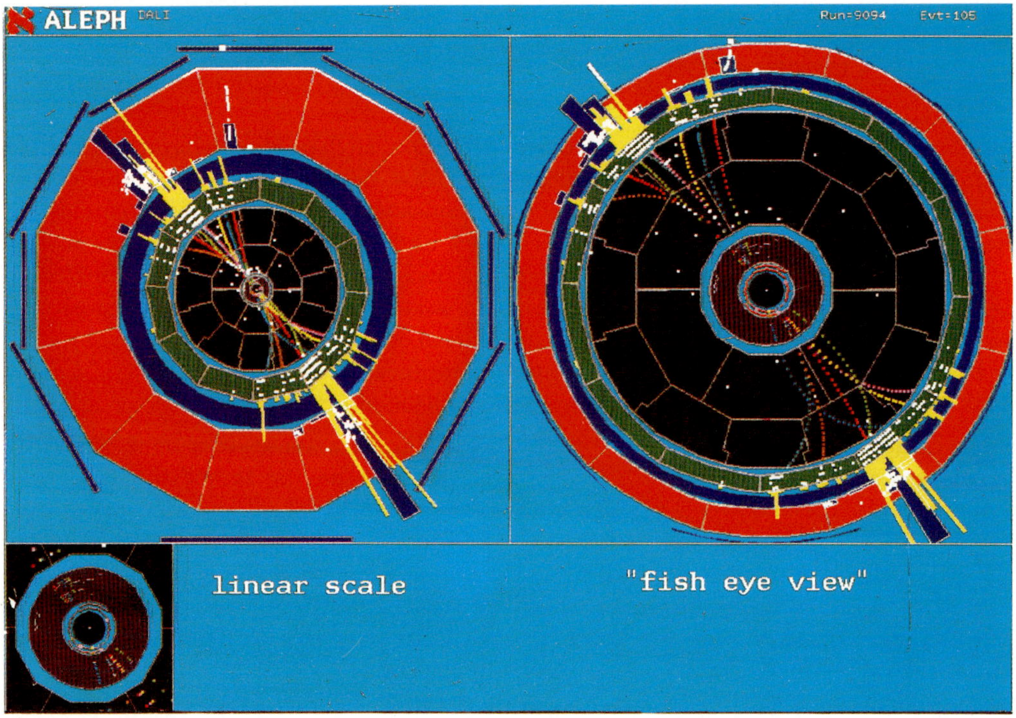

Fig. 2. Overall view of a typical two-jet event in the Aleph detector visualization program Dali, showing different scalings.

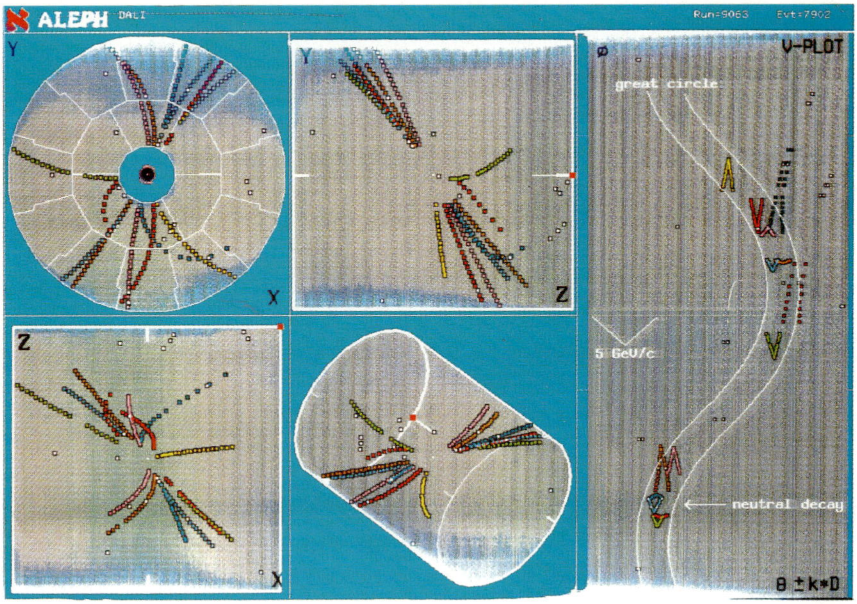

Fig. 3. The charged tracks of another Aleph event, showing only the main tracking detector data in different projections, including the V-plot.

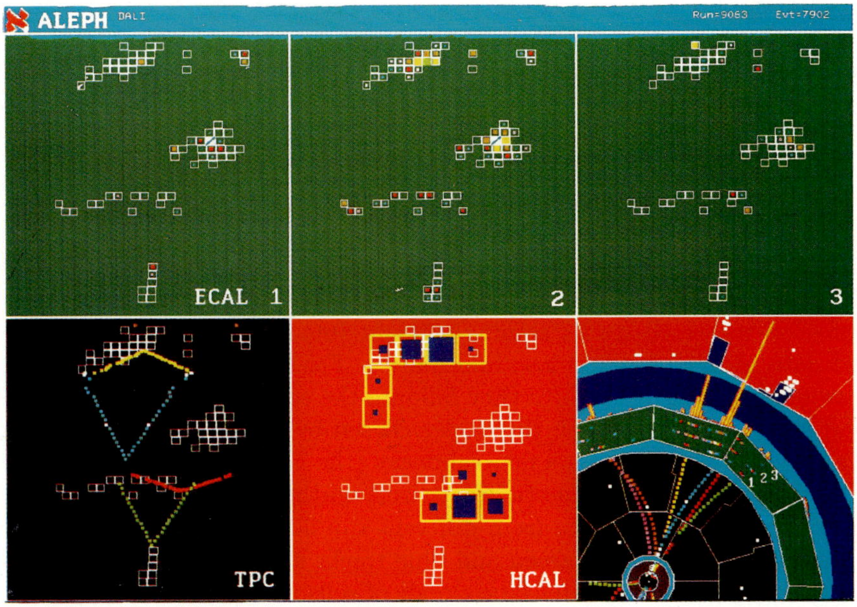

Fig. 4. Detail of one of the jets of the event of Figure 3, showing also the signals produced in the calorimeters.

the graphical display is to be able to represent as much of the available information as possible.

On a large modern experiment, there may be typically about a hundred physicists working on the computer programs, or software, of which a few will be experts in graphics techniques needed for event visualization and other visual aids. They will be responsible not only for the programming of graphics devices, but also for the study of the physiological effects of different graphics techniques in order to represent the largest amount of understandable information about the events for the physicist. They will make use of color, light intensity, contrast, magnification, linear and non-linear transformations, subset selection and suppression, and similar tools to help the physicist get the most out of the visualization. A description of these tools for non-specialists has recently appeared in Scientific American[4].

For example, the particle trajectories in the central tracking device will typically be represented by displaying the measured points in a given two-dimensional projection, with the color of the points indicating to which track the points were assigned by the pattern recognition software. Then a physicist who sees an apparently smooth track composed entirely of light blue points knows that the pattern recognition performed correctly on these points and assigned them all to the same track. On the other hand, if half the points are blue and the other half are red, he must look further to see if the assignments are right. He will then want to zoom in on that track to get a higher magnification and detect a possible anomaly. If it still looks like one track, he may choose another two-dimensional projection in order to see if the two halves separate in another view. Some systems show the third dimension by rotating the whole event slowly about some axis, and showing some projection of the event as it is rotating. This requires considerable computing resources in order to give the impression of smooth rotation, and the first systems to do this were extremely expensive, but nowadays an ordinary middle range workstation has enough compute power to handle such a technique.

Colour figures 2, 3 and 4 show some of the visualization possible with the Dali program used by the Aleph collaboration at LEP. Figure 2 shows how non-linear scaling is used to give more detail in the inner detectors than would be possible by a strictly proportional mapping of the apparatus. Additional zoom features allow even more detail to be seen, but at the expense of losing the rest of the view.

Figure 3 shows different ways of viewing just the charged tracks as recorded by the main tracking device, the TPC. The projections labelled X-Z and Y-Z are not simple projections, since tracks coming straight into or out of the plane would be projected onto small spots and lose information. For this reason, tracks are instead rotated around the beam axis until they fall into the corresponding plane. The fourth projection is a genuine one, with the angles chosen by the user. The fifth view is the V-plot (see below) which shows a neutral decay (the blue tracks) and shows that the whole event lies nearly in a plane (a great circle on the V-plot).

Figure 4 shows how the data from different sub-detectors are associated. For example, we see that the yellow, green and blue tracks all deposit quite a bit of energy in the electromagnetic calorimeter (the clouds of cells near the vertex in the V-plots), but only the yellow and red tracks deposit a lot of energy in the hadron calorimeter, indicating that they are indeed hadrons. In addition, a big electromagnetic shower appears just to the right of the blue track, where there is no charged track and no hadronic energy, so this is a high energy gamma-ray. A lower energy gamma appears in the upper right corner.

A good visualization program will allow the user to examine the data in all its aspects and using the full precision of the data.

5.1 The V-plot.

Occasionally it is necessary to invent a particularly radical graphical transformation of the data in order to transmit the maximum information in a form that the user can easily interpret. Such a transformation is the so-called *V-plot*, invented by Drevermann[3].

This transformation is useful when studying charged tracks in a high-precision cylindrical detector in a magnetic field, which is the case for most general-purpose detectors at colliders like LEP. It is best adapted to detectors which produce as raw data points in three dimensions (such as is the case for the Time Projection Chamber used by the Aleph experiment), but it can also be used for detectors in which the third dimension must be inferred by association of two or more two-dimensional measurements.

Under this transformation, each measured point is represented as a line whose length is the distance of the point from the outside detector boundary, whose x-coordinate at the centre of the line is the θ-coordinate of the point, and whose y-coordinate at the centre is the ϕ-coordinate of the point. Then in fact only the end-points of the lines are plotted. This produces, for a "normal" track with smooth curvature, a series of lines with constant θ (equal x) and slightly different ϕ (y), so that the end points of the lines form a $\vee$ for curvature of one sign, and a $\wedge$ for tracks of the opposite curvature. The opening angle of the $\vee$ or $\wedge$ will be small for slow, strongly curving tracks, and large for higher momentum tracks. An exactly straight track will produce a — (a degenerate $\vee$ with 180 degree opening angle.)

Such a plot has a number of great advantages over any simple two-dimensional projection of the tracks: First, it is easy to estimate the momentum of the tracks; second, the tracks are well separated in the V-plot, even though they all cross in space at the vertex (the transformation is not defined at the vertex); and thirdly, any anomaly (kink or change in sign) is easily spotted in the V-plot. Figure 5 shows an example of the usefulness of the V-plot in a fixed-terget experiment at rather high energy, where all the tracks appear to be straight in the usual projections, and are hopelessly confused. The V-plot shows clearly the separation of tracks and noise, and allows a good determination of the tracks' charges and momenta by the naked eye. For example, we see immediately that all tracks but three are positive, which is all but impossible to determine from any classical projection.

6 Detector Description Data Bases

An important part of most of the big programs in experimental HEP is the geometrical and physical description of the detector. Modern detectors may have many thousands of detector elements, each of which is defined by its shape and position in space and its physical properties. The physical properties can be separated into passive and

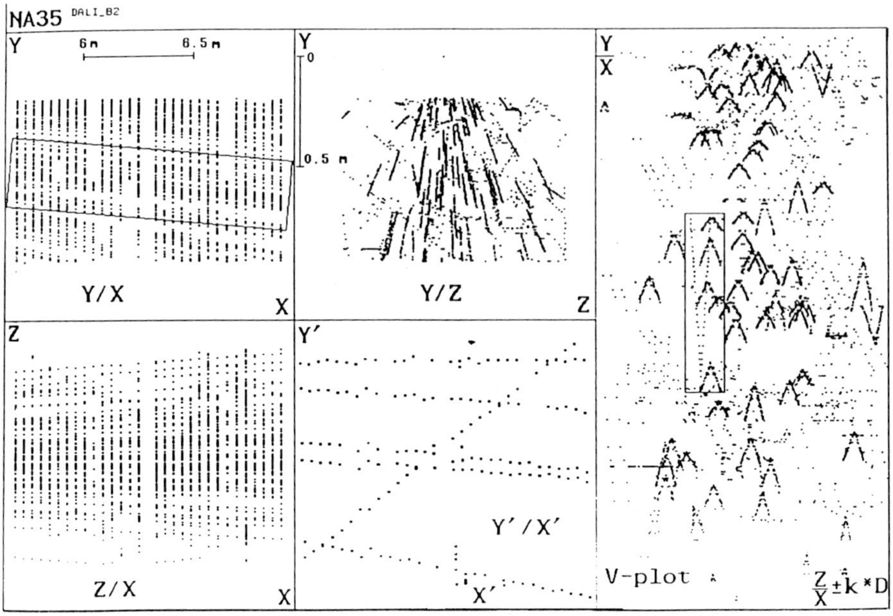

Fig. 5. High-energy tracks observed in the fixed-target experiment NA35, showing three classical projections and the V-plot.

active properties. The passive properties are determined by its chemical composition, mass density, magnetic field, etc.; they determine the probability that a particle of a given momentum and mass will interact or be deviated inside the element. The active properties determine what kind of signal will be produced when a given particle traverses the element. The total amount of information contained in the description of the detector may be very considerable, and it must be accessed very often by the reconstruction and simulation programs as well as the visualization systems. A typical question which must be answered very quickly by the data base system is: Given a particular (x, y, z)- or (r, θ, ϕ)-coordinate, in which element is it located and what is the composition of that element. This is not a trivial problem.

It is clearly essential that all programs which require the detector description should have access to the same data base. This seems obvious, but in practice is not so easy to implement. For very simple detectors it is usually easier to copy the description in different formats adapted specially to different programs; for the biggest detectors this is not only impractical, but also dangerous, since it is then very difficult to ensure that all programs are using equivalent descriptions. This means that the data base and its access routines must be organized so as to be usable by several different programs, often simultaneously.

The ultimate system for detector description would be to produce the data base before the detector is built, and to use that description as instructions to the builders of the detector apparatus. Then, assuming that they build it according to the instructions, the data base is automatically correct. This should also be the natural way to do things in the era of Computer Aided Design where the computer is in any case used in the construction of most detector elements. Unfortunately, the CAD programs tend to be commercial products with non-standard data base specifications, incompatible with the specifications used in the HEP programs. This situation is however likely to change in the near future as considerable work is now going into making the different systems able to talk to each other.

7 Monte Carlo Simulation

Probably over half of the considerable computer resources consumed by experimental HEP are devoted to simulation studies. This is probably true also of other areas of physics, both experimental and theoretical. Because of the great practical importance of the Monte Carlo Method, the balance of these lecture will be devoted to topics in Monte Carlo methodology, particularly the intelligent generation of random and better-than-random numbers.

7.1 Randomness

The theory of Monte Carlo tells us that Monte Carlo calculations are unbiased and converge with Gaussian-distributed statistical errors which decrease like $1/\sqrt{N}$ where N is the number of points or events simulated. This theory is based on the statistics of random variables through the Law of Large Numbers and the Central Limit Theorem. It requires that the "random numbers" used in the calculations be truly random (unpredictable), uniformly distributed (in probability) and independent. Two questions then arise:

- Is it possible to generate such numbers on a computer?
- Is it possible to generate numbers which are better than random?

We will address ourselves to these two questions.

7.2 Pseudorandom Numbers

Unfortunately, a computer alone, without contact to the outside world, is an entirely predictable machine incapable of acting randomly. It can, however, generate sequences of *pseudorandom* numbers which in some sense appear to be random, by making use of the phenomenon now known as numerical chaos.

I do not know of a rigorous mathematical test which can be applied to a given sequence of numbers — even a sequence of infinite length — to determine if it "appears to be random". The difficulty stems from the very definition of a truly random sequence: If a sequence of numbers is generated randomly, then all possible different sequences of values should be equally probable. This means that, given an already existing sequence, it should be just as probable as any other sequence, if it was generated truly randomly. This gives no way to determine the probability that it was indeed random. The way out of this paradox is to recognize that if a sequence is random, then all its subsequences must also be random, and therefore if certain subsequences appear with significantly different frequencies, that is evidence for non-randomness. This suggests a number of "tests of randomness" which increases superexponentially with the length of the sequence, but it still does not impose a unique recipe for measuring the randomness of a given sequence.

In practice, the methods for generating and testing pseudorandom number sequences have evolved with the need for longer and better sequences. An important step in this evolution has recently been observed as the traditional multiplicative congruential generators based on a single 32-bit integer became inadequate, not only for large calculations on supercomputers, but even for modest applications on personal workstations and PC's, due to the greatly increased computer power available. Even the best traditional generators have a period of $2^{30} \approx 10^9$, which is easily exhausted on a modern computer. Unfortunately, many unsuspecting physicists are still using the "standard" generators supplied traditionally by manufacturers and program libraries, even though new generators are now known which have much longer periods and other properties required for the most ambitious envisageable calculations.

It has been known for a long time how to improve a pseudorandom generator whose period is too short, and it is relatively easy to generalize from one seed to n seeds, which if they are correctly chosen lengthens the period by a power of n, but also takes n times as long to generate one number. The most modern methods use an arbitrary number of seeds n (usually of the order of ten to one hundred) in such a way that the period is lengthened by a little less than a power of n (typically the period goes from p to p^n/n) at an expense of only a factor of two in calculation time. A recent review article[5] gives details of these methods and gives portable Fortran code for a good 32-bit generator with a period of $\approx 10^{170}$ and nevertheless nearly as fast as the traditional manufacturer-supplied generators which have a period of only $\approx 10^9$.

8 Quasi-Monte Carlo

Even if the problem of pseudorandom number generation by a deterministic computer algorithm is now solved sufficiently well for the needs of the foreseeable future, we must still recognize that the convergence of the standard Monte Carlo method is extremely slow and is far from the performance which is theoretically possible according to more general principles. Let us now consider those more general principles and see that they lead to a different and very promising approach to the generation of well-distributed numbers.

8.1 Multidimensional Numerical Integration

Although the techniques of quasi-Monte Carlo will be applicable to many different Monte Carlo calculations, they will not be as universal as the traditional Monte Carlo methods. The theory of quasi-Monte Carlo is based on the theory of numerical integration over a hypercube, and we will have to investigate later to what extent we can take advantage of the acceleration of convergence also for other types of problems.

All numerical methods of integration of a function f over a hypercube Ω can be reduced to the formula:

$$\int_{\Omega} f\, d\underline{x} \approx \frac{1}{n}\sum_{i=1}^{n} \alpha_i f(\underline{x}_i) \qquad (1)$$

where the n points $\underline{x}_i$ and weights α_i must be specified by the method. The traditional Monte Carlo method for numerical integration specifies that all $\alpha_i = 1$, and the $\underline{x}_i$ are chosen randomly with uniform probability density over the hypercube.

We know that, under certain regularity conditions for the integrand f, it is possible to obtain much faster convergence, at least in a small number of dimensions, by using a "higher order" method, such as the Cartesian product of Simpson's Rules or of Gauss-Legendre Rules. The drawbacks of such *quadrature formulas* are well known:

1. Not all multidimensional quadrature rules have the desirable properties: $\alpha_i > 0$ and $\underline{x}_i \in \Omega$ for all i
2. The error estimate for such rules often depends on the average value of a higher derivative of f, which is not usually known, and may even be infinite.
3. If integration is over some other region than the hypercube, and if the transformation from the desired region to the hypercube is non-linear, then the convergence properties of the method are modified by the transformation, often drastically and to an extent which may be very hard to determine.
4. In a hypercube of d dimensions, the Cartesian product of one-dimensional rules converges much more slowly, like the power $1/d$ of the convergence of the one-dimensional rule, simply because the number of points increases like the power d.
5. For very high dimensions, it may simply be impossible to apply high-order rules because the number of points required is so large. For example, a 5-point rule in 20 dimensions requires $5^{20} \approx 10^{14}$ points.

These are precisely the problems which originally forced many physicists to turn to the Monte Carlo method. Now let us consider a generalization of the Monte Carlo method, namely the class of all numerical integrations based on equation 1 with all $\alpha_i = 1$. The $\underline{x}_i$ will not be chosen randomly, but instead will be calculated according to an algorithm designed to have better properties than random points. This class of methods is known as Quasi-Monte Carlo. The properties of such methods will depend only on:

- The function f through its *variation*, and
- The point set $\{\underline{x}_i\}$ through its *discrepancy*.

8.2 Variation of a Function

For traditional Monte Carlo, the important property of a function is its *variance*. In the more general theory of quasi-Monte Carlo, the relevant property will be the function's *variation*, which is not at all the same thing, although as its name implies, it is also a measure of how much the function deviates from a constant.

The variation of a function is not easy to define, but it can be thought of as a kind of average second derivative of the function. Functions of one dimension are a special case which does not especially interest us here, and in higher dimensions the definition is given in terms of the mixed second differences of the function. Let us define the variation for the two-dimensional case. We partition the x-axis by a series of n ordered x-values ξ_i, and similarly the y-axis is partitioned by the m ordered y-values η_j. Then take the double sum over the mixed second differences:

$$V_{nm} = \sum_{i=1}^{n} \sum_{j=1}^{m} |f(\xi_{i+1}, \eta_{j+1}) - f(\xi_{i+1}, \eta_j) - f(\xi_i, \eta_{j+1}) + f(\xi_i, \eta_j)| \qquad (2)$$

Then the *variation in the sense of Vitali* is the supremum of V_{nm} over all possible partitions (ξ, η). This is not generally considered sufficient since a function which is independent of one of the variables will then have a variation zero in the sense of Vitali, no matter how it varies with respect to the other, so it is more usual to add to this definition the one-dimensional variations of $f(x, 1)$ and $f(1, y)$, which produces the *variation in the sense of Hardy and Krause.*

The reader will have noticed that these definitions do not lend themselves to practical calculations. Indeed, unlike the variance, the variation of a function is not something for which one normally calculates approximate numerical values, which is unfortunate since as we shall see the error bound on quasi-Monte Carlo integration will be known in terms of the variation of the function. However it will still be of use in the theory, since the theorems about convergence will hold only for functions of *bounded variation,* so it will be of interest at least to know whether or not a variation is infinite. Continuous functions will as a rule have a bounded variation, but it can be seen rather easily that a multidimensional function which is discontinuous along a line or hyperplane of finite extent will have an infinite variation.

8.3 Discrepancy of a Point Set

The most important concept in quasi-Monte Carlo theory and practice is one which describes a set of points in the unit hypercube. This property is called *discrepancy* and is a measure of the non-uniformity of the density of these points. Given a set of points in the unit hypercube, let the function $N(\underline{x})$ defined at any position $\underline{x}$ in the hypercube give the number of points which fall in the hyper-interval $[0, \underline{x})$, and the function $V(\underline{x})$ give the volume of the same interval or hyper-rectangle from the origin to $\underline{x}$. The function $V(\underline{x})$ is therefore just the product of the coordinates of the position $\underline{x}$. Then the local discrepancy of this set of n points, at position $\underline{x}$ is:

$$D(\underline{x}) = \left| \frac{N(\underline{x})}{n} - V(\underline{x}) \right| \qquad (3)$$

That is, it is just the difference between fraction of points falling in the hyper-rectangle from the origin to $\underline{x}$, and the volume of that hyper-rectangle. Or, put another way, it is the difference between the number of points in this sub-volume and the expected number of points if the point set was distributed exactly uniformly. Clearly, for any finite point set, this difference cannot be zero everywhere.

But $D(\underline{x})$ is a local measure, and we are really interested in global measures of non-uniformity. Two such global measures are:

– *The mean square discrepancy:* $D_2 = \{\int_\Omega D^2(\underline{x})\, d\underline{x}\}^{1/2}$

– *The extreme discrepancy:* $D_\infty = \sup_x \{D(\underline{x})\}$

¿From here on, we will use the term ***local discrepancy*** for $D(\underline{x})$, and the unqualified term ***discrepancy*** will always refer to a global measure. In many contexts, any global measure may be used; when a particular one must be chosen, the default will be the *extreme discrepancy.*

8.4 The Koksma-Hlawka Inequality

The important result which allows us to set an upper bound on the error of a numerical integration, and which opened the door to the whole field of quasi-Monte Carlo, was first published by Koksma in 1942 for one dimension, and extended to higher dimensions in 1961 by Hlawka. The original papers (in Dutch and in German) are cited and summarized in a more accessible review of the subject in 1968 by Zaremba[6].

¿From equation 1 we define the error of any numerical integration as:

$$\delta I = \left| \int_\Omega f(\underline{x})\, d\underline{x} \; - \; \frac{1}{n} \sum_{i=1}^{n} \alpha_i f(\underline{x}_i) \right| \tag{4}$$

Restricting ourselves to quasi-Monte Carlo integration, that is all $\alpha_i = 1$, we can state the Koksma-Hlawka Inequality, which sets an upper limit on the error of such an approximation:

$$\delta I \; \leq \; V(f)\, D_\infty \tag{5}$$

where V is the variation of f in the sense of Hardy and Krause, and D_∞ is the extreme discrepancy of the point set $\underline{x}_i$. Since V is a constant property of the function f, this tells us that the way to obtain a good approximation to an integral is to find a point set with small discrepancy. Since the n-dependence of δi for traditional Monte Carlo is $1/\sqrt{n}$ we can see that the discrepancy of a set of truly random points must converge to zero like $1/\sqrt{n}$ in some appropriate expectation. The primary goal of quasi-Monte Carlo is then to find algorithms for generating point sets whose discrepancy is smaller than that of a truly random point set, and converges to zero faster than $1/\sqrt{n}$.

8.5 Finite Point Sets of Low Discrepancy

Since the results of Koksma and Hlawka there has been considerable study of the discrepancy of point sets with the goal of finding points with especially low discrepancy. The straightforward "experimental" approach is to start with points at arbitrary positions and then use numerical minimization methods to minimize the discrepancy by moving the points around, but this approach was out of the question until recently because of the enormous computational complexity involved. The major difficulty comes from the fact that the discrepancy is of interest only for relatively large sets of points in spaces of rather high number of dimensions: at the very least, let us say one hundred points in three dimensions. But this is already a minimization problem with three hundred free variables and no obvious starting point, and each evaluation of the discrepancy is a major calculation which itself requires finding a maximum value, although this is not quite as hard as it sounds since the maximum local discrepancy must lie at one of n^d positions, which is a large but finite number.

A more promising approach is to generate points according to a formula which contains some free parameters, and find (still numerically) the values of those parameters which minimize the discrepancy of the resulting point set. This amounts to finding the best point set of a given class of point sets, instead of aiming at finding the absolutely best point set. The result is that the number of free parameters is greatly reduced and the size of point sets can be increased accordingly. Another advantage is that the optimization can be done on a given set of points, but then the formula continued, with the optimized parameter values, to generate larger sets of points for the actual integration. Of course there is no guarantee that the parameter values which minimize the discrepancy for the smaller subset will also yield a larger set with low discrepancy, but it is certainly a good working hypothesis.

Probably the best known finite sequences found using the above method are those of Korobov. One such Korobov family is given by:

$$\underline{x}_k = \left.\frac{\beta_1 k}{n}\right|_{mod\ 1}, \left.\frac{\beta_2 k}{n}\right|_{mod\ 1}, \ldots, \left.\frac{\beta_d k}{n}\right|_{mod\ 1}, \quad k = 1, n \tag{6}$$

where the β_i are coefficients determined in order to minimize the discrepancy of n points in d dimensions. Discussion of Korobov sequences and references to the original Russian articles are given in [7] and [8]. The latter article describes a program for multiple integration using Korobov sequences. Extensive tables of optimal coefficients for Korobov sequences are given in [9].

8.6 Limits on the Discrepancy of Point Sets

The other approach to the study of discrepancy is to consider the asymptotic behavior of sequences generated according to algorithms capable of generating very long sequences. One thinks of these sequences as infinitely long, although the algorithms as implemented on a computer generally run up against an overflow somewhere if they are taken too far.

We first consider the theory, then go on in the next subsection to describe actual generating algorithms. A variety of different theorems have been proposed which set different limits on the asymptotic behavior of the discrepancy of a set of points. The earlier and most important results can be found in [10]. We summarize here the features common to most of these theorems.

The general idea is to show that it is possible to generate sets of points whose discrepancy, for large n, will tend to zero at least as fast as some function of n and the dimensionality d. In some cases, the proofs are constructive in the sense that they imply an actual algorithm for generating the points which have that discrepancy. The upper limits generally have the form:

$$D_\infty \leq k \frac{(\ln n)^{s(d)}}{n} \quad for\ n \to \infty \tag{7}$$

where k is a constant and $s(d)$ is some increasing function of the dimensionality. The earliest results, valid for the Halton (or van der Corput) generator given below, have the function $s(d)$ increasing very strongly, like the factorial, which makes these generators uninteresting for very high dimensions. However, for moderate dimensionalities the logarithmic dependence on n, even if raised to a small power, is not very serious, and the dominant n-dependence is still $1/n$ which is much faster than traditional Monte Carlo.

The results of Niederreiter[11] indicate that the function $s(d)$ can be made very slowly increasing, and even perhaps to decrease for large enough d. If this can be confirmed constructively, then it would mean that we could attain the $1/n$ behavior which seems to be the "speed limit". Considerable work is going on now to produce a quasirandom generator which obeys this theorem, and to investigate its behavior in practice.

8.7 Quasirandom Point Generators

Whereas pseudorandom number generators produce individual real numbers, quasirandom generators give us directly d-tuples of real numbers which are to be interpreted as the coordinates of points in a d-dimensional hypercube. Therefore, although we may speak loosely of quasirandom numbers, it is more appropriate to speak of quasirandom points or even better of point sets.

The earliest quasirandom generator is that of Richtmyer, for which the jth coordinate of the ith point is given simply by:

$$x_{ij} = iS_j \bmod 1 \tag{8}$$

where S_j is the square root of the jth prime number. This formula is very easy to program, and produces points which have good properties for small d. The van der Corput or Halton generator is however better.

A generator based on inverting fractions in base p (p prime) was proposed by van der Corput, and a clever algorithm to produce these numbers was published in 1960 by Halton. Unfortunately, Halton's algorithm turns out to be machine-dependent since it works only if the computer arithmetic is done with particular rules for rounding. A better algorithm is given here:

```
      SUBROUTINE RHALTON(RVEC)
C          Halton quasirandom number generator,
C              generates vector RVEC of ND numbers (dimensions)
C
C Usage:    dimension rvec(ndim)
C           call haltonin(ndim)      ! initialize for dimension ndim
C           do
C                call rhalton(rvec) ! generate one point
C           end do
C
      PARAMETER (NDIMEN=10)
C
C          NDIMEN is the maximum allowed value of NDIM
C   If you need more, increase NDIMEN and add more prime numbers
C             in the statement  DATA NBASE / ...      just below
```

```
C
      INTEGER NUMER(NDIMEN), NDENOM(NDIMEN), NBASE(NDIMEN)
      REAL RVEC(*),RDENOM(NDIMEN)
      SAVE NUMER,NDENOM,NBASE,ND,KOUNT
      DATA NBASE/2,3,5,7,11,13,17,19,23,29/
      DATA KOUNT/-10/
C ---------------------------------------------------------------
C          Normal entry to generate the quasirandom vector
      IF (KOUNT .LT. 0)  THEN
         WRITE (6,'(A)') ' RHALTON was not initialised.  Stop.'
         STOP
      ENDIF
C
      DO 500 I= 1, ND
         NXA = NDENOM(I) - NUMER(I)
         IF (NXA .EQ. 1)  THEN
            NUMER(I) = 1
            NDENOM(I) = NDENOM(I)*NBASE(I)
            RDENOM(I) = 1.0/REAL(NDENOM(I))
         ELSE
            NXB = NDENOM(I)/NBASE(I)
            DO 400 LOOP= 1,100
               IF (NXA .GT. NXB)  GO TO 410
               NXB = NXB/NBASE(I)
  400       CONTINUE
  410       CONTINUE
            NUMER(I) = (NBASE(I) + 1) * NXB - NXA
         ENDIF
         RVEC(I) = REAL(NUMER(I)) * RDENOM(I)
  500 CONTINUE
C
C--      KOUNT = KOUNT + 1
C--      IF (KOUNT .LT. 200)  THEN
C--         ND2 = MAX(ND,6)
C--         WRITE (6,'(12I6)') (NUMER(J),NDENOM(J),J=1,ND2)
C--      ENDIF
      RETURN
C ------------------------------------------------------------------
C          Initializing entry
      ENTRY HALTONIN(NDIM)
      IF (NDIM .GT. NDIMEN)  THEN
         WRITE (6,'(A,I4)')
     +     ' HALTONIN call exceeds maximum dimension', NDIMEN
         STOP
      ENDIF
      ND = NDIM
      KOUNT = 0
```

```
      DO 700 I= 1, ND
      NUMER(I) = 0
      NDENOM(I) = 1
  700 CONTINUE
      RETURN
      END
```

The above algorithm produces points with low discrepancy for smallish dimensionalities. Work is now in progress on other algorithms which promise much better discrepancy for higher dimensionalities.

8.8 The Current Status of Quasi-Monte Carlo

There is no doubt that the optimistic mathematical results concerning accelerated convergence using quasi-Monte Carlo are real. Berblinger and Schlier[12] have shown that even in cases where the mathematical results should not strictly apply, there is still an important gain by using quasirandom numbers rather than pseudorandom.

However a large number of problems still remains to be clarified before we can consider quasi-Monte Carlo to be as mature a discipline as the traditional variety:

1. We need to investigate the properties of the newest generators which have been programmed only recently and which it is hoped will attain the the low discrepancy bounds promised by the theory of Niederreiter.
2. We need to find a way to calculate approximately the variation of a function in order to have an error bound for the integral.
3. We need to understand how it is that the error is small even when the variation of the function is infinite.
4. We need to investigate the use of quasirandom numbers for problems other than multidimensional integration, in particular those involving random walks.

References

1. Kellner, G.: *Development of Software for Aleph using Structured Techniques,* Comp. Phys. Commun. **45** (1987) 229–243
2. Fisher, S.M., Palazzi, P.: *Using a Data Model from Software Design to Data Analysis: What Have We Learned?,* Comp. Phys. Commun. **57** (1989) 169–175
3. Drevermann, H., Grab, C.: *Graphical Concepts for the representation of Events in High Energy Physics,* Int. J. Mod. Physics C **1** (1990) 147–163
4. Breuker, H., Drevermann, H., Grab, C., Rademakers, A,Stone, H.: *Tracking and Imaging Elementary Particles,* Scientific American, Aug., 1991, 58–63
5. James, F.: *A Review of Pseudorandom Number Generators,* Comp. Phys. Commun., **60** (1990) 329–344
6. Zaremba, S.K.: *The Mathematical Basis of Monte Carlo and Quasi-Monte Carlo Methods,* SIAM Review **10** (1968) 303–314
7. Stroud, A.H.: *Approximate Calculation of Multiple Integrals,* Prentice-Hall, 1971
8. Zakrzewska, K., Dudek, J.,Nazarewicz, N.: *A Numerical Calculation of Multidimensional Integrals,* Comp. Phys. Commun. **14** (1978) 299–310

9. Keast, P.: *Technical Report 40,* Department of Computer Science, University of Toronto, Ontario, Canada
10. Kuipers, L. Niederreiter, H.: *Uniform Distribution of Sequences,* Wiley, 1974
11. Niederreiter, H.: *Point Sets and Sequences with Small Discrepancy,* Mh. Math. **104** (1987) 273–337
12. Berblinger, M. Schlier, Ch.: *Monte Carlo Integration with Quasirandom Numbers: Some Experience,* Comp. Phys. Commun. **66** (1991) 157–166

This article was processed using the LaTeX macro package with LMAMULT style

Seminars

Several participants contributed interesting seminar talks. It is not possible to include them here, but we want at least give the list of seminar speakers (alphabetically ordered) and topics covered by them.

J. Bohacik (SAS Bratislava):
Heavy Quarks Potential in the Continuum SU(2) Theory at Finite Temperature

W. Bürger (Techn. Univ. Wien):
Charge Polarization in Compact QED

C.R. Gattringer (Univ. Graz):
Study of Resonance Scattering in a Simple 2-d Lattice Model

E. Gerstenmayer (Techn. Univ. Wien):
Quantum Gravity on Simplicial Lattices

M. Grabenstein (Univ. Münster):
Acceptance Rates in Multigrid Monte Carlo

M. Hasenbusch (Univ Kaiserslautern):
High Precision Monte Carlo Study of Solid-on-solid Models

A. Hulsebos (Univ. Liverpool):
Multigrid Exploration of the XY-Model

E.M. Ilgenfritz (Univ. Leipzig):
Finite Temperature Higgs Transition at Weak Couplings

A.N. Ivanov (Univ. St. Petersburg) and **M. Nagy (SAS Bratislava):**
Problem of Bosonization of Local Four-quark Operators

R. Kenna (Univ. Graz):
Logarithmic Corrections to Scaling in Φ_4^4

C. Klimcik (Univ. Prague):
Grassmann Integral Topological Invariants

S. Lee (Univ. of Florida, Gainesville):
3-d Langevin

T. Lippert (Univ. Wuppertal):
Multi Scale Updating of U(1) Lattice Gauge Theory

P.V. Malyshev (Univ. Kiew):
Solution of the Cauchy Problem for BBGKY Hierarchy of Equations

E. Myers (Univ. of Texas, Austin):
Non Compact Sigma Models and their Relation to Quantum Gravity

W. Petersen (ETH Zürich):
New Methods to Solve Langevin Equations

H. J. Pirner (Univ. Heidelberg):
Color Dielectric Model of QCD

J. Przeszowski (PAS Warsaw):
Alpha Regularization for the Axial Gauge Condition - the Basis of Calculable Theory

H. Rosu (Univ. Bucharest):
On the Unruh Effect in Storage Rings

T. Trappenberg (RWTH Aachen):
Z(3) Instantons and Tunneling of Hot Gluons

F. Zimmermann (RWTH Aachen):
Finite Size Effects and Scaling Analysis of the O(4) Theory in the Broken Phase

Printing: Druckhaus Beltz, Hemsbach
Binding: Buchbinderei Schäffer, Grünstadt

Lecture Notes in Physics

For information about Vols. 1–374
please contact your bookseller or Springer-Verlag

Vol. 375: C. Bartocci, U. Bruzzo, R. Cianci (Eds.), Differential Geometric Methods in Theoretical Physics. Proceedings, 1990. XIX, 401 pages. 1991.

Vol. 376: D. Berényi, G. Hock (Eds.), High-Energy Ion-Atom Collisions. Proceedings, 1990. IX, 364 pages. 1991.

Vol. 377: W. J. Duschl, S.J. Wagner, M. Camenzind (Eds.), Variability of Active Galaxies. Proceedings, 1990. XII, 312 pages. 1991.

Vol. 378: C. Bendjaballah, 0. Hirota, S. Reynaud (Eds.), Quantum Aspects of Optical Communications. Proceedings 1990. VII, 389 pages. 1991.

Vol. 379: J. D. Hennig, W. Lücke, J. Tolar (Eds.), Differential Geometry, Group Representations, and Quantization. XI, 280 pages. 1991.

Vol. 380: I. Tuominen, D. Moss, G. Rüdiger (Eds.), The Sun and Cool Stars: activity, magnetism, dynamos. Proceedings, 1990. X, 530 pages. 1991.

Vol. 381: J. Casas-Vazquez, D. Jou (Eds.), Rheological Modelling: Thermodynamical and Statistical Approaches. Proceedings, 1990. VII, 378 pages. 1991.

Vol. 382: V.V. Dodonov, V. I. Man'ko (Eds.), Group Theoretical Methods in Physics. Proceedings, 1990. XVII, 601 pages. 1991.

Vol. 384: M. D. Smooke (Ed.), Reduced Kinetic Mechanisms and Asymptotic Approximations for Methane-Air Flames. V, 245 pages. 1991.

Vol. 385: A. Treves, G. C. Perola, L. Stella (Eds.), Iron Line Diagnostics in X-Ray Sources. Proceedings, Como, Italy 1990. IX, 312 pages. 1991.

Vol. 386: G. Pétré, A. Sanfeld (Eds.), Capillarity Today. Proceedings, Belgium 1990. XI, 384 pages. 1991.

Vol. 387: Y. Uchida, R. C. Canfield, T. Watanabe, E. Hiei (Eds.), Flare Physics in Solar Activity Maximum 22. Proceedings, 1990. X, 360 pages. 1991.

Vol. 388: D. Gough, J. Toomre (Eds.), Challenges to Theories of the Structure of Moderate-Mass Stars. Proceedings, 1990. VII, 414 pages. 1991.

Vol. 389: J. C. Miller, R. F. Haglund (Eds.), Laser Ablation-Mechanisms and Applications. Proceedings. IX, 362 pages, 1991.

Vol. 390: J. Heidmann, M. J. Klein (Eds.), Bioastronomy - The Search for Extraterrestrial Life. Proceedings, 1990. XVII, 413 pages. 1991.

Vol. 391: A. Zdziarski, M. Sikora (Eds.), Ralativistic Hadrons in Cosmic Compact Objects. Proceedings, 1990. XII, 182 pages. 1991.

Vol. 392: J.-D. Fournier, P.-L. Sulem (Eds.), Large-Scale Structures in Nonlinear Physics. Proceedings. VIII, 353 pages. 1991.

Vol. 393: M. Remoissenet, M.Peyrard (Eds.), Nonlinear Coherent Structures in Physics and Biology. Proceedings. XII, 398 pages. 1991.

Vol. 394: M. R. J. Hoch, R. H. Lemmer (Eds.), Low Temperature Physics. Proceedings. XXX,XXX pages. 1991.

Vol. 395: H. E. Trease, M. J. Fritts, W. P. Crowley (Eds.), Advances in the Free-Lagrange Method. Proceedings, 1990. XI, 327 pages. 1991.

Vol. 396: H. Mitter, H. Gausterer (Eds.), Recent Aspects of Quantum Fields. Proceedings. XIII, 332 pages. 1991.

Vol. 398: T. M. M. Verheggen (Ed.), Numerical Methods for the Simulation of Multi-Phase and Complex Flow. Proceedings, 1990. VI, 153 pages. 1992.

Vol. 399: Z. Švestka, B. V. Jackson, M. E. Machedo (Eds.), Eruptive Solar Flares. Proceedings, 1991. XIV, 409 pages. 1992.

Vol. 400: M. Dienes, M. Month, S. Turner (Eds.), Frontiers of Particle Beams: Intensity Limitations. Proceedings, 1990. IX, 610 pages. 1992.

Vol. 401: U. Heber, C. S. Jeffery (Eds.), The Atmospheres of Early-Type Stars. Proceedings, 1991. XIX, 450 pages. 1992.

Vol. 402: L. Boi, D. Flament, J.-M. Salanskis (Eds.), 1830-1930: A Century of Geometry. VIII, 304 pages. 1992.

Vol. 403: E. Balslev (Ed.), Schrödinger Operators. Proceedings, 1991. VIII, 264 pages. 1992.

Vol. 404: R. Schmidt, H. O. Lutz, R. Dreizler (Eds.), Nuclear Physics Concepts in the Study of Atomic Cluster Physics. Proceedings, 1991. XVIII, 363 pages. 1992.

Vol. 405: W. Hollik, R. Rückl, J. Wess (Eds.), Phenomenological Aspects of Supersymmetry. VII, 329 pages. 1992.

Vol. 406: R. Kayser, T. Schramm, L. Nieser (Eds.), Gravitational Lenses. Proceedings, 1991. XXII, 399 pages. 1992.

Vol. 407: P. L. Smith, W. L. Wiese (Eds.), Atomic and Molecular Data for Space Astronomy. VII, 158 pages. 1992.

Vol. 408: V. J. Martínez, M. Portilla, D. Sàez (Eds.), New Insights into the Universe. Proceedings, 1991. XI, 298 pages. 1992.

Vol. 409: H. Gausterer, C. B. Lang (Eds.), Computational Methods in Field Theory. Proceedings, 1992. XII, 274 pages. 1992.

New Series m: Monographs

Vol. m 1: H. Hora, Plasmas at High Temperature and Density. VIII, 442 pages. 1991.

Vol. m 2: P. Busch, P. J. Lahti, P. Mittelstaedt, The Quantum Theory of Measurement. XIII, 165 pages. 1991.

Vol. m 3: A. Heck, J. M. Perdang (Eds.), Applying Fractals in Astronomy. IX, 210 pages. 1991.

Vol. m 4: R. K. Zeytounian, Mécanique des fluides fondamentale. XV, 615 pages, 1991.

Vol. m 5: R. K. Zeytounian, Meteorological Fluid Dynamics. XI, 346 pages. 1991.

Vol. m 6: N. M. J. Woodhouse, Special Relativity. VIII, 86 pages. 1992.

Vol. m 7: G. Morandi, The Role of Topology in Classical and Quantum Physics. XIII, 239 pages. 1992.

Vol. m 8: D. Funaro, Polynomial Approximation of Differential Equations. X, 305 pages. 1992.

Vol. m 9: M. Namiki, Stochastic Quantization. X, 217 pages. 1992.

Vol. m 10: J. Hoppe, Lectures on Integrable Systems. VII, 111 pages. 1992.

Vol. m 11: A. D. Yaghjian, Relativistic Dynamics of a Charged Sphere. XII, 115 pages. 1992.

Vol. m 12: G. Esposito, Quantum Gravity, Quantum Cosmology and Lorentzian Geometries. XVI, 326 pages. 1992.